Oxford Applied and Engineering Mathematics

Oxford Applied and Engineering Mathematics

1 Alwyn Scott, *Nonlinear science: emergence and dynamics of coherent structures*
2 D.W. Jordan and P. Smith, *Nonlinear ordinary differential equations: An introduction to dynamical systems* Third edition
3 I.J. Sobey, *Introduction to interactive boundary layer theory*

Introduction to Interactive Boundary Layer Theory

I.J. Sobey
Computing Laboratory, Oxford University
Fellow in Engineering, St John's College

This book has been printed digitally and produced in a standard specification in order to ensure its continuing availability

OXFORD
UNIVERSITY PRESS

Great Clarendon Street, Oxford OX2 6DP
Oxford University Press is a department of the University of Oxford.
It furthers the University's objective of excellence in research, scholarship,
and education by publishing worldwide in
Oxford New York
Auckland Cape Town Dar es Salaam Hong Kong Karachi
Kuala Lumpur Madrid Melbourne Mexico City Nairobi
New Delhi Shanghai Taipei Toronto
With offices in
Argentina Austria Brazil Chile Czech Republic France Greece
Guatemala Hungary Italy Japan South Korea Poland Portugal
Singapore Switzerland Thailand Turkey Ukraine Vietnam

Oxford is a registered trade mark of Oxford University Press
in the UK and in certain other countries

Published in the United States
by Oxford University Press Inc., New York

Oxford is a registered trade mark of Oxford University Press
in the UK and in certain other countries

Published in the United States
by Oxford University Press Inc., New York

© I.J. Sobey, 2000

The moral rights of the author have been asserted

Database right Oxford University Press (maker)

Reprinted 2005

All rights reserved. No part of this publication may be reproduced,
stored in a retrieval system, or transmitted, in any form or by any means,
without the prior permission in writing of Oxford University Press,
or as expressly permitted by law, or under terms agreed with the appropriate
reprographics rights organization. Enquiries concerning reproduction
outside the scope of the above should be sent to the Rights Department,
Oxford University Press, at the address above

You must not circulate this book in any other binding or cover
And you must impose this same condition on any acquirer

ISBN 0-19-850675-9

PREFACE

It is now a little over a quarter of a century since I started as a graduate student in the Department of Applied Mathematics and Theoretical Physics in Cambridge. DAMTP was a fascinating department and contained an unusual concentration of the leading fluid dynamicists and indeed physical applied mathematicians in the world. There were many disparate research activities: DAMTP was mostly composed of fluid mechanists under the leadership of George Batchelor and James Lighthill, theoretical physicists lead by Polkinghorne and solid mechanists under Hill. In the fluid mechanics section of the department, in common with the other sections, we felt at the forefront of research throughout the world. There was a steady stream of distinguished visitors who were given a sometimes fair and sometimes robust reception at seminars. Above all there was a feeling of excitement about the department. Amongst the last duties of a graduate student was to give a main seminar. A rite of passage which in my case was complicated by my being due to row in the May bumps shortly before my talk. I think we rowed over and I cycled furiously back to DAMTP to give my talk. Perhaps my obvious physical exhaustion meant I was given a relatively easy time. My talk was on an analysis of flow very near a slot, an attempt to model some features of flow near the junction of an intercostal artery with the descending aorta. It was the second half of my dissertation and was only marginally successful in showing the type of flow field which might occur in steady or quasi-steady flow. Nevertheless it began my interest in interactive boundary layer theory; not that it was a true use of an interactive theory since the conditions were so severe that the flow region was embedded within the region where an interactive theory would have been needed but nevertheless it required use of some of the elements of that theory.

Even from before I had started at DAMTP, a research area of growing importance was the recently discovered triple deck theory and from after I arrived, I can remember being bemused by debates swirling around me about whether these ideas were right, were they relevant; who could believe a theory which used the one-eighth power of the inverse Reynolds number as a small parameter. I do not think some of the scepticism I perceived was just due to the fact that the fundamental discoveries occurred elsewhere. Rather I think it was because at that time, any analysis of fluid flow dealt with either very low Reynolds number flow, very high Reynolds number turbulent flow or inviscid flow. The area of intermediate Reynolds number flow was just too intractable and there were only a small number of experimental works and a smaller number of computed results available to give insight or intuition. The idea that a theory based on very large Reynolds number *laminar* flow could be used to understand laminar flows at more modest Reynolds numbers was naturally difficult to accept although the possibility of this happening had been noted many years before in 1956 by George Batchelor. Added to that, there were predictions from the theory such as those relating to upstream influence which at the time seemed counter intuitive to some.

Since that time, interactive boundary layer theory has proven to be a remarkable advance in theoretical fluid mechanics. Yet the theory is scattered through the literature and its background fading. The theory can be very difficult to follow with different authors using different notation. There is also an inherent complexity when using multiple matched asymptotic expansions. This book has come from a desire to have a graduate level text which draws the basic theory together to provide a starting point for research using interactive boundary layer theory or to give a readable (albeit mathematical) description for readers who are not boundary layer specialists. It is not intended as an advanced or indeed complete description of triple deck theory and many important developments such as those relating to compressible flow are not described at all.

In writing this book I have tried to provide the historical context for the development of the theory since it seems to me that there is much work which led up to interactive boundary layer theory which is now fading from general knowledge but without an understanding of which, it can be difficult to see the importance and place of interactive theory; hence the very long second chapter dealing with flow around a flat plate and the long historical chapter on separation from a cylinder. I also have a strong belief that the great expansion in the use of computational fluid mechanics to 'solve' flow problems has not eliminated the need for theoretical work, quite the contrary, it has made even more significant asymptotic solutions which can be used to validate or be validated by computation of flows. The need for asymptotic methods in fluid mechanics to be understood by workers in computational fluid mechanics is very great. Particularly the strengths and weaknesses of asymptotic solutions and the extent to which a computed solution should or should not agree with an approximate asymptotic solution. It is also the case that many applications of computational fluid mechanics are carried out in physics, chemistry and engineering by people who have not studied fluid mechanics in depth and I hope this book will help their understanding of fluid mechanics. There is even a hope that this book will appeal to any scientist interested in some really good applied mathematical analysis.

Deciding the content of this book has been quite difficult. I wanted to include such material as I thought would help in understanding the present position of interactive boundary layer theory but not to repeat easily available first order boundary layer theory. As I have said, I also did not intend to write a lot about some of the more complex multi-zone models for separation. I have restricted matters to two-dimensional incompressible steady laminar flow. I have excluded both three-dimensional and unsteady flow partially because these are still evolving areas and partly because the success so far of asymptotic methods has mostly been for two-dimensional steady flow.

The first chapter is intended as an introduction for readers who do not have a background in traditional fluid mechanics. Other than setting out some notation all the material in chapter one would most likely be covered in basic fluid mechanics courses. My original plan was to break the material only into chapters but when I was someway through writing it seemed much more sensible to think of the material in three sections, one on the triple deck, one on separation and one on channel flows, a structure which is similar to the first two parts in the survey paper by Smith (1982).

In the first section I have tried to set out the pre-triple deck understanding of flow about a plate aligned with the flow. This starts with a semi-infinite plate and continues with the case of a finite plate which brings a wake into consideration. This leads to Goldstein's near wake structure and a singularity in the approximation at the trailing edge, and it is the resolution of that singularity by Messiter and Stewartson, which was a key development in interactive boundary layer theory (along with application to separation in a compressible boundary layer by Neiland and Stewartson & Williams). It seems to me essential to show both why interactive boundary layer theory was needed and to show its place in the hierarchy of approximation. The third chapter covers the independent resolution by Messiter and Stewartson of the trailing edge flow structure. I have tried to present this very difficult material in a clear and consistent manner, starting with a heuristic explanation of the different length scales which arise but following with more formal description of the expansion procedure and then a computed solution of the interaction equations. In the fourth chapter it seems appropriate to discuss briefly the numerical solution of the incompressible triple deck equations and so conclude the first section.

In the second section separation is described, beginning with an extended view of how separation was approached within the context of conventional boundary layer theory and moving on to the ideas of Sychev and Smith and comparing predictions from interactive theory with computed solutions of the full Navier–Stokes equations. The main thrust of the section is separation from a cylinder and there is a concentration on the local flow picture near the cylinder. There is some discussion of the global flow field but that is limited to an introductory level.

In the third section the theory is continued with the development of interactive theory by Smith to channel flows. The interactive problem is formulated and developed for small asymmetric and symmetric channel boundaries where the pressure is constant across the channel. Then in considering upstream influence, the consequences of indentations so large that the pressure varies across the channel are studied. This leads to predictions from interactive theory which can be compared to computed solutions of the Navier–Stokes equations. In the last chapter we consider loss of symmetry and bifurcation in symmetric channels with a sudden expansion (the Coanda effect) as an example problem which is not yet fully described and where I believe more fruitful work will emerge.

Associated with the book is a web site where material described in the appendix is available. In particular programs used to recompute previous results are available there as are other codes I have written for a time dependent solution of the two-dimensional Navier–Stokes equations. Problems suggested in the appendix will be updated and extended as time passes. It is intended that material in the web site should provide a suitable resource for graduate level teaching.

Oxford
July 2000

I.J.S.

ACKNOWLEDGEMENT

This book has been mostly a solitary effort. However, I have been fortunate to have had at critical moments support from unlooked for quarters. The first suggestion that I might write a book was from Professor J.R. Blake as long ago as 1995 and I unreservedly thank him for his initial push which started me writing. I have also to thank Professor L.C. Woods, FRS, who although retired was willing to talk to me about free streamline theory and from whose book I have learned much. I thank Robert McLachlan for letting me have a copy of his Doctoral Thesis. I have tried to see first hand all the works referred to in the bibliography and I thank the Computing Laboratory Librarian, Gordon Riddell, for helping me find copies of older papers, as well as all those anonymous people who have helped to make the Radcliffe Science Library one of the great collections of older scientific journals.

The diagrams in this book are the result of computations by myself. In doing this I have tried to faithfully implement methods of previous workers but where I have referred to 'so-and-so's' method it needs to be clear that it is the spirit of the method which I have implemented rather than an exact reproduction. If small differences or even small errors have crept in during my calculations it should also be clear that differences are intended to extend previous calculations and errors, if there are any, are unintentional on my part and should not be considered a part of the original author's work. Many of the ideas described in this book were completed long before I was even aware of fluid mechanics. I have tried to describe some of the history of my subject based on many hours reading in libraries but if I have omitted some relevant work or not reported some other correctly then that is entirely unintentional.

Before going to OUP this book was written using $\text{\LaTeX}\,2_\varepsilon$ running under Linux. Program development has been with gcc (both C and fortran programs) and many graphs produced by xvgr and others using MATLAB and xfig.

I thank Mahua Nandi and Elizabeth Johnston of OUP for their help and encouragement.

My greatest thanks, however, go to my wife, Wendy and my sons, Jack and Thomas for their unwavering support and encouragement throughout the time I have been writing this book.

CONTENTS

1 Mathematical and Fluid Mechanical Introduction 1
  1.1 Introduction 1
  1.2 The Navier–Stokes equations 3
  1.3 Boundary conditions 5
  1.4 Asymptotic methods 5
  1.5 The Euler equations and potential flow 9
  1.6 Stokes flow 10
  1.7 Oseen's approximation 11
  1.8 Basic boundary layer theory 13
  1.9 Drag 17
  1.10 Summary and overview 20

I THE TRIPLE DECK

2 The Boundary Layer on a Flat Plate 25
  2.1 Introduction 25
  2.2 Semi-infinite plate – Rectangular coordinates 26
  2.3 Semi-infinite plate - Parabolic coordinates 36
  2.4 The drag on a section of semi-infinite plate 45
  2.5 The wake behind a finite length plate 49
  2.6 Near wake region 50
  2.7 Far wake expansion 59
  2.8 The drag on a finite plate 69
  2.9 Summary 74

3 The Triple Deck 76
  3.1 Introduction 76
  3.2 Formulation 82
  3.3 The middle deck 83
  3.4 The outer deck 85
  3.5 The inner deck 86
  3.6 Computed results 88
  3.7 Drag 90
  3.8 Numerical solution of the Navier–Stokes equations 91
  3.9 Summary 96

4 Numerical Solution of Triple Deck Equations 97
  4.1 Introduction 97
  4.2 Numerical solution in rectangular coordinates 98
  4.3 Solution using sublayer coordinates 103

4.4	A spectral method	104
4.5	Channel flow	106

II SEPARATION

5 Introduction to Separation — 111

6 Separated Flow about a Cylinder — 115
- 6.1 Observation at moderate Reynolds number — 115
- 6.2 Free streamline theory — 122
- 6.3 Boundary layer with a variable pressure gradient — 149
- 6.4 Combined boundary layer – free streamline models — 164
- 6.5 Goldstein's hypothesis of a boundary layer singularity — 169
- 6.6 Direct numerical solution of boundary layer equations — 176
- 6.7 Reprise — 183
- 6.8 Numerical solution of Navier–Stokes equations — 184
- 6.9 Attempts to resolve Goldstein's singularity — 194
- 6.10 Summary — 198

7 Prediction of Separation from a Cylinder — 199
- 7.1 Introduction — 199
- 7.2 Sychev's hypothesis for separation — 204
- 7.3 Smith's solution near separation — 206
- 7.4 Separation from a cylinder — 208
- 7.5 Comparison with numerical solutions — 210
- 7.6 Prandtl–Batchelor flow — 212
- 7.7 Summary — 218

III CHANNEL FLOW

8 Introduction to Channel Flow — 223
- 8.1 Introduction — 223
- 8.2 Asymmetric channels: $R^{-1} \ll \epsilon \ll R^{-1/7}$ — 228
- 8.3 Symmetric channels: $R^{-1} \ll \epsilon \ll 1$ — 233
- 8.4 Free streamline theory — 234
- 8.5 Computed examples — 246
- 8.6 Numerical solution of the Navier–Stokes equations — 250
- 8.7 Flow near a corner — 252
- 8.8 Summary — 261

9 Upstream Influence — 263
- 9.1 Introduction — 263
- 9.2 Asymmetric channels: $\epsilon \sim R^{-1/7}$ — 263
- 9.3 Upstream influence — 266
- 9.4 A numerical example — 276
- 9.5 Symmetric channels — 277

9.6	Prandtl–Batchelor flow in channels	282
9.7	Summary	282

10 Coanda Effect   284
    10.1 Introduction   284
    10.2 Symmetry and bifurcation   284
    10.3 Bifurcation solutions from Navier–Stokes equations   290
    10.4 Application of interactive boundary layer theory   292
    10.5 Summary   298

Appendix A Problems and Computer Programs   299
    A.1 Chapter 1 – Introduction   299
    A.2 Chapter 2 – Flat plate   300
    A.3 Chapter 3 & 4 – Triple deck   300
    A.4 Chapter 5 & 6 – Separation   301
    A.5 Chapter 7 – Prediction of separation from a cylinder   303
    A.6 Chapter 8 – Channel flow   303
    A.7 Chapter 9 – Upstream influence   304
    A.8 Chapter 10 – Coanda effect   306

Bibliography   307

Author Index   323

Subject Index   327

1
MATHEMATICAL AND FLUID MECHANICAL INTRODUCTION

1.1 Introduction

We live in a world which is largely fluid. Air, oceans, rivers and so on are all fluids whose behaviour is mostly described by a simple set of continuum equations, the Navier–Stokes equations. There are many fluids which are not described by these equations: visco-elastic fluids, visco-plastic fluids, thixotropic fluids and others all show more complicated behaviour, but for most fluids we encounter, the Navier–Stokes equations are adequate to model their motion. These equations themselves are only a continuum expression of Newton's second law of motion: rate of change of momentum equals applied force. Thus, given the assumption of a linear relation between stress and strain rate, the equations are simple to derive and simple to write down but at their centre lies an almost insurmountable complication: they are non-linear. When we calculate the rate of change of momentum at a point in a fluid we observe different fluid particles at different times, so to determine the correct value we have to add to the instantaneous rate of change of momentum with respect to time any spatial difference in the rate at which momentum arrives at and leaves the neighbourhood of the point. Momentum is of course linear in the velocity of a particle; the rate of arrival of momentum involves products of velocity components and so we have a non-linearity at the heart of fluid motion. A most important consequence is that there are very few exact solutions for fluid flow. Most that we do have are ones where the non-linearity vanishes exactly; for instance, in parallel flow the rate of arrival of momentum at a point will exactly equal the rate of departure so the non-linearity has no effect. The very small number of exact solutions mean that approximate solutions play a key role in our understanding of fluid motion. If a flow is steady then the Navier–Stokes equations, when non-dimensionalised, depend on only one parameter: a Reynolds number which is proportional to a length scale, a velocity scale and inversely proportional to the kinematic viscosity. The vast array of different steady fluid motions must then occur for particular combinations of Reynolds number, boundary geometry and flow constraints.

When fluid flows around a body, if the flow rate is low enough then fluid arrives at the leading (relative to the fluid motion) part of the body at a stagnation or attachment point, moves around the body and leaves from a trailing point. In moving around the body the fluid velocity will vary and so will the local fluid pressure. Such flows are described as attached to the body since near

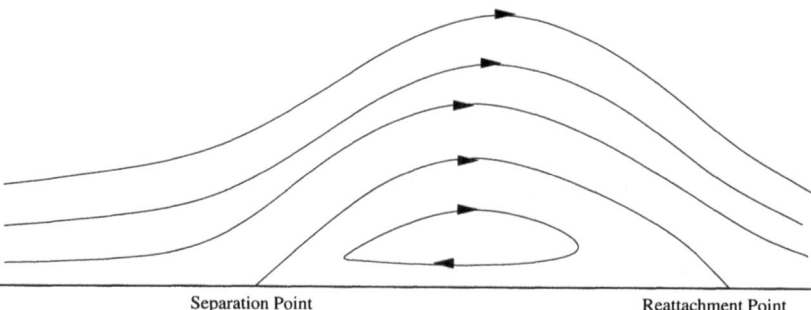

FIG. 1.1. Sketch of flow field near a separated region on a boundary.

the body, fluid particles follow the body shape. If the flow rate about the body increases (or the body speed in an otherwise stationary fluid increases) then a significant change in the nature of the flow can occur. It is observed that at some point on the body surface, fluid does not continue to follow the body shape but rather moves almost directly away from the surface. Further to the rear of the body fluid comes *towards* the oncoming flow and that too moves away from the surface at a *separation* point. Fluid which moves away from the body can only move directly away for a short distance before being affected by the oncoming flow and swept rearwards. Since the fluid is incompressible, the fluid moving away from the body has to either reattach to the body or join fluid from the other side of the body in the wake behind the body. This is illustrated in Figure 1.1. The region within which the flow has separated from the body and then reattached is called a vortex, and is fluid which the body is "carrying" through the oncoming flow. This separation region is a part of fluid mechanics for which we have had virtually no analytic description for most of the twentieth century and before that. Its occurrence is also a significant event, for instance indicating stall on an aircraft wing. Gross physical quantities such as lift on a wing or heat transfer from a body can vary substantially between a separated and an unseparated flow so the prediction of separation is vital in engineering design involving fluid motion. It is also the case that steady laminar separation is only the first of a sequence of flow events which dominate practical flows. As the flow rate increases, it is observed that a single separated region will itself have further internal separated parts, vortices within vortices. Such flows become unstable and can result in periodic shedding of vortices. The range of flow rates for which periodic shedding occurs is limited, at even higher flow rates the vortices become disorganised and the flow is eventually (with increasing flow rate) called turbulent. An important goal in the study of fluid mechanics is to be able to quantify this cascade of events. Boundary layer theory provides the beginning of such understanding and at present, it is the only rational analytic tool we have to probe the structure of flows within this cascade.

Although fluid flow becomes turbulent at high Reynolds number, to try to understand such a significant event as laminar separation we have to consider high Reynolds number asymptotic solutions of the equations of motion, ignoring that real flows may be turbulent and examining laminar flow high Reynolds number solutions of the equations of motion. Even when we consider high Reynolds number asymptotics we are still extremely restricted in our ability to get to the heart of how fluids behave near separation. The simplest asymptotic formulation for viscous flow near a boundary is adequate to describe flow which is attached to a body but if a adverse pressure gradient is specified, a singularity develops in the resulting boundary layer equations as the separation point is approached. It is of course not a singularity in the fluid but a singularity in the solutions of the approximate equations which have replaced the full non-linear equations of motion. What was faced was whether these approximate equations could be modified in some way so that they were still able to provide analytic asymptotic solutions and to describe separation (or even just highly modified unseparated flows) or whether such attempts were all in vain and one could only resort to numerical solutions of the full equations of motion. There is nothing wrong with numerical solutions and we are certainly able to calculate most two-dimensional laminar flows, steady or unsteady but what we do not get from a numerical solution is the understanding which comes from analytic solutions of how the flow depends on governing parameters.

1.2 The Navier–Stokes equations

Laminar viscous flow starts with the Navier–Stokes equations of motion of a Newtonian fluid. There are many texts which could be consulted for their derivation, for instance Panton (1996) but a recommended text is Batchelor (1967). If the velocity is non-dimensionalised with respect to a scale U, length with respect to a scale L, and time with respect to a frequency Ω, there are two non-dimensional parameters which are conventionally used, a Reynolds number,

$$R = \frac{UL}{\nu}, \qquad (1.1)$$

where ν is the kinematic viscosity, and a Strouhal number,

$$St = \frac{\Omega L}{U}. \qquad (1.2)$$

The Navier–Stokes equations are a restriction of a simple continuum model of a fluid for the case where the shear tensor is proportional to the strain rate tensor, the constant of proportionality being the viscosity, μ. The kinematic viscosity is $\nu = \mu/\rho$ where μ is the viscosity and ρ the density.

Let **u** be a non-dimensional velocity. If the pressure is scaled by ρU^2 to give a non-dimensional pressure p, then in the absence of body forces and gravity, conservation of momentum gives

$$\boxed{St\frac{\partial \mathbf{u}}{\partial t} + \mathbf{u}\cdot\nabla\mathbf{u} = -\nabla p + \frac{1}{R}\nabla^2\mathbf{u}.} \tag{1.3}$$

To this must be added the continuity equation, which for an incompressible fluid is just

$$\nabla\cdot\mathbf{u} = 0. \tag{1.4}$$

If the flow is steady, then the first term of (1.3) is missing as $St = 0$. This deceptively simple set of equations has very few exact solutions, mostly those occur when the non-linear term $\mathbf{u}\cdot\nabla\mathbf{u}$ vanishes identically.

One important class of flows is flows where the vorticity, $\nabla\times\mathbf{u}$, vanishes throughout the flow region. An old symbol for curl was *rot*, a shortened form of rotation which has the same meaning in German and English and flows with vanishing vorticity are called irrotational. In which case the velocity can be represented by a potential,

$$\mathbf{u} = \nabla\phi. \tag{1.5}$$

To satisfy the continuity equation (1.4), the potential ϕ must satisfy a Laplace equation,

$$\nabla^2\phi = 0. \tag{1.6}$$

In this case the momentum equation, (1.3), determines the pressure. In steady flow there will be a simple Bernouilli equation, $p + \frac{1}{2}\mathbf{u}\cdot\mathbf{u}$ being constant on streamlines. In unsteady flow the Bernouilli equation is $St\frac{\partial\phi}{\partial t} + p + \frac{1}{2}\mathbf{u}\cdot\mathbf{u}$ constant on a streamline.

In any two-dimensional flow, with $\mathbf{u} = (u,v)$ and coordinates (x,y), it is possible to define a stream-function, ψ, by

$$u = \frac{\partial\psi}{\partial y}, \quad v = -\frac{\partial\psi}{\partial x}. \tag{1.7}$$

A two-dimensional flow defined this way will automatically satisfy the continuity equation and differences in stream-function give the fluid flux between points. The vorticity is $\omega = \frac{\partial v}{\partial x} - \frac{\partial u}{\partial y}$, and the stream-function is determined by

$$\nabla^2\psi = -\omega, \tag{1.8}$$

while the vorticity satisfies a convection–diffusion equation,

$$\boxed{St\frac{\partial\omega}{\partial t} + \mathbf{u}\cdot\nabla\omega = \frac{1}{R}\nabla^2\omega.} \tag{1.9}$$

The vorticity can be thought of as twice the local rate of angular rotation at a point. This can be seen from considering a rotating flow field, $(u,v) = \Omega(-y,x)$ whence $\omega = 2\Omega$.

The stream-function vorticity formulation is extremely important for numerical solution of steady and unsteady two-dimensional viscous flow. Since differences in the stream-function determine differences in the flux of fluid between points, by specifying the stream-function on boundaries it is possible to automatically satisfy the continuity equation in a numerical scheme. The stream-function is also important at a theoretical level, since if we have the stream-function then we have essentially a full description of the flow – the pressure still has to be determined by integration but that is straightforward. The bulk of this book is oriented towards determining the stream-function

$$\psi = \psi(x,y;R), \qquad (1.10)$$

as an asymptotic sequence for $R \to \infty$ and we try to keep that as a unifying theme.

1.3 Boundary conditions

If there are solid boundaries around which fluid flows or which move through the fluid then it is clear from mass conservation that the normal velocity of the fluid must match the normal velocity of the boundary. Observation shows that if the fluid is viscous, it is not just the fluid velocity normal to the boundary which must coincide with the normal boundary velocity, the fluid velocity must also coincide with the boundary velocity. This is called a no-slip condition. This is not a universal condition but it is (or appears to be) satisfied by the vast majority of fluids we encounter. To complete the mathematical specification of a flow field, the Navier–Stokes equations and the continuity equation must have added this boundary condition on solid boundaries and suitable boundary conditions far from any boundaries. There is an essential difference between imposing one component of velocity, the normal component and imposing both components of velocity at a boundary.

In inviscid incompressible flow, there needs only be one boundary condition specified on a body, the normal velocity of the fluid must match that of the body: two boundary conditions cannot be arbitrarily specified. Hence for inviscid flow there can be a difference in value between the tangential fluid velocity and the tangential boundary velocity, described as a slip velocity.

1.4 Asymptotic methods

As the Navier–Stokes equations are mostly too difficult to solve exactly we are thrown back onto approximate analysis. Many approximate methods are ad-hoc studies of the important influences in a given physical problem and often provide solutions for major engineering questions as well as physical insight into problems. However, if a problem is to be studied deeply it is necessary to be very careful in formulation and approximation. The core of this is asymptotic analysis. There are many good texts which set out the mathematics of asymptotic analysis. Here we shall only define the notation which will be used throughout this book and give some simple heuristic comments on the motivation for asymptotic methods. Two excellent introductions to this subject are Erdelyi (1955) and Copson (1965): each is concise and readable while retaining mathematical rigour without it becoming an overbearing objective in itself. Two good recent texts are Hinch (1991) and Georgescu (1995). Chronologically between these two pairs of texts are Bender and Orszag (1978) and Eckhaus (1973). My own favourites are Copson and Bender & Orszag. Van Dyke (1964) is not a particularly mathematical introduction to asymptotic analysis, being much more concerned with physical understanding in fluid mechanics, but Chapter 3 of that text is quite an adequate introduction to asymptotic analysis. Much of our present view of matched asymptotic analysis in fluid mechanics comes from Kaplun (1967).

Behind all the work described later in this book is the idea that solutions of a differential system which are too complicated to write down analytically can nevertheless be approximated by other, less complicated functions, at least in parts of their region of definition. As we shall see, we will find expressions such as 'close to a boundary', 'far from a boundary' providing a heuristic framework for us to approximate a flow field. So we need some way to compare functions in the vicinity of a limit point. Here it should be obvious that if the functions are continuous, there are only two possibilities, either one function is much smaller than the other as the limit point is approached or else they are roughly the same size! If $f(x)$ and $g(x)$ are functions defined near a point $x = a$ then we write $f(x) = o(g(x))$ as $x \to a$ if $f(x)/g(x) \to 0$ as $x \to a$.

Having decided on a symbol to denote that one function is asymptotically smaller than another (and assuming $g \neq o(f)$ as $x \to a$), then anything else is covered by writing $f = O(g)$ as $x \to a$, and this will be true provided that f/g is bounded as $x \to a$. The two order symbols o, O are usually given equal prominence but it should be clear that o is the more powerful statement about the comparative behaviour of two functions near a limit point.

A sequence of functions $\{f_n\}$ is defined to be asymptotically ordered as $x \to a$ if $f_{n+1} = o(f_n)$ as $x \to a$ for $n = 0, 1, 2, \cdots$. Given such as asymptotic sequence and a function $f(x)$ defined near $x = a$, there is an asymptotic expansion of the function f if numbers $a_0, a_1, \cdots$, can be found such that

$$f(x) - \sum_{m=0}^{n} a_m f_m(x) = o(f_n(x)), \qquad (1.11)$$

as $x \to a$ for $n = 0, 1, 2, \cdots$; in which case we write

$$f \sim \sum_{m} a_m f_m \quad \text{as} \quad x \to a. \qquad (1.12)$$

If such an asymptotic expansion exists then the coefficients can be constructed from $a_0, a_1, \cdots$, from

$$a_n = \lim_{x \to a} \frac{f(x) - \sum_{m=0}^{m=n-1} a_m f_m(x)}{f_n(x)}. \tag{1.13}$$

These simple definitions become more alive when the function f has two or more arguments, for instance $f = f(x, \epsilon)$, $a \leq x \leq b$, $\epsilon \geq 0$, and an asymptotic expansion is sought for f when ϵ is small compared with one. Now, given an asymptotic sequence $\{f_n(\epsilon)\}$ as $\epsilon \to 0$, can functions $g_n(x)$ be found such that

$$f(x, \epsilon) \sim \sum_m f_m(\epsilon) g_m(x) \quad \text{as} \quad \epsilon \to 0? \tag{1.14}$$

At this point what has been deliberately omitted is the simple condition $a \leq x \leq b$. If the asymptotic expansion is valid over the whole range of x then the expansion is called uniform but unfortunately, in boundary layer theory it may be necessary to consider different asymptotic expansions in different regions of the flow.

This brings us to matched asymptotic expansions, a subject which is usually included in texts on asymptotic methods and indeed is nowadays part of many undergraduate mathematics courses. Anyone interested in the application of matched asymptotics in fluid mechanics should regard Van Dyke (1964) as essential reading. Rather than repeat easily available mathematical examples here, all I want to do is give an illustrative physical example of how matched asymptotic expansions can come about. It is essentially the explanation taught to me by Prof E.O.Tuck when I was an undergraduate at Adelaide University in the early 1970s (see also Tuck (1971)).

Consider a breakwater of infinite depth located at $x = 0$ against which waves are coming from $x = -\infty$. Suppose there is a small hole of size h a distance H down the breakwater and let $\epsilon = h/H$ be small. Tuck sought to determine the proportion of incoming wave energy which was transmitted through the hole and the proportion reflected. It would have been natural to try to solve this problem by supposing $\epsilon \to 0$ and trying to determine an asymptotic expansion of the solution in powers of ϵ. What Tuck observed was that ϵ was the ratio of two length scales so that as $\epsilon \to 0$ he could consider two hypothetical observers. An observer of scale H saw only the hole shrink until the flow near to it resembles an unsteady point source/sink. This observer knew the rate at which energy was coming from $x = -\infty$ but could not determine how great or small the source/sink strength would be. Next consider a second hypothetical observer of scale h. As $\epsilon \to 0$ this observer saw the free surface go off to infinity and so knew nothing about the rate at which energy was arriving from $x = -\infty$. This observer could, however, calculate the flow through the hole in an infinite plate. On their own, neither observer could produce an asymptotic expansion for their 'local' flow problem

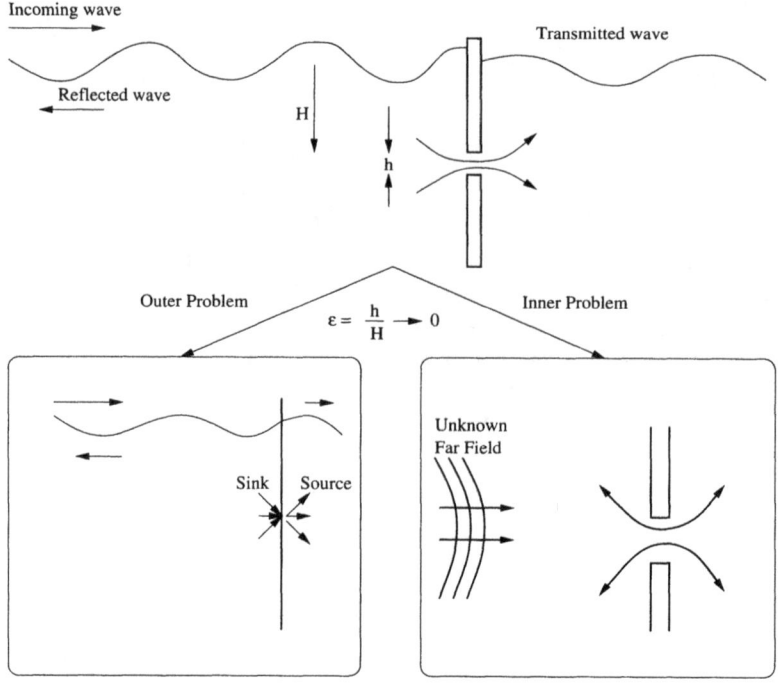

FIG. 1.2. Model matched asymptotic problem, after Tuck (1971).

which was free of unknown constants. Matched asymptotic methods provided a systematic way for the two observers to compare asymptotic expansions and so resolve unknown constants. A vital part of this process was that the two expansions should had a region of overlap so that they could be matched to each other to resolve the unknown constants. In Tuck's problem, observers on both scales saw a region where the flow was essentially like a source/sink combination. The outer observer saw a time-varying sink on one side of the wall and source on the other side while an observer on the inner scale saw the far field as being that of a sink on one side of the wall and a source on the other. This provided the vital overlap region which allowed the two expansions to be matched and the unknown constants resolved.

Once expansions at each scale are known, it is then in principle possible to develop a composite expansion which is asymptotically equal to the expansion of each observer while also being defined over the whole spatial region. The composite expansion is a convenient method to calculate an approximate solution for practical use, it is not though a proper asymptotic solution of the original differential system. It will however be indistinguishable from the true asymptotic solutions as the expansion parameter tends to its limit value. As observed by Van Dyke, problems where an asymptotic expansion is sought in terms of a parameter which is the ratio of two physical scales invariably lead to matched asymptotic expansions. In fluid mechanics it is common to find the formulation of a flow problem leads to or is based on a comparison of length scales. Just the expressions 'near field' and 'far field' imply solutions for different length scales. Thus many recent theoretical developments in fluid mechanics are based on the formalism and the understanding which comes from matched asymptotic analysis.

1.5 The Euler equations and potential flow

If a fluid has no viscosity then application of conservation of mass and momentum leads to a set of equations called the Euler equations. These are usually associated with compressible flow but can be reduced to apply to an incompressible fluid. In steady incompressible flow which neglects gravity or other body forces the Euler equations are a continuity equation,

$$\nabla \cdot \mathbf{u} = 0, \tag{1.15}$$

and momentum equations,

$$\mathbf{u} \cdot \nabla \mathbf{u} = -\nabla p. \tag{1.16}$$

The infinite Reynolds number limit for the Navier–Stokes equations reduces to these Euler equations. This is generally thought of as a limit where viscous effects are small compared with inertial effects but alternately it is a limit where the ratio of a body length scale to a viscous length scale, ν/U becomes large. Since the Reynolds number it is the ratio of two length scales, asymptotics involving $R \to \infty$ will satisfy Van Dyke's criterion for the application of matched asymptotics.

At this point is is worth recalling the physical origin of viscosity. If we imagine two streams of fluid moving at slightly different velocity then lateral diffusion or Brownian motion will result in an exchange of fluid particles. Those from the slower stream will have to speed up, requiring momentum which will provide a drag force on the other particles in the faster stream. Particles from the faster stream moving to the slower stream will have to be slowed down providing an acceleration force on the slower stream. A simple argument then provides that the tangential stress between two layers of fluid should be proportional to the local velocity gradient across the direction of flow, the constant of proportionality being the viscosity. Indeed viscosity is often measured by observing the force required to maintain a uniform velocity gradient, for instance in a thin layer between two walls moving tangentially. Central to our understanding of viscosity is that the velocity should be continuous. However, the Euler equations allow solutions with discontinuous velocity. Thus the limit process of $R \to \infty$ has to be singular in some sense since we cannot have a true discontinuity in the Navier–Stokes solutions, only layers which become infinitesimally thin but within which the velocity remains continuous (that is, using a continuum hypothesis and neglecting any molecular limit to the length scale). Both the Navier–Stokes equations and the Euler equations admit sets of non-unique solutions, many of which we do not yet know. As a result are still open questions about the relation between solutions of the Navier–Stokes equations in the limit $R \to \infty$ and solutions of the Euler equations; see for instance Lagerstrom (1975) for an attempt to codify relations between solutions of the two sets of equations.

If the flow is irrotational then a velocity potential can be used with $\mathbf{u} = \nabla \phi$. In two-dimensions the velocity potential, ϕ, and the stream-function, ψ, can be combined in a complex potential,

$$w = \phi + i\psi, \qquad (1.17)$$

so that using a complex coordinate $z = x + iy$,

$$\frac{dw}{dz} = u - iv. \qquad (1.18)$$

In regions where the flow is irrotational both ϕ and ψ are harmonic functions so *any* analytic function $w(z)$ will define a flow field of some sort. A simple example is

$$w \sim z^\beta = r^\beta e^{i\beta\theta} \quad \text{in polar coordinates,} \qquad (1.19)$$

or

$$\phi = r^\beta \cos \beta\theta \quad \psi = r^\beta \sin \beta\theta. \qquad (1.20)$$

Thus this flow has ψ constant when $\sin\beta\theta = 0$ or $\theta = 0$, $\theta = \pm\frac{\pi}{\beta}$. Depending on whether $\beta < 1$ or $\beta > 1$ this defines flow around an obtuse or acute corner. If $\beta > 1$ and depending on the sign of w this models flow around the leading or trailing edge of a wedge.

There are many texts which describe potential flow, Milne-Thompson (1938) is particularly good.

1.6 Stokes flow

The low Reynolds number limit for the Navier–Stokes equations occurs when inertial effects are small compared to viscous effects and is called Stokes flow. This limit too is a comparison of a viscous length scale, ν/U, with a body length scale, L, and so a limit where matched asymptotics should be expected to play an important role.

If $R \to 0$ in the Navier–Stokes equations, (1.3), the scaling we have used for the pressure fails with the non-dimensional pressure becoming infinite. In that case we need to rescale the pressure and instead of scaling pressure with ρU^2, we should use μUL where $\mu = \rho\nu$ is the viscosity. Using p_s for pressure non-dimensionalised in this way for Stokes flow, the Navier–Stokes equations are

$$RSt\frac{\partial \mathbf{u}}{\partial t} + R\mathbf{u} \cdot \nabla \mathbf{u} = -\nabla p_s + \nabla^2 \mathbf{u}. \qquad (1.21)$$

Then in steady flow, if $R \to 0$ and the velocity and pressure are expanded

$$\mathbf{u} \sim \mathbf{u}_0 + o(1), \quad p_s \sim p_0 + o(1) \text{ as } R \to 0, \qquad (1.22)$$

the leading order terms should satisfy

$$\nabla^2 \mathbf{u}_0 = \nabla p_0. \qquad (1.23)$$

In unsteady flow it is convenient to use $\alpha^2 = RSt = \Omega L^2/\nu$ and to consider α^2 varying independently of $R \to 0$. In the Stokes limit of $R = 0$ and for steady flow, since $\nabla \times \nabla p_0 = 0$, the vorticity ω is harmonic and in two-dimensional flow, the stream-function ψ is biharmonic

$$\nabla^2 \omega = 0, \quad \nabla^4 \psi = 0.$$

In three-dimensional flow the velocity is biharmonic.

Thus at the centre of studies for Stokes flow are solutions of the biharmonic equation. The implication is that in two-dimensions when determining the stream-function we should require four boundary conditions. We have only three obvious conditions: two velocities on the body and the velocity in the far field. For the Navier–Stokes equations conditions on the body and the far flow field can be augmented by requiring that the vorticity become exponentially small outside of a body-wake region. It is more difficult to apply this directly to Stokes flow where unfortunately the solutions can in effect be singular at infinity. The limit $R \to 0$ can require more complicated expansions than a simple power series in Reynolds number and needs to be matched to a different form of approximation far from a body.

1.7 Oseen's approximation

The need for a more complicated asymptotic expansion than a power series in R when $R \to 0$ arises for example for flow around a cylinder. If the equations for steady Stokes flow about a body such as a cylinder are solved in two dimensions then the solution has unacceptable behaviour far from the body: the solution does not tend to parallel flow and thus is not physically plausible. Essentially the solution of the steady Stokes equations for a cylinder moving in a fluid is one where viscosity having had an infinite time to act, results in all the fluid at infinity moving with the body (see also Stokes (1850) and Lamb (1932) article 343). This difficulty was studied by Oseen (1910) who treated the far field as a perturbation of uniform flow and included the effect of low Reynolds number inertial terms. Later this was formalised using matched asymptotic methods as a low Reynolds number expansion of the Navier–Stokes equations. Although it is not directly a part of boundary layer theory, reference to Oseen's approximation runs throughout fluid mechanics so we shall consider a simple outline here.

Consider flow around a cylinder at low Reynolds numbers. A simple expansion of the stream-function in terms of powers of Reynolds number gives

$$\psi \sim \psi_0(x,y) + R\psi_1(x,y) + O(R^2), \qquad (1.24)$$

and the leading order term ψ_0 satisfies

$$\nabla^4 \psi_0 = 0, \qquad (1.25)$$

with solution in polar coordinates

$$\psi_0 = A(\frac{1}{r} - r + r\log r)\sin\theta, \qquad (1.26)$$

where the unknown constant A cannot be found by trying to satisfy the condition of uniform flow upstream (where $\psi \sim r\sin\theta$) because of the presence of the $r\log r$ term. Thus the expansion appears to fail at the first step. This is called Stokes paradox.

Oseen considered an approximation

$$\psi \sim y + \Psi_1, \qquad (1.27)$$

which gave after linearisation in the Navier–Stokes equations,

$$\boxed{\nabla^4 \Psi_1 = R\frac{\partial}{\partial x}\nabla^2 \Psi_1.} \qquad (1.28)$$

Unfortunately this expansion would not be valid near the cylinder where the flow could not be described as a perturbation to uniform flow. This pragmatic approximation was put into a matched asymptotic framework by Proudman and Pearson (1957). A very good description of the process has been given by Van Dyke (1964), Chapter 8.

The resolution of Stokes paradox came from observing that the expansion (1.24) was not asymptotic when $Rr = O(1)$ for $R \to 0$ and that a scaled radius $\tilde{r} = Rr$ should be introduced.

The Stokes expansion which was taken to be valid near the cylinder was when considered far from the cylinder

$$\psi \sim (-\frac{A\log R}{R}\tilde{r} + \frac{A}{R}\tilde{r} + \frac{A}{R}\tilde{r}\log\tilde{r} - AR\frac{1}{\tilde{r}})\sin\theta, \qquad (1.29)$$

while Oseen's expansion began

$$\psi \sim \frac{1}{R}\tilde{r}\sin\theta + \Psi. \qquad (1.30)$$

The leading order terms as $R \to 0$ would match if the constant A was chosen to be

$$A = -\frac{1}{\log R}, \tag{1.31}$$

(the last three terms in (1.29) then being asymptotically smaller than the first).

As Van Dyke has observed, even (1.31) was insufficiently general although it does encapsulate how Oseen's approximation could be matched to a solution near the cylinder so as to resolve Stokes paradox. Further details can be found in Van Dyke's book.

Since Oseen's approximation linearises the Navier–Stokes equations it was possible to determine the solution for a point force at the origin with stationary fluid at infinity. The velocity components for that were given by Lamb (1932) and the stream-function for that case was (in polar coordinates)

$$\Psi = \frac{r}{\pi} \int_0^\theta e^{\frac{1}{2}rR\cos\alpha}[K_0(\frac{rR}{2})\cos\alpha + K_1(\frac{rR}{2})]d\alpha, \tag{1.32}$$

where K_0 and K_1 were appropriate Bessel functions.

This solution, called a doublet, was applied to various problems by assuming a distribution of doublets and attempting to solve an integral equation for distribution strength which would lead to specified velocity, for instance uniform velocity on a circle to model flow around a cylinder or uniform flow on a segment of line to model flow past a flat plate (Bairstow et al. (1923), Piercy and Winny (1933)). The drag on the body was then given by a simple integral of the doublet distribution. This example will be discussed further in section 2.8.1.

1.8 Basic boundary layer theory

Having looked briefly at the low Reynolds number limit for the Navier–Stokes equations we can now turn to the main area of interest in this book. The large Reynolds number limit is one where the effects of viscosity occur on a length scale which becomes smaller and smaller in comparison with a body length scale. Layers where viscous effect continue to be important are called a boundary layers. Note that there is no requirement that a boundary layer must be attached to a solid boundary.

Although this is a book introducing advanced boundary layer concepts, we will spend a little time here setting out the basics of boundary layer theory. There are many books which deal with these ideas and nearly all modern 'fluid mechanics' texts cover basic boundary layer theory in some detail. Prandtl's ideas were given in later editions of Lamb's 'Hydrodynamics' [Lamb (1932)], other older books are Prandtl and Tietjens (1934), Goldstein (1938), more recent books specifically on boundary layers are Rosenhead (1963), Meksyn (1961), Curle (1962), Stewartson (1964) Schlichting (1979), Young (1989), Rogers (1992) and Oleinik and Samokhin (1999). There are also two papers on the history of boundary layers, Tani (1977) and Van Dyke (1994) which are well worth reading.

Armed with the tools of asymptotic analysis, a leading order description of a boundary layer is nowadays a routine task. Suppose the Reynolds number is large and that the effect of viscosity is confined to a layer near the wall where the velocity changes from the wall velocity to the main fluid velocity. As the layer is thin, derivatives across the flow direction might be expected to be larger than derivatives in the flow direction. The continuity equation indicates that the transverse velocity must then be small compared to the streamwise velocity in the same ratio. This can be quantified by supposing a body of size L will have a boundary layer of thickness $L\delta$ where δ is small compared with one. Then if the streamwise velocity is of size U the transverse velocity in a boundary layer will be of size $U\delta$. Now consider the streamwise and transverse momentum equations for steady two-dimensional flow. If we non-dimensionalise on the scale of an observer who remains comparable to the boundary layer thickness the streamwise coordinate by $\hat{x} = Lx$ and the transverse coordinate by $\hat{y} = L\delta Y$ then the non-dimensional form of the Navier–Stokes equations becomes, for x-momentum,

$$u\frac{\partial u}{\partial x} + v\frac{\partial u}{\partial Y} = -\frac{\partial p}{\partial x} + \frac{1}{R\delta^2}\frac{\partial^2 u}{\partial Y^2} + \frac{1}{R}\frac{\partial^2 u}{\partial x^2}, \qquad (1.33)$$

and for y-momentum,

$$\delta^2(u\frac{\partial v}{\partial x} + v\frac{\partial v}{\partial Y}) = -\frac{\partial p}{\partial y} + \frac{1}{R}\frac{\partial^2 v}{\partial Y^2} + \frac{\delta^2}{R}\frac{\partial^2 v}{\partial x^2}. \qquad (1.34)$$

If on the other hand we were an observer of height comparable with the body size and used a transverse coordinate $\hat{y} = Ly$ then we would use the Navier–Stokes equations as given in (1.3) and as $R \to \infty$ these would have as their leading approximation just potential flow. If we were to calculate potential flow we would not be able to satisfy all the boundary conditions on a body since for solutions of a Laplace equation we can specify only one boundary condition around the boundary of the domain. We recognise this as a matched asymptotic expansion problem: in the momentum equations 'seen' by an observer of scale $L\delta$. As $\delta \ll 1$ and $R \gg 1$ the sensible combination in the x-momentum equation, (1.33), is to take $R\delta^2 = 1$ as $R \to \infty$. This will mean that the last term will be absent from a leading order expansion of the x-momentum equation and heuristically, the velocities should be approximately given by

$$u\frac{\partial u}{\partial x} + v\frac{\partial u}{\partial Y} = -\frac{\partial p}{\partial x} + \frac{\partial^2 u}{\partial Y^2}, \qquad (1.35)$$

with the boundary conditions $u = v = 0$ on $Y = 0$ and u specified as $Y \to \infty$. Consider now the y-momentum equation, (1.34), using the scaling $\delta = R^{-1/2}$, the leading order form of this equation is simply

$$\frac{\partial p}{\partial y} \sim 0, \qquad (1.36)$$

so to leading order, the pressure is uniform *across* the boundary layer. Using this simplification means that for some forms of pressure, we can attempt to determine a similarity solution of the continuity and x-momentum equations, using a similarity variable of the form $x^b Y$. This then is what might be called basic boundary layer theory. There are two regions in which we must obtain asymptotic expansions, an outer region where to leading order the flow is potential, and an inner region where the flow is given by the leading order terms of (1.33), (1.34) and continuity and they are matched to the outer flow by considering $y \to 0$ in the outer flow and $Y \to \infty$ in the boundary layer. If the pressure throughout the boundary is constant then the boundary layer is usually called a Blasius boundary layer, if the pressure gradient on the boundary is a power of x (we have rather cavalierly taken rectangular coordinates, a closer analysis would allow x to be distance along a curved boundary) then the boundary layer is called a Falkner–Skan boundary layer. We will examine the matching process in detail in the next chapter. For the present, consider the leading order terms of (1.33), for the case of zero pressure gradient and in stream-function form,

$$\psi_Y \psi_{xY} - \psi_x \psi_{YY} = \psi_{YYY}, \tag{1.37}$$

then if $s = Y x^b$ and $\psi = x^{1+b} F(s)$ (we will derive this form properly in the next chapter) then

$$F''' + (1+b) F' F'' - (1+2b) F'^2 = 0. \tag{1.38}$$

If the velocity is zero at $Y = 0$ then two boundary conditions are $F(0) = F'(0) = 0$, a third still has to come from matching the behaviour of ψ as $Y \to \infty$ with the stream-function in the outer flow, matching which will also determine b.

If the external streamwise velocity varies with x, say $u = u_e(x)$, then taking the limit of (1.35) as $Y \to \infty$ gives $p'(x) = -u_e u_e'$ and (1.37) has to be replaced by

$$\psi_Y \psi_{xY} - \psi_x \psi_{YY} - \psi_{YYY} = u_e(x) u_e'(x). \tag{1.39}$$

This equation has similarity solutions for only a restricted set of external velocity forms $u_e(x)$.

This brief glimpse at the material we shall shortly examine in detail, is the essence of boundary layer theory from its inception by Prandtl in 1904. There have been other attempts to model flow in the boundary layer, for example Smith (1938) tried to use an energy minimisation approach but Prandtl's approach has endured. Depending on the far field and depending on the applied pressure gradient, equation (1.37) or an inhomogeneous form of it, has to be solved with appropriate conditions as $Y \to \infty$.

Two avenues have been explored, one associated originally with Blasius attempted a series representation using polynomials in the similarity variable s and an equation such as (1.38) to determine coefficients. We shall see more on this in the next chapter.

Another method, associated with Karman, considered integrals of flow quantities across the boundary layer and so obtained an approximate differential equation for the streamwise development the integral quantities along the boundary layer, Karman (1921). The use of an integral approach for this type of boundary layer was crystallised by Pohlhausen (1921) and we shall look into that later.

Nowadays it is a simple matter to integrate equations such as (1.38) numerically but for dealing with models of turbulent boundary layers, the integral approach has considerable significance.

One important integral is the displacement thickness, $\delta_T(x)$ defined by

$$u_e \delta_T(x) = \int_0^\infty u_e(x) - u(x,y) \, dy, \tag{1.40}$$

which can be interpreted as the local increase in body thickness which will be 'seen' by the outer potential flow.

A second integral is the momentum thickness, $\theta(x)$ given by

$$u_e^2 \theta(x) = \int_0^\infty u(x,y)(u_e(x) - u(x,y)) dy, \tag{1.41}$$

and is interpreted as the reduction in momentum flux caused by the boundary layer.

With these two definitions, an integral form of the momentum equation for the boundary layer can be derived as

$$\boxed{\frac{d}{dx}[\theta u_e^2] + \delta_T u_e \frac{du_e}{dx} = \tau_W,} \tag{1.42}$$

where $\rho U^2 \tau_W$ is the dimensional wall stress. This is a very general statement about a boundary layer and has been and continues to be an important engineering tool.

The approximate method for dealing with the integral equation, (1.42), proposed by Pohlhausen revolved around realising that a similarity solution for the velocity profile used the same geometric shape for the boundary layer as it evolved and changed thickness in the streamwise direction. If we suppose the local boundary layer thickness is $\delta(x)$ and let $u(x,y) \sim u_e(x) f(y/\delta(x))$, when $y \leq \delta$, $u(x,y) \sim u_e(x)$ when $y > \delta$, then the wall stress is

$$\tau_W = \frac{u_e f'(0)}{R\delta}, \tag{1.43}$$

and the displacement and momentum thicknesses are

$$\delta_T = \delta \int_0^1 [1 - f(\eta)] d\eta, \text{ and } \theta = \delta \int_0^1 f(\eta)[1 - f(\eta)] d\eta. \tag{1.44}$$

Eliminating δ,

$$\delta_T = H\theta, \qquad (1.45)$$

where $H = \int_0^1 [1 - f(\eta)]d\eta / \int_0^1 f(\eta)[1 - f(\eta)]d\eta$, is a form factor which in this case, relies only on the geometric shape of the velocity profile. If the velocity profile is more complicated, for instance in a wake where the geometric shape varies with distance from the body, a form factor can still be defined but it will no longer be a constant. Using the form factor H, (1.42), can be viewed as an ordinary differential equation for θ,

$$\boxed{\frac{d}{dx}[\theta u_e^2] + H\theta u_e \frac{du_e}{dx} = \frac{1}{R}\frac{f'(0)u_e A}{\theta},} \qquad (1.46)$$

where $A = \int_0^1 f(\eta)[1 - f(\eta)]d\eta$ is known. We shall see shortly that there is a good reason for choosing θ as dependent variable because it is intimately connected to the drag of body.

1.9 Drag

It is one of our most common experiences to stand on a windy day and feel the force of the air as it passes by. The streamwise component of the force on a body as a fluid moves past it is called the drag. Any discussion about drag is slightly complicated by the way the drag has traditionally been broken down into different components. If there is a non-uniform pressure field around the body we can calculate a pressure drag from the integral around the body of the streamwise component of pressure force (that is, pressure times normal area with the normal area having a vector property). We can also calculate a viscous drag by integrating the streamwise component of the viscous stresses on the body. If the body moves in a fluid with a free surface we can calculate a wave drag from the kinetic energy under the waves in the wake. Whereas pressure and viscous drag are easily defined, other 'drag' components can sometimes seem more elusive. As an example, the term wave drag is also associated with losses in shocks in compressible flow. Form drag, induced drag, self-induced drag and others are all used variously to describe components of the drag. Our main interest in this work is in the pressure and viscous drag terms and their contribution to the total drag.

Our ideas concerning scaling of drag in fluid flow can be traced back to at least Newton (1686) (*Principia*, Book II, Proposition XXXIII, Theorem XXVI) who argued that drag was the result of particles in the fluid impacting against the body and changing momentum due to these collisions. Thus the rate at which collisions would occur would be given by the fluid velocity and the body front area, the change in momentum would depend on the density and whether the collision were elastic or inelastic; but in any case changes in momentum would scale with the main flow velocity. Thus he reasoned that drag should scale with density, velocity squared and frontal area of the body. Although he did not consider any additional changes in momentum due to further collisions between fluid particles, the scaling he proposed is still used. Hence the physical drag $\hat{D}$ experienced by a body of length scale L moving with velocity U in a fluid of density ρ which is at rest far from the body or for a body which is at rest in a fluid moving with the same velocity U far from the body might be written

$$\hat{D} = \frac{1}{2}\rho U^2 L^2 C_D. \tag{1.47}$$

The non-dimensional constant of proportionality, C_D is called the drag coefficient. Pressure drag, viscous drag and so on are all components of the drag and can be associated with suitable drag coefficients. Using the definition of drag, a major goal in fluid mechanics is to be able to predict

$$C_D = C_D(R), \tag{1.48}$$

for a given body shape as the Reynolds number R varies since the drag $\hat{D}$ gives the power $U\hat{D}$ needed to maintain steady motion. Unfortunately, $C_D(R)$ is a complicated function. It can be non-unique in transition flows where there may be no stable steady solution, and it may only be defined by an average in turbulent flow. The complicated nature of $C_D(R)$ is also a reason why separation is such an important fluid mechanical 'event', since the drag for flow around a streamlined body and the drag around a similar sized body with separation can be very different.

If we turn to the calculation of drag, then again we have to look back to Newton. Using a large control volume (from $y = -BL$ to $y = BL$ and $z = -BL$ to $z = BL$) about the body we can calculate the difference between oncoming and leaving momentum flux (if the pressure is p, the flow rate Q and normal velocity to an area A is u, then the momentum flux across A is $pA + \rho Qu$) to determine the drag on a body.

In the case of an external flow where the upstream pressure is recovered far downstream, if Q_B is the fluid flux across the upstream side of our control volume and $Uu(x, y, z)$ is the fluid velocity across the downstream side of the control volume, then the drag will be given by

$$\hat{D} = \lim_{x \to \infty} \lim_{B \to \infty} [\rho Q_B U - \int_{-B}^{B}\int_{-B}^{B} \rho U^2 L^2 u^2(x,y,z) dy dz], \tag{1.49}$$

and as

$$U = \lim_{B \to \infty} \frac{Q_B}{4B^2 L^2} = \lim_{B \to \infty} \frac{U}{4B^2} \int_{-B}^{B}\int_{-B}^{B} u(x,y) dy dz, \tag{1.50}$$

to satisfy continuity, so it must be the case that

$$\hat{D} = \lim_{x \to \infty} \rho U^2 L^2 \int_{-\infty}^{\infty}\int_{-\infty}^{\infty} u(1-u) dy dz. \tag{1.51}$$

We see that in two-dimensional flow, where we calculate the drag per unit width, the integral in (1.51) is just (1.41) with $u_e = 1$, so that under those circumstances, the momentum thickness far downstream is twice the drag coefficient,

$$C_D = \frac{1}{2}\theta_\infty. \tag{1.52}$$

When we consider viscous flow over a body, with say Ly as coordinate normal to the body and Uu the velocity tangential to the body then the local viscous stress tangential to the body is given by

$$\tau = \frac{\mu U}{L}\frac{\partial u}{\partial y}, \tag{1.53}$$

so we can define a local friction coefficient,

$$c_f = \frac{\tau}{\frac{1}{2}\rho U^2} = \frac{2}{R}\frac{\partial u}{\partial y}, \tag{1.54}$$

The viscous drag on the body will then be an integral over the body of the component of c_f in the direction of the far flow field.

If an incompressible flow is irrotational and inviscid then the force on a body can be found from integrating the momentum flux over the surface of an arbitrarily large circle (in two-dimensions) enclosing the body. Let $\mathbf{ds} = \mathbf{n}ds$ be an element of the circle ds with normal $\mathbf{n}$ directed outwards. Then the force on the body is

$$\mathbf{F} = -\rho U^2 [\oint_C p\,\mathbf{ds} + \oint_C \mathbf{u}(\mathbf{u}\cdot\mathbf{ds})]. \tag{1.55}$$

The pressure can be replaced using Bernouilli's equation

$$p = p_\infty + \frac{1}{2}(1 - q^2), \tag{1.56}$$

where the speed is $q = |\mathbf{u}|$. Suppose that far from the body the velocity is written

$$\mathbf{u} = \mathbf{U} + \mathbf{u}_1,$$

with $\mathbf{U}$ is a constant vector satisfying $|\mathbf{U}| = 1$ and $|\mathbf{u}_1| \ll 1$. Strictly speaking, $\mathbf{u}_1$ should decay such that the integrals over C which follow and which are linear in $\mathbf{u}_1$ all exist and the neglected integrals over C which are quadratic in $\mathbf{u}_1$, vanish.

Then neglecting those integrals which are quadratic in $\mathbf{u}_1$, the force becomes

$$\frac{\mathbf{F}}{\rho U^2} = -\oint_C \mathbf{u}_1(\mathbf{U}\cdot\mathbf{ds}) - \oint_C \mathbf{U}(\mathbf{u}_1\cdot\mathbf{ds}) + \oint_C (\mathbf{U}\cdot\mathbf{u}_1)\mathbf{ds}. \tag{1.57}$$

The second term vanishes since the fluid is incompressible so that

$$\frac{\mathbf{F}}{\rho U^2} = \oint_C (\mathbf{U} \cdot \mathbf{u}_1)\mathbf{ds} - \oint_C \mathbf{u}_1(\mathbf{U} \cdot \mathbf{ds}), \qquad (1.58)$$

or equivalently,

$$\frac{\mathbf{F}}{\rho U^2} = \oint_C (\mathbf{U} \cdot \mathbf{u})\mathbf{ds} - \oint_C \mathbf{u}(\mathbf{U} \cdot \mathbf{ds}). \qquad (1.59)$$

This is a famous result since $\mathbf{F} \cdot \mathbf{U} = 0$ and there is no force in the direction of the flow, known as d'Alembert's paradox. The only force which can arise in an inviscid irrotational flow about a finite body must be at right angles to the flow direction and is called a lift force.

1.10 Summary and overview

In this brief description of both asymptotic notation and fluid mechanics we have been concerned primarily with setting out some of the notation which will be used in the rest of this book and secondarily with describing material taken for granted through the rest of this work. We have indicated the central role which the stream-function plays in two-dimensional flow and that a fundamental objective is to determine asymptotic expansions for the stream-function for large Reynolds number. Although it is the physical velocities and pressure which are the true objectives in fluid mechanics by calculating the stream-function we can obtain the velocities by differentiation and the pressure by a further integration. Thus it is reasonable to think that determining the stream-function is the key to solving for a two-dimensional flow.

The material we shall study is separated into three parts. In the first part we consider the flow around a flat plate. This is the simplest and best understood boundary layer flow and the simplification is not just that the geometry is straightforward, but also that separation does not occur. Consequently it is also the flow for which higher order boundary layer theory is most developed. In this part we look at higher order boundary layer theory for a semi-infinite plate and see that there are unresolved technical difficulties in developing that theory. Then a finite plate is studied and a difficulty arises at the trailing edge of the plate. That difficulty is largely resolved by the introduction of a triple deck structure at the trailing edge which does not assume that the pressure is specified but rather the pressure emerges through an interaction between flow in the various parts of the flow. We examine some of the detail of that structure and numerical methods for solving the resultant non-linear interactive boundary layer equations.

In the second part we consider a steady separated external flow, focusing on a particular example of flow about a circular cylinder. Here the range of technical difficulties is greater than for flow around a flat plate and we begin by examining how the flow around the cylinder and near separation was modelled before turning to attempts to provide a solution using interactive boundary layer theory. There have been many different attempts to model separated flow and a substantial objective in this section is to bring out an historical view of this problem, showing why it was so difficult and why it occupied some of the most able theoretical fluid dynamicists for most of the twentieth century. Even now there remain some unsatisfactory loose ends in our understanding and we try to draw those out too.

In the third part we consider channel flows where there is quite a wide range of understanding. The range of problems which can be studied is broad and of engineering importance. It will emerge that there are significant differences between flow in symmetric channels and flow in asymmetric channels and some surprising predictions come out of interactive boundary layer theory. In this area too there remain important unresolved problems.

Since this book is intended only to be an introduction to interactive boundary layer theory and because the theory can be so complicated, the depth of treatment has been limited to an introductory level. Older, more historical material is in places studied in some depth, not just because of its own intrinsic interest but in order to focus as clearly as possible on the significance of the results from interactive boundary layer theory in advancing theoretical fluid mechanics.

Part I

The Triple Deck

2

THE BOUNDARY LAYER ON A FLAT PLATE

2.1 Introduction

This chapter and the next are concerned with a simple boundary layer problem of flow past a flat plate aligned with the stream. In the present chapter we will follow developments up to 1969. Boundary layer theory development had followed two avenues, on one hand seeking to describe finite Reynolds number flows over bodies such as cylinders and airfoils, flows which involved separation, whilst on the other hand finding greater analytic success with the geometrically simpler flat plate where separation was absent. Even with the simplification of geometry, the development of an analysis for a flat plate boundary layer saw two periods of activity before 1969.

The first began with Prandtl in 1904 and finished with Goldstein's description during the 1930s of the boundary layer over a semi-infinite and finite flat plate (Prandtl (1904), Blasius (1908), Goldstein (1933a,b)). This period saw the completion of the leading order asymptotic analysis of the boundary layer but without the techniques of matched asymptotic analysis, the development of higher order analysis could not proceed. Equally worrying was the singularity in the leading order solution for the wake behind a finite plate. Here, in order to balance the sudden change in $\frac{\partial u}{\partial y}$ as the flow moved past the trailing edge, a singularity with infinite transverse velocities appeared at the trailing edge.

The second period of development occurred during the 1950s and followed from the greater understanding of asymptotic analysis gained with the development of matched asymptotic analysis. Even so there were false trails before a proper analysis of higher order theory was given by Goldstein (1955), Goldstein (1960) and Imai (1957). Whilst this seemed to complete the analysis for a semi-infinite flat plate, the analysis had reached another stumbling block. Whereas the form of the solution was now understood, an unknown constant appeared in the second order solution. This constant multiplied an eigenfunction of a linearised boundary layer equation and could be attributed to unknown details of the flow field at the leading edge of the plate. Despite the efforts of Van Dyke(1964) and despite the knowledge of the form of the asymptotic expansion in both the boundary layer and the outer potential region, we do not yet know how to determine the unknown constant at this order (or the other unknowns which will appear in higher order terms) and so the analysis cannot proceed further. Of equal concern is that the second phase of understanding was concerned with a semi-infinite plate and the difficulty associated with the trailing edge was not overcome.

In the next chapter we will follow the work of Stewartson (1969) and Messiter (1970) who simultaneously resolved the asymptotic structure of flow near the trailing edge by introducing a "triple deck" expansion; in this chapter we will follow work which lead up to the discovery of the triple deck structure at the trailing edge.

2.2 Semi-infinite plate – Rectangular coordinates

2.2.1 Formulation

The analysis began with the non-dimensional Navier–Stokes equations for velocity (u,v) in the (x,y) directions and the pressure p,

$$u_x + v_y = 0, \tag{2.1}$$

$$uu_x + vu_y = -p_x + R^{-1}\nabla^2 u, \tag{2.2}$$

$$uv_x + vv_y = -p_y + R^{-1}\nabla^2 v. \tag{2.3}$$

To these equations was added the boundary conditions,

$$u \to 1 \text{ as } x \to -\infty, \tag{2.4}$$

and

$$u = v = 0 \text{ on } y = 0, \; x \geq 0. \tag{2.5}$$

To analyse these equations define a stream-function $\psi(x,y)$ such that

$$u = \psi_y, \quad v = -\psi_x, \tag{2.6}$$

when the boundary conditions become

$$\psi \to y \text{ as } x \to -\infty, \tag{2.7}$$

$$\psi = \psi_y = 0 \text{ on } y = 0, \; x \geq 0. \tag{2.8}$$

In order to fix the stream-function take $\psi = 0$ as $x \to -\infty$ on $y = 0$. An asymptotic expansion of (2.1) to (2.3) was sought for $R \to \infty$ by expanding the stream-function,

$$\boxed{\psi \sim \epsilon_0(R)\psi_0(x,y) + \epsilon_1(R)\psi_1(x,y) + \cdots ,} \tag{2.9}$$

where $\{\epsilon_j(R)\}$ was an unknown asymptotic sequence as $R \to \infty$. In order to have $\epsilon_0(R)\psi_0 \sim y$ as $x \to -\infty$ choose $\epsilon_0(R) = 1$ and immediately see that $\psi_0(x,y) = y$ satisfies the equations of motion and the upstream inflow condition but not the boundary condition (2.8) on the plate. At this stage, if (2.9) is substituted into the Navier–Stokes equations then further terms in the expansion will be zero and the failure to satisfy the boundary conditions on the plate apparently cannot be removed.

To overcome this difficulty suppose that the expansion (2.9) holds only far away from the plate, near the plate it is necessary to seek a different expansion. To do this "near the plate" has to be defined more carefully by choosing a length scale $\delta_0(R)$ and an inner coordinate Y with $y = \delta_0(R)Y$ so that "near the plate" can be interpreted as y small but $Y = O(1)$.

Then expand

$$\psi \sim \delta_0(R)\Psi_0(x,Y) + \delta_1(R)\Psi_1(x,y) + \cdots, \qquad (2.10)$$

$$p \sim g_0(R)P_0(x,Y) + g_1(R)P_1(x,Y) + \cdots, \qquad (2.11)$$

where $\{\delta_j(R)\}$ and $\{g_j(R)\}$ are asymptotic sequences as $R \to \infty$. We have slightly jumped ahead in choosing the inner length scaling to be the same as the leading order scaling for the stream-function, $\delta_0(R)$. This has been done so that the expansion (2.10) gives a velocity whose leading term is $O(1)$ as $R \to \infty$. It is also necessary to bear in mind that an expansion such as (2.10) will still fail if the sequence $\{\delta_j \Psi_j\}$ fails to be asymptotic as $R \to \infty$ regardless that the sequence $\{\delta_j\}$ is asymptotic, as for instance is the case when $x \to 0$.

The expansions (2.10) and (2.11) are continued so that the velocities are given by

$$u \sim \frac{\partial \Psi_0}{\partial Y} + \frac{\delta_1}{\delta_0}\frac{\partial \Psi_1}{\partial Y} + \cdots, \qquad (2.12)$$

$$v \sim -\delta_0 \frac{\partial \Psi_0}{\partial x} - \delta_1 \frac{\partial \Psi_1}{\partial x} + \cdots. \qquad (2.13)$$

Substituting these in the Navier–Stokes equations, the x-momentum equation gives (multiplied by $R\delta_0^2(R)$),

$$R\delta_0^2[\Psi_{0Y}\Psi_{0xY} - \Psi_{0x}\Psi_{0YY} + g_0 P_{0x}] - \Psi_{0YYY} + \cdots = 0. \qquad (2.14)$$

The y-momentum equation, multiplied by $R\delta_0$, becomes

$$Rg_0 P_{0Y} - R\delta_0^2[\Psi_{0Y}\Psi_{0xx} - \Psi_{0x}\Psi_{0xY}] + \Psi_{0xYY} + \cdots = 0. \qquad (2.15)$$

The natural scaling which emerges is $R\delta_0^2 = 1$ and $g_0 = 1$ giving the leading order equations

$$\boxed{\Psi_{0Y}\Psi_{0xY} - \Psi_{0x}\Psi_{0YY} + P_{0x} - \Psi_{0YYY} = 0,} \qquad (2.16)$$

and

$$\boxed{P_{0Y} = 0.} \qquad (2.17)$$

The second of these equations shows that the leading order contribution to the pressure is constant across the boundary layer so that $P_0(x,Y) \equiv P_0(x)$.

This initial phase of the expansion can be summarised by: as $R \to \infty$,

$$\psi \sim y + \epsilon_1(R)\psi_1(x,y) + \cdots \quad \text{with} \quad y >> R^{-1/2}, \qquad (2.18)$$

and

$$\psi \sim R^{-1/2}\Psi_0(x,Y) + \delta_1(R)\Psi_1(x,Y) + \cdots \quad \text{with} \quad y = R^{-1/2}Y \sim O(R^{-1/2}). \qquad (2.19)$$

What shortly emerges is that the conditions on y in (2.18) and (2.19) need to be tightened by including an x-dependence. The expansion (2.18) as $y \to 0$ has to be matched to the expansion (2.19) as $Y \to \infty$. In that case if $y = R^{-1/2}Y$ is substituted into the outer expansion (2.18),

$$\psi \sim R^{-1/2}Y + \epsilon_1(R)\psi_1(x, R^{-1/2}Y) + \cdots, \qquad (2.20)$$

or

$$\psi \sim R^{-1/2}Y + \epsilon_1\psi_1(x,0) + R^{-1/2}\epsilon_1 Y \psi_{1Y}(x,0) + \cdots. \qquad (2.21)$$

The first two terms of this expression cannot be distinguished from an asymptotic viewpoint, but the third and subsequent terms can be discarded as being asymptotically smaller than either of the first two terms. The expansion (2.21) has now to be compared with the expansion of the inner approximation (2.19) as $Y \to \infty$. For the inner expansion to match will require

$$\Psi_0(x,Y) \sim Y \quad \text{as} \quad Y \to \infty. \qquad (2.22)$$

It is not possible to say anything further about the size of ϵ_1, later this will be determined by matching the outer solution to the inner solution for Ψ_0.

Equation (2.16) as $Y \to \infty$ shows that the pressure gradient will vanish so there is no loss of generality in taking $P_0(x) = 0$.

2.2.2 Leading order boundary layer solution

We have thus come to a full statement of the problem for Ψ_0:

$$\Psi_{0Y}\Psi_{0xY} - \Psi_{0x}\Psi_{0YY} - \Psi_{0YYY} = 0, \qquad (2.23)$$

together with the boundary conditions $\Psi_0(x,0) = \Psi_{0Y}(x,0) = 0$, and $\Psi_0(x,Y) \sim Y$ as $Y \to \infty$, the conditions being for $x > 0$.

We can try to find solutions of equation (2.23) which are of the form

$$\Psi_0(x,Y) = x^a F(Y x^b) \equiv x^a F(s), \qquad (2.24)$$

where $s = Y x^b$. Using this form of solution it is easy to see that the non-linear or inertia terms in (2.23) will be $x^{2a+2b-1} \times \text{function}(s)$ whilst the viscous term will be $x^{a+3b} \times$ function(s) so the partial differential equation (2.23) will reduce to an ordinary differential equation in s provided $a - b = 1$. Hence there are many combinations of a and b which could be used, in this case since $\Psi_0 \sim Y$ as $Y \to \infty$, we will need $F(s) \sim s$ as $s \to \infty$ and this implies $a + b = 0$ in addition to $a - b = 1$, so that we must choose $a = -b = 1/2$. Thus we write the stream-function as

$$\Psi_0(x, Y) = \sqrt{2x} F_0\left(\frac{Y}{\sqrt{2x}}\right) \equiv \sqrt{2x} F_0(s), \qquad (2.25)$$

where we have $s = Y/\sqrt{2x}$ and F_0 satisfies

$$\boxed{F_0''' + F_0 F_0'' = 0,} \qquad (2.26)$$

together with boundary conditions $F_0(0) = F_0'(0) = 0$ and $F_0'(s) \to s$ as $s \to \infty$. The factors of $\sqrt{2}$ which appear in (2.25) are somewhat arbitrary, different authors have used slightly differing notation, in this case we are following Van Dyke (1964).

This equation is usually solved numerically, the division of boundary conditions between $s = 0$ and $s \to \infty$ being dealt with by a simple transformation so as to have a problem with all the boundary conditions at $s = 0$.

If the function G_0 satisfies

$$G_0''' + G_0 G_0'' = 0, \qquad (2.27)$$

and boundary conditions $G_0(0) = G_0'(0) = 0$ and $G_0''(0) = 1$, then if $G_0'(s) \to \alpha_0$ as $s \to \infty$, the required function F_0 is given by

$$F_0(s) = \frac{1}{\sqrt{\alpha_0}} G_0\left(\frac{s}{\sqrt{\alpha_0}}\right). \qquad (2.28)$$

It is straightforward to show that for small s,

$$G_0(s) \sim \frac{1}{2} s^2 - \frac{1}{5!} s^5 + \frac{11}{8!} s^8 + \cdots, \qquad (2.29)$$

so that

$$\boxed{F_0(s) \sim \frac{1}{2} \alpha_0^{-3/2} s^2 - \frac{1}{5!} \alpha_0^{-3} s^5 + \frac{11}{8!} \alpha_0^{-9/2} s^8 + \cdots \quad \text{as} \quad s \to 0.} \qquad (2.30)$$

Weyl (1942) considered the series solution of equations of the form

$$G^{(\nu+1)} + 2GG^{(\nu)} = 0, \qquad (2.31)$$

with boundary conditions $G(0) = G'(0) = \ldots = G^{(\nu-1)} = 0$, $G^{(\nu)} = 1$ and showed that the coefficients of a power series solution

$$G(s) = \frac{s^\nu}{\nu!} \sum_0^\infty (-1)^n c_n s^{n(\nu+1)}, \qquad (2.32)$$

satisfy

$$\left\{\frac{2\nu!}{(2\nu+1)!}\right\}^n \leq c_n \leq \left\{\frac{2}{(\nu+1)(\nu+1)!}\right\}^n. \qquad (2.33)$$

This series converges if $c_n s^{n(\nu+1)} \to 0$ as $n \to \infty$. Thus convergence is assured if

$$\frac{2|s|^{\nu+1}}{(\nu+1)(\nu+1)!} < 1, \qquad (2.34)$$

and divergence will occur if

$$\frac{2|s|^{\nu+1}}{(2\nu+1)!} > 1. \qquad (2.35)$$

Hence the radius of convergence ρ of the series satisfies

$$\frac{1}{2}(\nu+1)(\nu+1)! \leq \rho^{\nu+1} < \frac{1}{2}(\nu+1)\ldots(2\nu+1). \qquad (2.36)$$

In the case where $\nu = 2$ this gives

$$9^{1/3} \leq \rho < 30^{1/3} \qquad (2.37)$$

and for (2.26) with $G = 2^{-1/3} G_0(2^{2/3} s)$, the radius of convergence, ρ_0 of a series solution for G_0 will satisfy

$$\frac{9^{\frac{1}{3}}}{4} \leq \rho_0 < \frac{30^{\frac{1}{3}}}{4}. \qquad (2.38)$$

This range does not cover the region of the boundary layer for which a solution is needed (which extends closer to $s = 5$) so that the series solution is a useful analytic tool but is not suitable for practical computations.

The asymptotic behaviour of F_0 for large s is also of great importance and has been widely studied, starting with Blasius (1908). If we write $F_0 \sim s - g_0(s)$ where $g_0 = o(s)$ as $s \to \infty$, then the equation for g_0 is

$$g_0''' + s g_0'' - g_0 g_0'' = 0, \qquad (2.39)$$

with boundary conditions $g_0(0) = g_0'(0) = 0$. Examination of asymptotic balances in the differential equation (2.39) when s is large shows that

$$g_0''' + s g_0'' \sim 0, \quad \text{as } s \to \infty, \qquad (2.40)$$

so that

$$g_0'' \sim \exp(-s^2/2) \quad \text{as } s \to \infty, \qquad (2.41)$$

where there will be a constant multiplier attached to the solution for g_0''. As $g_0 = o(s)$ for large s, we must have $g_0(s) \sim$ constant+exponentially small terms.

Semi-infinite plate – Rectangular coordinates

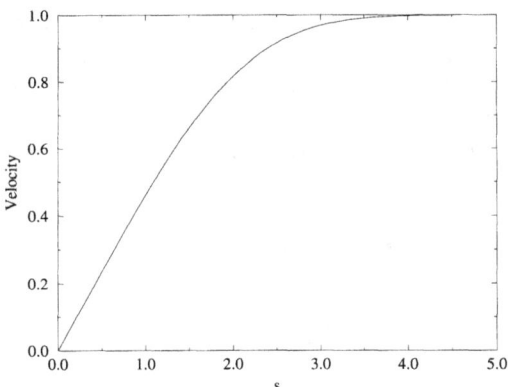

FIG. 2.1. Calculated boundary layer velocity across a boundary layer.

The most illuminating way to look at the limit as $s \to \infty$ of the function g_0 is to follow Kaplun (1956) and write

$$g_0(s) = \int_0^s [1 - F_0'(s)]\,ds, \qquad (2.42)$$

so that as $s \to \infty$,

$$g_0(s) \sim \beta_0 \equiv \int_0^\infty [1 - F_0'(s)]\,ds \qquad (2.43)$$

The constant β_0 must be evaluated numerically. Finally we can write

$$\boxed{F_0(s) \sim s - \beta_0 + O(\exp(-s^2/2)) \quad \text{as} \quad s \to \infty.} \qquad (2.44)$$

The solution to (2.27) obtained from a fourth order Runge–Kutta scheme between $s = 0$ and $s = 5$ with step size $\Delta s = .01$ gives

$$\alpha_0 = 1.65519\cdots. \qquad (2.45)$$

Using this solution to obtain F_0 and calculate β_0 in (2.43) gives

$$\beta_0 = 1.21649\cdots. \qquad (2.46)$$

The non-dimensional velocity in the boundary layer, F_0' is illustrated in Figure 2.1.

2.2.3 Second order outer flow

Having determined the leading order term the ideas of matched asymptotic analysis can be used to evaluate further terms in the expansions (2.18) and (2.19).

The leading order outer term was just uniform flow. This term failed to satisfy the boundary conditions on the plate and so it was necessary to calculate the leading order term in an expansion valid near the plate. Now an investigation of how the inner expansion develops far from the plate can be used to match that behaviour to an improved outer expansion,

$$\psi \sim y + \epsilon_1(R)\psi_1(x,y), \qquad (2.47)$$

where $\epsilon_1(R)$ is not known. We know that $\psi_1 \to 0$ at infinity but we do not have *inviscid* boundary conditions on the plate. In the two term inner expansion,

$$\psi \sim R^{-1/2}\sqrt{2x}F_0(\frac{Y}{\sqrt{2x}}) + \delta_1(R)\Psi_1(x,Y), \qquad (2.48)$$

where $\delta_1 = o(R^{-1/2})$, allow $Y \to \infty$ and use the expansion for F_0, (2.44), then

$$\psi \sim \sqrt{\frac{2x}{R}}(\frac{Y}{\sqrt{2x}} - \beta_0) + O(\delta_1), \qquad (2.49)$$

or with $Y \sim R^{1/2}y$,

$$\psi \sim y - R^{-1/2}\beta_0\sqrt{2x} + o(R^{-1/2}) \text{ as } Y \to \infty, \qquad (2.50)$$

where exponentially small terms have been neglected. This will match with the outer solution (2.47) when $y \to 0$ if ϵ_1 is chosen by $\epsilon_1(R) = R^{-1/2}$ and the stream-function satisfies an *inviscid* boundary condition

$$\psi_1(x,0) = -\beta_0\sqrt{2x}. \qquad (2.51)$$

Substituting (2.47) into the Navier–Stokes equations shows that ψ_1 also satisfies a potential equation and so the solution (which comes most easily from complex variable theory) is

$$\psi_1(x,y) = -\beta_0\Re\sqrt{2(x+iy)}. \qquad (2.52)$$

Note that to the outer flow, the boundary layer to this order has exactly the same effect as inviscid flow about a parabolic body $y \sim \beta_0\sqrt{2x/R}$. In Figure 2.2 we show the body detail for three Reynolds numbers, $R = 500, 1000, 10000$. Note that on the scale shown detail near the leading edge is lost, a much magnified view of the case $R = 500$ is shown in Figure 2.3 where it is clear that the 'nose' of the parabolic body lies upstream of the leading edge: the complex potential of the leading terms in the outer flow is

$$w = z - i\beta_0\sqrt{\frac{2}{R}}z^{1/2}. \qquad (2.53)$$

Hence the complex velocity $u - iv$ is

$$\frac{dw}{dz} = 1 - \frac{i\beta_0}{\sqrt{2R}}z^{-1/2}, \qquad (2.54)$$

and the stagnation point is at $z = -\beta_0^2/(2R)$: in the case where $R = 500$, this is $z = -0.00148$.

Semi-infinite plate – Rectangular coordinates

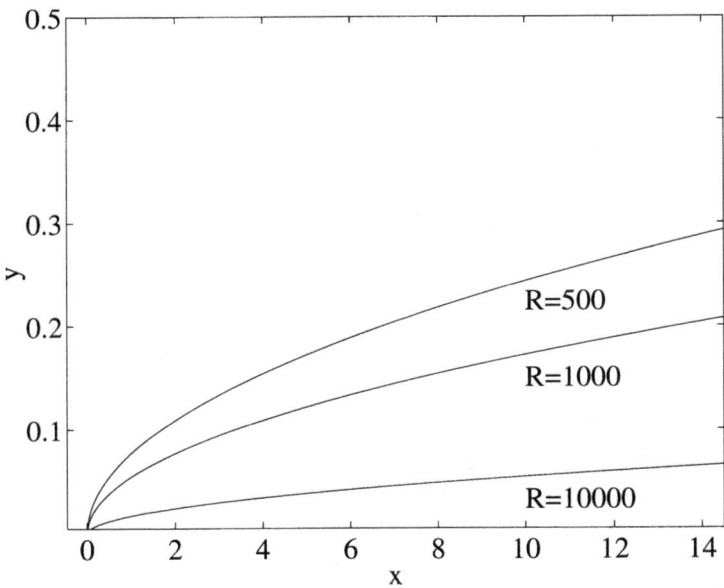

FIG. 2.2. Curves showing the parabolic body which is 'seen' by the outer flow for $R = 500, 1000, 10000$.

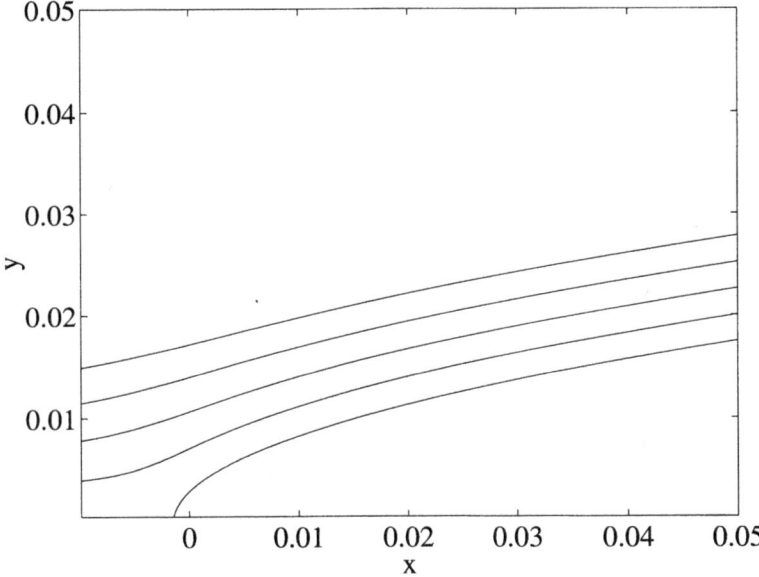

FIG. 2.3. Detail of outer potential flow field near the leading edge of plate for $R = 500$.

Having obtained two terms of the outer expansion the next step is to examine whether the limit of the outer expansion as $y \to 0$ can be used to improve the inner expansion.

2.2.4 Second order inner expansion

Here the analysis starts with the three term outer expansion and as each term of the outer expansion will represent some form of potential flow, the individual functions will be analytic as $y \to 0$, so that formally expanding for small $y = R^{-1/2}Y$,

$$\psi \sim R^{-1/2}Y + R^{-1/2}\psi_1(x,0) + R^{-1}Y\psi_{1y}(x,0) + \epsilon_2 \psi_2(x,0) + \cdots . \quad (2.55)$$

At this point the size of the second order outer term is not known, only that $\epsilon_2 = o(R^{-1/2})$. The first two terms in (2.55) are the ones we have just studied, the third suggests a term in the inner expansion of order R^{-1} and the fourth term may be of the same order as the third term or it may be different, that cannot be judged at this point.

The expansion (2.55) implies that the inner expansion should be continued to a second term by

$$\psi \sim R^{-1/2}\Psi_0(x,Y) + R^{-1}\Psi_1(x,Y). \quad (2.56)$$

If this expansion is substituted into the Navier–Stokes equations and higher order terms neglected, then Ψ_1 must satisfy

$$\Psi_{1YYYY} + \Psi_{0x}\Psi_{1YYY} + \Psi_{0YYY}\Psi_{1x} - \Psi_{0Y}\Psi_{1xYY} - \Psi_{0xYY}\Psi_{1Y} = 0. \quad (2.57)$$

This equation can be integrated once with respect to Y. The constant of integration (in this case a function of x) can be seen to vanish because of the properties of Ψ_0 and Ψ_1 as $Y \to \infty$, so that

$$\Psi_{1YYY} + \Psi_{0x}\Psi_{1YY} + \Psi_{0YY}\Psi_{1x} - \Psi_{0Y}\Psi_{1xY} - \Psi_{0xY}\Psi_{1Y} = 0. \quad (2.58)$$

If we now substitute $\Psi_0 = \sqrt{2x} F_0(s)$ and $\Psi_1 = x^a F_1(s)$ where a is unknown, then

$$F_1''' + F_0 F_1'' + (1 - 2a) F_0' F_1' + 2a F_0'' F_1 = 0. \quad (2.59)$$

The boundary conditions for F_1 on the plate are the same as for F_0, namely $F_1(0) = F_1'(0) = 0$ in order that the velocities will vanish on the plate. If we return to the outer expansion (2.55) and use (2.52) then in the case of a flat plate $\psi_{1y}(x,0) \equiv 0$, so even allowing $\epsilon_2 = O(R^{-1})$ it is still necessary that $F_1 = o(Y)$ as $Y \to \infty$ (in fact to be consistent with (2.55) F_1 needs to tend to at most a constant as $Y \to \infty$).

Solutions of equations of the form (2.59) satisfying homogeneous boundary conditions at $s = 0$ and with values $o(Y)$ as $Y \to \infty$ have been calculated as an eigenvalue problem by Libby and Fox (1963) and Libby (1965). Generally the eigenvalues a and eigensolutions have to be found numerically but the first (in descending values of a) can be determined analytically and occurs for $a = -1/2$ when

$$F_1(s) = F_0(s) - sF_0'(s). \tag{2.60}$$

Is it the case that ϵ_2 should be chosen as $\epsilon_2 = R^{-1}$ and an unknown multiple of F_1 allowed to enter the expansion at this stage? The answer to this is quite subtle and to understand how this question is answered we have to go back to the Navier–Stokes equations, written in a compact form,

$$R^{-1}\nabla^4\psi + \frac{\partial(\psi, \nabla^2\psi)}{\partial(x,y)} = 0. \tag{2.61}$$

If magnified coordinates $(\mathcal{X}, \mathcal{Y}) = R(x, y)$ are defined and a magnified stream-function $\Upsilon = R\psi$ then the Reynolds number disappears from the stream-function equation which becomes,

$$\nabla_1^4\Upsilon + \frac{\partial(\Upsilon, \nabla_1^2\Upsilon)}{\partial(\mathcal{X}, \mathcal{Y})} = 0, \tag{2.62}$$

where ∇_1^2 is the Laplace operator in the magnified coordinates. In the case of an arbitrary flow about a general body, the Reynolds number would reappear in the boundary conditions on the body as well as in the far field conditions, but for the case of uniform flow about a semi-infinite flat plate, the boundary conditions are $\Upsilon_\mathcal{X} = \Upsilon_\mathcal{Y} = 0$ on the plate $\mathcal{Y} = 0$, $\mathcal{X} > 0$ and $\Upsilon \to \mathcal{Y}$ far from the plate. Thus the solution of the original flow field must be able to be found from the solution of this new parameter-less problem when $\mathcal{X}$ and $\mathcal{Y}$ are large. Hence we should expect the asymptotic expansion which we have been developing to be consistent with this new description whereby $\Upsilon = \Upsilon(\mathcal{X}, \mathcal{Y})$.

If we take the terms already found thus far in the outer and inner expansions, they can be written

$$R\psi_{\text{outer}} \sim \mathcal{Y} - \beta_0 \Re\sqrt{2(\mathcal{X} + i\mathcal{Y})}, \tag{2.63}$$

and

$$R\psi_{\text{inner}} \sim \sqrt{2\mathcal{X}} F_0(\frac{\mathcal{Y}}{\sqrt{2\mathcal{X}}}), \tag{2.64}$$

and so they each fall into this new scheme. If a term proportional to the first eigen-solution of (2.59) was allowed it would lead to an expansion,

$$R\psi_{\text{inner}} \sim \sqrt{2\mathcal{X}} F_0(\frac{\mathcal{Y}}{\sqrt{2\mathcal{X}}}) + \frac{R^{1/2}}{\sqrt{2\mathcal{X}}} F_1(\frac{\mathcal{Y}}{\sqrt{2\mathcal{X}}}), \tag{2.65}$$

it could only be accepted if F_1 itself were $O(R^{-1/2})$ to remove the factor of $R^{1/2}$ in the last term of (2.65). This of course would make the eigen-solution part of a higher order contribution to the expansion, of order $R^{-3/2}$. Thus it was deduced that there could be no term in the inner expansion of order R^{-1} and that in the outer expansion, $\epsilon_2 = o(R^{-1})$. This brought matters to something of a crossroad in seeking an expansion of this boundary layer flow. Two terms had been found for both the inner and the outer expansion (the second inner term is zero by virtue of $\psi_{1y}(x,0) \equiv 0$) using rectangular coordinates. However, the form of the outer perturbation (2.52), was a very strong indicator that the use of rectangular coordinates was perhaps misguided and that better progress might be made using parabolic coordinates. This was indeed the case and the next term in the expansion was more easily calculated in parabolic coordinates although it would have been perfectly possible to soldier on in rectangular coordinates. Thus to some extent the analysis had to be taken a step back and began again, having understood a great deal about the structure of the expansion. In particular it is a key point that instead of expanding (2.61) for large Reynolds number, we can equivalently expand (2.62) for large $\mathcal{X}$ and large Υ and expect to see a boundary layer structure develop. The importance of Υ being large is that if $\mathcal{X}$ and $\mathcal{Y}$ are the same size, then the right-hand side of (2.62) dominates and essentially each term in an expansion for Υ will satisfy potential flow (assuming an irrotational upstream flow), whereas if $\mathcal{Y}$ is asymptotically of order $\sqrt{\mathcal{X}}$ then (2.62) will simplify to a boundary layer form of equation.

2.3 Semi-infinite plate - Parabolic coordinates

2.3.1 Definition

Define parabolic coordinates by

$$\mathcal{Z} = \mathcal{X} + i\mathcal{Y} = \frac{1}{2}(\xi + i\eta)^2 = \frac{1}{2}\zeta^2. \qquad (2.66)$$

so that in our original Cartesian coordinates, $2Rx = \xi^2 - \eta^2$ and $Ry = \xi\eta$. Thus for large x we have the approximation

$$\xi \sim \sqrt{2Rx}, \qquad (2.67)$$

$$\eta \sim \sqrt{R}y/(\sqrt{2x}) \equiv Y/(\sqrt{2x}) = s. \qquad (2.68)$$

We need to define a branch cut in the $\mathcal{Z}$ plane so that $0 \leq \arg\mathcal{Z} \leq 2\pi$ and $0 \leq \arg\zeta \leq \pi$. This has the effect of opening the plate up so that the top and bottom sides of the plate become $\eta = 0$ and $\xi > 0$ and $\xi < 0$ respectively (see diagram). For the purposes of the outer field we can define a complex potential w whose imaginary part is the stream-function, $\Upsilon = \Im w$. Each term of an outer flow expansion will be harmonic and so w will be an analytic function of ζ. The upstream flow is given by $w = \zeta^2/2$.

Next we must consider the transformed Navier–Stokes equations; written in parabolic coordinates the Laplace operator becomes

$$\nabla_1^2 = \frac{1}{\xi^2 + \eta^2}\left(\frac{\partial^2}{\partial\xi^2} + \frac{\partial^2}{\partial\eta^2}\right) \equiv \frac{1}{\xi^2 + \eta^2}\nabla_2^2, \qquad (2.69)$$

Semi-infinite plate - Parabolic coordinates

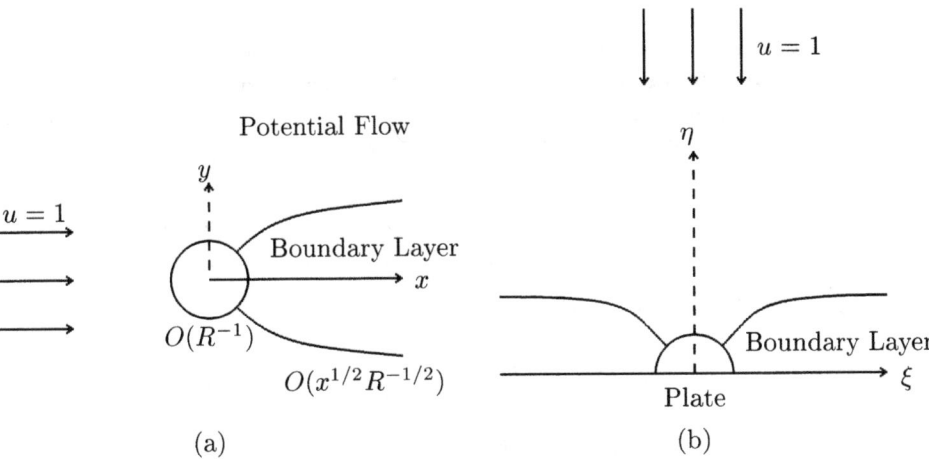

FIG. 2.4. Rectangular (a) and parabolic (b) coordinate systems. Note that in (b), to leading order, the boundary layer has constant thickness for large ξ.

where we use ∇_2^2 for the Laplace operator in parabolic coordinates. Accordingly

$$\nabla_1^4 = \frac{1}{(\xi^2+\eta^2)^2}\nabla_2^4 + \frac{1}{(\xi^2+\eta^2)^3}\left(4 - 2\xi\frac{\partial}{\partial\xi} - 2\eta\frac{\partial}{\partial\eta}\right)\nabla_2^2. \tag{2.70}$$

If we now turn to the non-linear terms, we can write

$$\frac{\partial(\Upsilon, \nabla_1^2\Upsilon)}{\partial(\mathcal{X},\mathcal{Y})} = \frac{\partial(\Upsilon, \nabla_1^2\Upsilon)}{\partial(\xi,\eta)}\frac{\partial(\xi,\eta)}{\partial(\mathcal{X},\mathcal{Y})} \tag{2.71}$$

$$= \frac{1}{\xi^2+\eta^2}\frac{\partial(\Upsilon, \nabla_1^2\Upsilon)}{\partial(\xi,\eta)} \tag{2.72}$$

$$= \frac{1}{(\xi^2+\eta^2)^2}\frac{\partial(\Upsilon, \nabla_2^2\Upsilon)}{\partial(\xi,\eta)} + \frac{2}{(\xi^2+\eta^2)^3}\left(\xi\frac{\partial\Upsilon}{\partial\eta} - \eta\frac{\partial\Upsilon}{\partial\xi}\right)\nabla_2^2\Upsilon. \tag{2.73}$$

so that the Navier–Stokes equations become

$$\nabla_2^4\Upsilon + \frac{\partial(\Upsilon, \nabla_2^2\Upsilon)}{\partial(\xi,\eta)} + V(\Upsilon,\xi,\eta)\nabla_2^2\Upsilon = 0, \tag{2.74}$$

where

$$V = \frac{1}{\xi^2+\eta^2}\left(4(1 - \xi\frac{\partial}{\partial\xi} - \eta\frac{\partial}{\partial\eta}) + 2(\xi\frac{\partial\Upsilon}{\partial\eta} - \eta\frac{\partial\Upsilon}{\partial\xi})\right). \tag{2.75}$$

2.3.2 Power series expansion for large ξ

Now expand (2.74) for large Υ and ξ, assuming that $\frac{\partial}{\partial\eta} \gg \frac{\partial}{\partial\xi}$ when we are close to the plate. This of course leads to a very similar problem to that already studied in rectangular coordinates. The question which needs most thought is the nature of the expansion of Υ for large ξ. Symmetry dictates that Υ should be an odd function of ξ and the far field is to leading order $\Upsilon \sim \xi\eta$. It is perhaps most natural to seek an expansion in descending powers of ξ,

$$\Upsilon \sim \xi f_0(\eta) + \text{sgn}\xi f_1(\eta) + \frac{1}{\xi}f_2(\eta) + \cdots, \quad \xi \gg 1. \tag{2.76}$$

If this is substituted into (2.74) we obtain successively,

$$(f_0''' + f_0 f_0'')' = 0, \tag{2.77}$$

$$(f_1''' + f_0 f_1'' + f_0' f_1')' = 0, \tag{2.78}$$

$$(f_2''' + f_0 f_2'' + 2 f_0' f_2' - f_0'' f_2)' = \frac{\mathrm{d}}{\mathrm{d}\eta}(f_0 - \eta f_0')^2, \tag{2.79}$$

where the boundary conditions on the plate are $f_j(0) = f_j'(0) = 0$ for $j = 0, 1, 2, \ldots$. As in the case of rectangular coordinates, $f_0(\eta) \sim \eta$ as $\eta \to \infty$, so that f_0 is identical to F_0 and has the same properties, $f_0 \sim \eta - \beta_0 +$ exponentially small terms as $\eta \to \infty$.

It is worth recalling that when we tried to find a term in the stream-function expansion of order $O(R^{-1})$ in rectangular coordinates, we found that no such term should be present in the expansion because it appeared to lead to an inconsistency in the asymptotic structure of the solution. In this case with parabolic coordinates, the equivalent term is embodied by f_1. Do we have grounds to conclude that f_1 should be zero? If we briefly return to the far field condition $\Upsilon \sim \xi\eta$ as $\eta \to \infty$, then using the first two terms of (2.76), the far field condition can be written

$$\frac{\xi f_0(\eta) + \text{sgn}\xi f_1(\eta) - \xi\eta}{\xi\eta} \to 0 \text{ as } \eta \to \infty. \tag{2.80}$$

This shows that what is needed is $f_1 = o(\eta)$ as $\eta \to \infty$. Turning to the equation (2.77) and integrating twice (introducing constants a_1 and b_1) then

$$f_1'' + f_0 f_1' = a_1\eta + b_1, \tag{2.81}$$

so that for $\eta \to \infty$

$$f_1'' + \eta f_1' \sim a_1\eta + b_1. \tag{2.82}$$

The homogeneous solutions will be $f_1 \sim \beta_1$, a constant, and $f_1 \sim \exp(-\eta^2/2)$. The particular integral terms must vanish by virtue of $f_1 = o(\eta)$ so that it is necessary for $a_1 = b_1 = 0$. The question now arises as to whether it is permissable for the constant β_1 to be non-zero. Certainly a non-zero value will satisfy the far field condition $\Upsilon \sim \xi\eta$. Fortunately this "radiation type" condition is not sufficient to define the outer flow field, and we have to recall that the outer flow field is a potential flow. So for large ζ, the stream function Υ must be the imaginary part of an analytic function $w(\zeta)$ (so that we are guaranteed that Υ is harmonic). In this case we cannot find any such complex potential. Thus we have to conclude that $f_1 \equiv 0$ and come to the same conclusion as before, that the expansion should start

$$\Upsilon \sim \xi f_0(\eta) + \frac{1}{\xi} f_2(\eta) + \cdots . \tag{2.83}$$

Of course it has not been shown that this is the correct expansion, only that there is no $O(1)$ term. We will shortly find that in fact this is not the correct expansion. We can also note that the behaviour of f_0 as $\eta \to \infty$ and (2.83) imply that the outer potential flow must have an expansion beginning

$$w(\zeta) \sim \frac{1}{2}\zeta^2 - i\beta_0\zeta + \cdots \quad \text{as } |\zeta| \to \infty. \tag{2.84}$$

2.3.3 Failure of power series expansion

To understand the inadequacy of (2.83) it is necessary understand more about the solution of (2.79) which defines the function f_2. If (2.79) is integrated once, introducing a constant a_2 then

$$f_2''' + f_0 f_2'' + 2f_0' f_2' - f_0'' f_2 = a_2 + (f_0 - \eta f_0')^2. \tag{2.85}$$

Consider solutions when $\eta \ll 1$. If we use $f_0 \sim \alpha_0^{-3/2}\eta^2/2$ and try $f_2 \sim k\eta^p$, then

$$kp(p-1)(p-2)\eta^{p-3} + \frac{k\alpha_0^{-3/2}}{2}(p(p-1) + 4p - 2)\eta^p \sim a_2, \tag{2.86}$$

and of course only the first term on the left-hand side is relevant to an asymptotic balance, showing that the particular integral will have a power series beginning $a_2\eta^3/6$, whilst the three solutions of the homogeneous equation will be proportional to power series beginning 1, η and η^2. Following Goldstein, the four solutions can be labelled $f_0^{(1)},...,f_0^{(4)}$ such that

$$f_2^{(1)} \sim 1, \quad f_2^{(2)} \sim \eta, \quad f_2^{(3)} \sim \eta^2, \quad f_2^{(4)} \sim a_2\eta^3/3, \quad \eta \ll 1. \tag{2.87}$$

Clearly only the last two solutions can satisfy the boundary conditions $f_2(0) = f_2'(0) = 0$ on the plate. Thus the solution will involve two unknown constants, a_2 and an unknown multiplier to $f_2^{(3)}$, which has to be determined from the conditions as $\eta \to \infty$. It is also straight-forward to show that $\eta f_0' - f_0$ is a solution of the homogeneous form of (2.85) with value and derivative zero at $\eta = 0$, so that

$$f_2^{(3)} = \eta f_0' - f_0. \tag{2.88}$$

Next consider the behaviour of solutions for large η. From the asymptotic behaviour of f_0 it is necessary that

$$f_2''' + \eta f_2'' + 2f_2' \sim a_2 + \beta_0^2. \tag{2.89}$$

Four types of asymptotic behaviour can be recognised, not necessarily corresponding to $f_2^{(1)}$ to $f_2^{(4)}$, rather each of the solutions $f_2^{(j)}$ will have an asymptotic expansion which could be a mixture of the four solutions to (2.89). Examination of homogeneous solutions to (2.89) shows that there can be balances

$$\text{(a)} \qquad f_2''' + \eta f_2 \sim 0, \tag{2.90}$$

with solution $f'' \sim \exp(-\eta^2/2)$,

$$\text{(b)} \qquad \eta f_2'' + 2f_2' \sim 0, \tag{2.91}$$

with solution $f_2' \sim \eta^{-1}$,

$$\text{(c)} \qquad f_2'' + 2f_2' \sim 0, \tag{2.92}$$

with a constant as solution.

The particular integral, $f_2^{(4)}$, will have $f_2^{(4)} \sim \frac{1}{2}(a_2+\beta_0^2)\eta$ [from (2.89)]. Hence the most general asymptotic behaviour would be

$$f_2^{(3)} \sim c_{3,1} + c_{3,2}\eta^{-1} + c_{3,3}\exp(-\eta^2/2), \tag{2.93}$$

and

$$f_2^{(4)} \sim \frac{1}{2}(a_2 + \beta_0^2)\eta + c_{4,1} + c_{4,2}\eta^{-1} + c_{4,3}\exp(-\eta^2/2), \tag{2.94}$$

where constants $c_{i,j}$ can be regarded as known from numerical solution of (2.85) and its homogeneous form given the initial behaviour (2.87). The general solution for f_2 is then an arbitrary multiple of $f_2^{(3)}$, say A_2, added to $f_2^{(4)}$, so that the solution has two unknowns, a_2 and A_2, satisfies the boundary conditions at $\eta = 0$ but has yet to be matched to the far field when $\eta \to \infty$. It has to be decided whether it is possible to determine the constants a_2 and A_2 so as to match the boundary layer to a potential flow. The answer is that it cannot be done, but it is interesting to see how this comes about.

Since $f_2^{(3)} = \eta f_0' - f_0$, it must be the case that $c_{3,1} = \beta_0$ and $c_{3,2} = 0$. It appears not possible to determine closed expressions for $c_{4,1}$ and $c_{4,2}$ in terms of

a_2 and A_2 but for the present they can be regarded as computable functions of a_2 and A_2.

Next observe that it is possible to apply various forms for the far field boundary condition, both

$$f_2 = o(\eta) \text{ as } \eta \to \infty, \tag{2.95}$$

and

$$f_2 \to 0 \text{ as } \eta \to \infty, \tag{2.96}$$

seeming equally plausible. In either case $a_2 + \beta_0^2 = 0$ is needed to remove the linear term in η so that a_2 would be fixed and only A_2 remains to be determined.

It is straight-forward to evaluate f_2 numerically using Runge–Kutta integration with initial conditions $f_2(0) = f_2'(0) = 0$ and $f_2''(0) = \alpha_0^{-3/2} A_2$. To calculate $c_{4,2}$ the same idea can be applied as Kaplun used to calculate β_0, (2.43).

Suppose $f_2(\eta) = f_2(\infty) + B(\eta)/\eta$, where $B(0) = 0$ (the constraint on B as $\eta \to 0$ is $B(\eta) \sim -f_2(\infty)\eta + O(\eta^3)$ to satisfy the boundary conditions on f_2). Then it is simple to see that $B'(\eta) = \eta f_2'(\eta) + f_2(\eta) - f_2(\infty)$ and so that

$$c_{4,2} = B(\infty) = \int_0^\infty [\eta f_2'(\eta) + f_2(\eta) - f_2(\infty)] d\eta. \tag{2.97}$$

Using this method with A_2 varied between 3 and 7 then the graphs shown in Figure 2.3 are obtained for the limiting values $f_2 \sim \beta_0 A_2 + c_{4,1}$ and $c_{4,2}$. It can be seen that the condition $f_2 \to 0$ can be satisfied for a suitable choice of A_2 but that it is not possible to simultaneously enforce $c_{4,2} = 0$. This presents a difficulty because it is not physically or mathematically plausible for the boundary layer far field to have this algebraic form of decay.

On the mathematical side, this decay would give a contribution to Υ like $(\xi\eta)^{-1} = \mathcal{Y}^{-1}$ as the solution passes from the boundary layer to the potential flow region and there is no potential function which behaves like $\mathcal{Y}^{-1}$ as the positive real axis is approached. Thus this form of decay cannot be matched to any form of potential outer flow.

On the physical side, there is an assumption that the vorticity must decay exponentially in the potential flow region. This is because vorticity changes are transmitted **across** streamlines by a diffusive mechanism and once the potential flow region is reached the vorticity should decay as a solution of a diffusion equation: that is exponentially.

Thus if $c_{4,2}$ cannot be eliminated then the expansion (2.83) must fail. Before accepting this conclusion Goldstein tried one further idea. Could he abandon far field conditions such as (2.95) or (2.96) and instead determine a_2 so as to eliminate $c_{4,2}$? Leaving aside the computational question of whether there is a value of a_2 for which $c_{4,2} = 0$, suppose it is possible to have

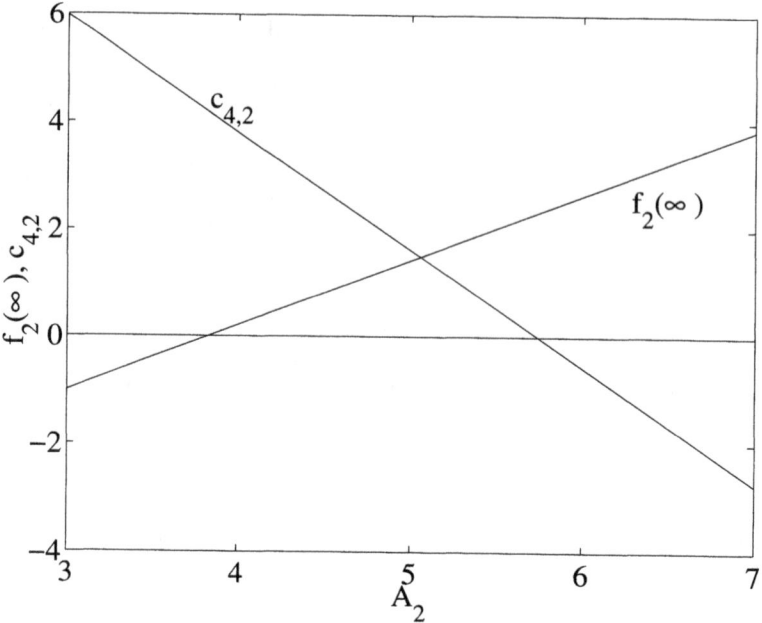

FIG. 2.5. Limiting values for f_2 and $c_{4,2}$ calculated from solutions of (2.85) as the parameter A_2 varies.

$$f_2 \sim \alpha_2 \eta - \beta_2 + \text{exponentially small terms, as } \eta \to \infty. \qquad (2.98)$$

Could this be matched to a potential outer flow? In that case a complex potential $w(\zeta)$ would be needed whose imaginary part was asymptotic to

$$\frac{\alpha_2 \eta}{\xi} - \frac{\beta_2}{\xi}, \qquad (2.99)$$

as the plate was approached. The second term could be accommodated by a stream-function $\beta_2 \xi/(\xi^2 + \eta^2)$ corresponding to a complex potential $i\beta_2 \zeta^{-1}$. The first term was more difficult. A natural potential would be $\alpha_2 \tan^{-1}(\eta/\xi)$ (with complex potential $\alpha_2 \log \zeta$) but that lead to an unacceptable discontinuity $\pi \alpha_2$ for $\eta \to \infty$ with $\xi = 0^+$ and $\xi = 0^-$. One can try to move the discontinuity around the $\xi - \eta$ plane but it remains a fundamental difficulty and Goldstein had to conclude that the expansion (2.83) failed. Alden (1948) had derived a second order expansion in parabolic coordinates but used the solution with incorrect far field behaviour.

2.3.4 Goldstein's resolution of the large ξ expansion

Goldstein (1960) came to the conclusion that to some extent the problem had been approached the wrong way around by beginning with an expansion in the boundary layer. What should have been more carefully thought about was the form of the outer flow field. In particular, he reasoned that the nature of the complex potential $w(\zeta)$ for large ζ was more complicated than had been hitherto thought and that the *least* complicated expansion for w would be

$$w \sim \frac{1}{2}\zeta^2 - i\beta_0\zeta + i\frac{a_{21} + a_{22}\log\zeta'}{\zeta} + \frac{a_{31} + a_{32}\log\zeta' + a_{33}\log^2\zeta'}{\zeta^2} + \cdots, \quad (2.100)$$

where $\zeta' = \eta - i\xi$ so that

$$\Im\log\zeta' = -\tan^{-1}\frac{\xi}{\eta} = \frac{\pi}{2}\text{sgn}\xi - \tan^{-1}\frac{\eta}{\xi}, \quad (2.101)$$

to remove the difficulty observed in using an inverse tan function in the outer flow. As observed by Murray (1964, 1965), even this expansion may prove too simple for higher order terms. One reason for thinking that the expansion (2.100) may prove inadequate is that the eigen-solutions to (2.57) conveniently start with a term proportional to ξ^{-1}, but subsequent eigen-solutions correspond to non-integral powers of ξ so that the whole form of (2.100) beyond the third term may be incorrect.

If we neglect terms of order $o(\zeta^{-1})$ in the outer expansion, then the corresponding expansion in the boundary layer will be

$$\Upsilon \sim \xi f_0(\eta) + \frac{1}{\xi}[f_2(\eta) + g_2(\eta)\log|\xi|], \quad (2.102)$$

with far field conditions $f_2 \to a_{21}$ and $g_2 \to a_{22}$. The use of the logarithmic term and the appearance of the function g_2 provides the flexibility needed to remove any algebraic decay in the boundary layer far field but because (2.102) has come from (2.100), we will be guaranteed matching to a potential outer flow. We can also note that the expansion (2.102) will become invalid when $\xi \sim 1$ or $x = O(R^{-1})$. Thus there will be a region close to the leading edge where the boundary layer expansion will break down and where it may be necessary to use the full Navier–Stokes equations to describe the flow.

To calculate f_2 and g_2 substitute (2.102) into the Navier–Stokes equations (2.74) to obtain

$$(f_2''' + f_0 f_2'' + 2f_0' f_2' - f_0'' f_2)' = \frac{d}{d\eta}(f_0 - \eta f_0')^2 + f_0' g_2'' - f_0''' g_2, \quad (2.103)$$

and

$$(g_2''' + f_0 g_2'' + 2f_0' g_2' - f_0'' g_2)' = 0, \quad (2.104)$$

and boundary conditions $f_2(0) = f_2'(0) = g_2(0) = g_2'(0) = 0$. The far field boundary conditions are that f_2 and g_2 tend to constants exponentially. The function g_2 can be determined analytically since $\eta f_0' - f_0$ satisfies the conditions at $\eta = 0$ and tends to a constant(β_0) exponentially as $\eta \to \infty$. Hence we can put

$$g_2 = B_2(\eta f_0' - f_0), \qquad (2.105)$$

where B_2 is an undetermined constant.

Now return to (2.103) and let

$$G_2(\eta) = \int_0^\eta [f_0'(\eta f_0' - f_0)'' - f_0'''(\eta f_0' - f_0)]\,\mathrm{d}\eta, \qquad (2.106)$$

so that using our previous notation,

$$f_2''' + f_0 f_2'' + 2 f_0' f_2' - f_0'' f_2 = a_2 + (f_0 - \eta f_0')^2 + B_2 G_2. \qquad (2.107)$$

Again we follow Goldstein and use the notation $f_2^{(5)}$ for a particular integral for the last term of (2.107) when $B_2 = 1$, which satisfies the boundary conditions at $\eta = 0$ and has asymptotic behaviour

$$f_2^{(5)} \sim c_{5,1} + c_{5,2}\eta^{-1} + c_{5,3}\exp(-\eta^2/2). \qquad (2.108)$$

Using the functions $f_2^{(3)}$ and $f_2^{(4)}$ described above and again choosing $a_2 = -\beta_0^2$ to eliminate $O(\eta)$ behaviour in $f_2^{(4)}$ then the solution of (2.100) is

$$f_2 = A_2 f_2^{(3)} + f_2^{(4)} + B_2 f_2^{(5)}, \qquad (2.109)$$

with asymptotic behaviour

$$f_2 \sim A_2 c_{3,1} + c_{4,1} + B_2 c_{5,1} + (c_{4,2} + B_2 c_{5,2})\eta^{-1} + O(\exp(-\eta^2/2)). \qquad (2.110)$$

Goldstein conjectured that B_2 could be determined by $c_{4,2} + B_2 c_{5,2} = 0$ so as to remove the unacceptable algebraic decay while still allowing the boundary layer expansion to be matched to an outer potential flow. This was later shown to be possible by Murray (1965) who computed solutions to the differential equation (2.103) and determined coefficients equivalent to $c_{i,j}$. However, we have no way to determine A_2 [remember that $f_2^{(3)}$ is the first eigen-solution of (2.103) satisfying the boundary conditions at $\eta = 0$ and with $o(\eta)$ behaviour as $\eta \to \infty$]. The view of Goldstein and subsequent workers was that the constant A_2 could only be found by knowing more about the solution near the leading edge. Murray calculated solutions to differential equations associated with higher order terms in (2.100) and additional unknowns enter the solution as the order of the expansion increases. Thus for practical purposes we cannot continue the analysis and the question of how A_2 is determined remains unknown.

2.3.5 Return to rectangular coordinates

To use the asymptotic expansion, (2.102) for $\Upsilon = R\psi$, we need to be able to move back to rectangular coordinates. To do that we have to use the expansion of parabolic coordinates in terms of rectangular coordinates for large Rx,

$$\xi \sim \sqrt{2xR}\{1 + \frac{1}{8}\frac{y^2}{x^2}\} \equiv \sqrt{2xR}(1 + \frac{s^2}{4Rx}) \qquad (2.111)$$

and

$$\eta \sim \frac{\sqrt{Ry}}{\sqrt{2x}}\{1 - \frac{1}{8}\frac{y^2}{x^2}\} \equiv s(1 - \frac{s^2}{4Rx}) \qquad (2.112)$$

where $s = \sqrt{R}y/\sqrt{2x}$ is the boundary layer coordinate defined previously. Using these expansions we can write, for example

$$f_0(\eta) \sim f_0(s) - \frac{1}{4}x^{-1}R^{-1}s^3 f_0'(s), \quad Rx \gg 1, \qquad (2.113)$$

so that in the boundary layer when $x > 0$,

$$\psi \sim R^{-1/2}\sqrt{2x}f_0(s) + \frac{R^{-3/2}}{\sqrt{(2x)}}[f_2(s) + \frac{1}{4}s^2 f_0(s) - \frac{1}{8}s^3 f_0'(s) + g_2(s)\log\sqrt{2xR}]. \qquad (2.114)$$

This again emphasises that the correct form of the boundary layer expansion in rectangular coordinates is neither (2.56) nor (2.65) but rather

$$\psi \sim R^{-1/2}\sqrt{2x}F_0(s) + x^{-1/2}R^{-3/2}[F_2(s) + g_2(s)\log\sqrt{2xR}], \qquad (2.115)$$

where we have already observed that $F_0 \equiv f_0$; we can now add $F_2 = f_2/\sqrt{2} + s^2 f_0/4 - s^3 f_0'/4$.

2.4 The drag on a section of semi-infinite plate

Once the stream-function expansion, (2.115) has been determined the viscous drag on a section of plate should be able to be determined from integrating the local viscous stress. Unfortunately this cannot be carried out directly since the local velocity gradient at the wall will be given by

$$\frac{\partial u}{\partial y} \sim \frac{R^{1/2}}{\sqrt{2x}}F_0''(0) + \frac{R^{-1/2}}{\sqrt{2x^3}}[F_2''(0) + g_2''(0)\log\sqrt{2xR}], \qquad (2.116)$$

and the second term is not integrable at the leading edge of the plate. Of course the expansion (2.115) is not valid near the leading edge and the non-integrability of (2.116) there is only a reflection that a different expansion would be necessary to calculate the leading edge drag correctly. Imai (1957) was able nevertheless to calculate the leading order drag terms by considering an integral of the momentum flux over a large circle.

Imai (1951) had considered general results about the far field of a general body in a uniform stream. His main concern had been to show that a consistent expansion structure existed using Oseen's approximation for the far field. Earlier Filon (1926) had thought that the moment on a body could not apparently be found using Oseen's approximation but Imai was able to show that this was because the number of terms needed in an Oseen expansion to calculate the moment to a given order was greater than Filon had appreciated. Part of his work though, was to generalise the Blasius formula for the drag on a body by including the thin layer of non-zero vorticity which would be generated by viscous flow around the body. Imai considered the stream-function in two parts,

$$\psi = \psi_\omega + \Im w, \qquad (2.117)$$

where w was the complex potential for an irrotational outer flow and ψ_ω was the rotational field with vorticity $\omega = -\nabla^2 \psi_\omega$. If the drag and lift coefficients, scaled according to (1.47), are c_D and c_L then Imai's result was that

$$c_D - ic_L = i\oint \left(\frac{dw}{dz}\right)^2 dz - i\int_W K dz - 2i\int_W \omega \bar{z} d\psi_\omega - \frac{4}{R}\int_W \frac{\partial \omega}{\partial \bar{z}} \bar{z} d\bar{z}, \qquad (2.118)$$

with W representing an integral over the wake and

$$K(\psi_\omega, w) = 4\left[\left(\frac{\partial \psi_\omega}{\partial z}\right)^2 - i\frac{\partial \psi_\omega}{\partial z}\frac{dw}{dz}\right]. \qquad (2.119)$$

The first term in the integral (2.118) was the result of Blasius, the remaining terms generalised that result to include vorticity in the wake. This enabled Imai to determine the leading order terms for the drag without needing to know the details of the flow near the leading edge.

Imai's result (2.118) has to be modified a little to obtain the drag on the leading section of length x of a semi-infinite plate. Defining the drag coefficient by

$$c_D = \frac{\hat{D}}{\frac{1}{2}\rho U^2 \times 2x}, \qquad (2.120)$$

then the first term of (2.118), the Blasius result will give a contribution to the drag

$$\frac{i}{2x}\oint \left(\frac{dw}{dz}\right)^2 dz,$$

and using only the first two terms of the outer potential flow, (2.53)), a pole at $z = 0$ gives a contribution to the drag

$$\frac{\pi \beta_0^2}{2Rx} = \frac{2.3245...}{Rx}. \qquad (2.121)$$

The other parts of (2.118) are quite difficult to evaluate, it is easier to consider a momentum flux integral

$$\int_{\text{boundary layer}} u(1-u)dy,$$

which with velocity from (2.115),

$$u \sim f_0'(s) + \frac{1}{\sqrt{2Rx}}(F_2'(s) + g_2'(s)\log\sqrt{2Rx}),$$

gives a viscous drag coefficient (using both sides of the plate)

$$\frac{2\sqrt{2}}{\sqrt{Rx}}\{\int_0^\infty f_0'(1-f_0')ds + \frac{1}{\sqrt{2Rx}}\int_0^\infty (1-2f_0')(F_2' + g_2'\log\sqrt{2Rx})ds\}.$$

The first integral can be integrated by parts and the differential equation for f_0 used to obtain

$$\int_0^\infty f_0'(1-f_0')ds = f_0''(0) = \alpha_0^{-3/2},$$

and g_2 is known but F_2 involves an unknown constant. The leading viscous drag term is therefore

$$\frac{2\sqrt{2}}{\sqrt{Rx}}\alpha_0^{-3/2} = \frac{1.328...}{\sqrt{Rx}},$$

and combining the viscous and pressure drag terms, Imai found the overall drag coefficient for the section $[0, x]$ of the plate to be

$$\boxed{c_D \sim \frac{1.33}{(Rx)^{1/2}} + \frac{2.32}{Rx} - \frac{2.20}{(Rx)^{3/2}}\log\sqrt{Rx} + O(R^{-3/2})} \qquad (2.122)$$

Imai attempted an ad-hoc resolution for the unknown constant which arose in the term of order $R^{-3/2}$ by patching (2.122) to a low Reynolds number approximation for the drag but it is still the formal position that how the expansion (2.122) might be determined beyond these terms is unknown.

There have been two other attempts to determine the drag coefficient for a section of a semi-infinite plate which are of interest. Dean (1954) used what seemed a non-rational approximation for the boundary layer problem in parabolic coordinates but which is possibly amenable to more formal treatment.

Suppose (2.74) is written in the symbolic form

$$\nabla_2^4 \Upsilon = \mathcal{L}[\Upsilon], \qquad (2.123)$$

where the non-linear operator $\mathcal{L}$ represents the right-hand side of (2.74).

If Υ_A is a first approximation to the solution of (2.123) then Dean showed that as before,

$$\Upsilon_A \sim \xi f_0(\eta), \qquad (2.124)$$

where f_0 was given by (2.77).

Dean then considered a second approximation, Υ_B where we might write

$$\Upsilon_B = \Upsilon_A + \delta\Upsilon. \qquad (2.125)$$

Assuming that the perturbation $\delta\Upsilon$ is in some sense small compared to Υ_A then

$$\nabla_2^4 \Upsilon_B = \nabla_2^4 \Upsilon_A + \nabla_2^4 \delta\Upsilon = \mathcal{L}[\Upsilon_A] + \nabla_2^4 \delta\Upsilon. \qquad (2.126)$$

Dean assumed that the second term in the right-hand side of (2.126) was small compared to the first so that the second approximation satisfied an inhomogeneous **linear** equation,

$$\nabla_2^4 \Upsilon_B = \mathcal{L}[\Upsilon_A], \qquad (2.127)$$

for which the Green's function was known and that allowed the solution to be written as a convolution integral of the Green's function with $\mathcal{L}[\Upsilon_A]$. Examination of his solution shows that the assumption that $\delta\Upsilon$ can be neglected is good near the plate although formally, not near the origin.

The result of some tedious algebra gave the second approximation and Dean was able to show that the ratio Υ_B/Υ_A was around 1.2 and deduced that the drag coefficient should be increased by an amount of this size. Dean's second approximation can however be used to calculate the local wall shear or local drag coefficient and surprisingly, a formal estimate of the integrated drag coefficient.

The local drag coefficient, c_f, is given in terms of Υ and the parabolic coordinates we are using by

$$c_f(x) = \frac{1}{Rx} \frac{\partial^2 \Upsilon}{\partial \eta^2}\bigg|_{\eta=0}, \qquad (2.128)$$

and using $\xi = \sqrt{2Rx}$ on the plate, Dean's second approximation resulted in

$$c_f(x) \approx \frac{\sqrt{2} f_0''(0)}{\sqrt{Rx}} + \frac{1}{2\sqrt{2Rx}} \int_0^\infty \frac{H(\eta)}{Rx + 2\eta^2} \eta \, d\eta, \qquad (2.129)$$

where $H(\eta) = 2\eta^2 f_0'''(f_0 - \eta f_0') - 2\eta^3 [f_0'']^2$. Now the remarkable point about this result is that whereas most expansions have only the leading term being integrable from the leading edge of the plate (see for instance the work needed by Imai to determine the second term of the expansion); this expansion has both terms integrable from the leading edge.

Using $c_D = \int_0^x c_f(x)\mathrm{d}x/x$, this gives

$$c_D = \frac{2\sqrt{2}f_0''(0)}{\sqrt{Rx}} + \frac{1}{2Rx}\int_0^\infty H(\eta)\tan^{-1}\left(\frac{\sqrt{Rx}}{\sqrt{2}\eta}\right)\mathrm{d}\eta, \qquad (2.130)$$

and for large Reynolds number,

$$c_D \sim \frac{1.33}{\sqrt{Rx}} + \frac{2.32}{Rx} + O((Rx)^{-3/2}). \qquad (2.131)$$

It seems unexpected that the second term is identical to that found by Imai. Davis (1967) attributes this observation about Dean's approximation to Van Dyke. It seems that Dean's method has not yet been put on a firm rational footing and it may be that further work formalising the approximation would be rewarding. It bears considerable similarity with an iterative method which treats the non-linear terms of the Navier–Stokes equations as a source term for a biharmonic equation excepting it does not start with a Stokes flow as the initial solution, rather (2.124). It is not clear if Dean's method could be applied successfully to problems apart from flow about a semi-infinite plate because of its reliance on a special first approximation.

Another method to develop higher order expansions was due to Davis (1967) who applied a series truncation method of Van Dyke to flow about a flat plate. The stream-function Υ was expanded

$$\Upsilon \sim \xi\{g_1(\eta) + (\xi - \sqrt{2Rx})g_2(\eta) + (\xi - \sqrt{2Rx})^2 g_3(\eta) + \cdots\}, \qquad (2.132)$$

and powers of $\xi - \sqrt{2Rx}$ were equated in (2.123) to give ordinary differential equations for $g_1, g_2, \ldots$. Of course this is a pragmatic and indeed somewhat inconsistent method rather than a formal rational expansion method, but in computing g_1 and g_2 Davis showed that the approximation seemed to be leading towards very similar results to those of Dean.

2.5 The wake behind a finite length plate

We now turn to the flow about a plate of finite length. The results we shall describe are mainly from Filon (1926), Tollmien (1931) and Goldstein (1930, 1933a, 1933b). Filon considered general properties of the wake behind a body. Goldstein's first paper studied the wake immediately behind the trailing edge of a finite length flat plate. Tollmien and Goldstein each considered the wake far behind the plate, where the velocity was a perturbation of uniform flow and Goldstein also related his results back to the general results of Filon.

The parabolic nature of boundary layer equations meant that it was possible to use the velocity profile are a given distance from the leading edge of a plate as an upstream boundary condition for the evolution of a wake. This lead to a description using two expansions for the near wake. One expansion corresponded to a layer near the centreline of the plate where the velocity had to rise rapidly from zero on the plate to a finite value; we will follow Stewartson and call this expansion a Goldstein inner wake. Further from the centreline, the inner wake expansion had to be matched to a different expansion, valid near the edge of the oncoming boundary layer and we will call that a Goldstein outer wake.

The work of Goldstein appeared to complete the lowest order expansion for flow about a finite plate but carried with it a number of difficulties: the transition from plate boundary layer to inner wake is so abrupt that it was only achieved by having singular transverse velocities at the trailing edge, the wake had no influence on the oncoming boundary layer and the far wake expansion failed when continued to higher order.

Consider a plate of length $\tilde{L}$ in the non-dimensional (x,y) coordinate system defined above. Does the value of $\tilde{L}$ matter? In fact it is not a central issue so long as the plate is greater in length than R^{-1}, since as we have already seen, a boundary layer expansion will exist so long as $\xi \gg 1$, which translates into $x \gg R^{-1}$. Since the value of $\tilde{L}$ is largely immaterial we will choose $\tilde{L} = 1$; this corresponds to an original non-dimensionalisation based on the physical length of the plate rather than the scale of an observer.

2.6 Near wake region

Suppose that the plate lies in $-1 \leq x \leq 0$, $y = 0$ and that a solution for the stream-function is sought in the wake region $x > 0$. A simple view of the flow is that a boundary layer of the type described in section 2.3 develops on the plate and that the wake is the evolution of the velocity profile passing the trailing edge. There is one problem with this view. We have seen that Goldstein's resolution of a higher order boundary layer expansion revolved critically around having the correct outer potential expansion. The potential problem is elliptic and so what happens in the wake must be able to influence the upstream development of the boundary layer on the plate. It is the application of this idea which finally allowed Stewartson and Messiter independently to resolve the wake near the trailing edge; for the present assume that at least to leading order, the wake does not alter the boundary layer on the plate.

The leading order longitudinal velocity in a boundary layer a unit distance from the leading edge will be defined by (2.25) so that

$$u(0,Y) \sim F_0'(Y/\sqrt{2}), \qquad (2.133)$$

where $y = R^{-1/2}Y/\sqrt{2}$ is the boundary layer coordinate at the trailing edge. As detailed above, this was known to Goldstein as a series solution with leading term

$$F_0'(Y) \sim \alpha_0^{-3/2} Y/\sqrt{2} + O(Y^4), \qquad (2.134)$$

so it was natural that Goldstein should investigate the evolution of flows with (2.133) as an upstream boundary condition within the context of boundary layer theory (so that the transverse coordinate is stretched by $R^{1/2}$). In doing this, the lowest order term of the stream-function expansion for large Reynolds number must still satisfy (2.16) and (2.17). Thus solutions of the form $x^a F(Y x^b)$ will still be appropriate only now, in addition to the condition $a - b = 1$, we will need $a + 2b = 0$ in order to satisfy (2.134) as $x \to 0$ (recall that for the plate boundary layer, the expansion far from the plate gave $a + b = 0$). Hence we should have a leading order term in the expansion which behaves like $x^{2/3} \times \text{function}(Y x^{-1/3})$.

2.6.1 Goldstein inner wake

In order to develop an expansion near the centreline and immediately downstream of the trailing edge, we need to carefully keep track of exactly how the asymptotic expansion of the stream-function at large Reynolds number is proceeding. In terms of the coordinate system centred at the trailing edge, the oncoming boundary layer is described by

$$\psi \sim R^{-1/2}\sqrt{2(1+x)} F_0\left(\frac{Y}{\sqrt{2(1+x)}}\right) + o(R^{-1/2}), \quad -1 + O(R^{-1}) < x \le 0, \tag{2.135}$$

provided of course that the upstream influence of the wake on the developing boundary layer is neglected. It is this expansion which gives (2.133) as $x \to 0^-$. In order to match an expansion in the wake to this as $x \to 0^+$, we expand the stream-function in the wake as

$$\psi \sim R^{-1/2} \psi_0(x, Y) + o(R^{-1/2}). \tag{2.136}$$

It is convenient at this point to use an inner wake similarity variable $s = Y/x^{1/3}$, see also for instance Hakkinen and Rott (1965). Then suppose that for small x, that the function ψ_0 is expanded by

$$\psi_0 \sim x^{2/3} \{ H_0(s) + x H_1(s) + x^2 H_2(s) + \cdots \}, \tag{2.137}$$

so that the stream-function can be matched to the boundary layer leaving the plate,

$$\psi \sim \frac{\sqrt{2}}{4} \alpha_0^{-3/2} Y^2 - \frac{1}{4\,5!} \alpha_0^{-3} Y^5 + \frac{11\sqrt{2}}{16\,8!} \alpha_0^{-9/2} Y^8 + \cdots. \tag{2.138}$$

Since $x \to 0^+$ corresponds to $s \to \infty$ this gives boundary conditions

$$H_0 \sim \sqrt{2} \alpha_0^{-3/2} s^2 / 4, \quad H_1 \sim -\frac{1}{4\,5!} \alpha_0^{-3} s^5, \quad \text{as } s \to \infty, \tag{2.139}$$

and so on. The velocities will be

$$u \sim x^{1/3} \left(H_0'(s) + x H_1'(s) + \cdots \right), \tag{2.140}$$

$$v = \frac{1}{3}x^{-1/3}\left(sH_0'(s) - 2H_0(s)\right) + \frac{1}{3}x^{2/3}\left(sH_1'(s) - 5H_1(s)\right) + \cdots, \quad (2.141)$$

so that the boundary conditions $v = 0$ and $\frac{\partial u}{\partial y} = 0$ on $Y = 0$ become $H_r(0) = H_r''(0) = 0$ for $r = 0, 1, 2, \cdots$. That a singularity could occur at the trailing edge was apparent in (2.141) unless of course $sH_0' - 2H_0 \to 0$ as $s \to \infty$.

If we continue with the assumption that there is no pressure gradient in the wake then substitution of (2.136) into the Navier–Stokes equations of course leads to the leading order boundary layer equation (2.23) and the expansion (2.137) gives for H_0 and H_1,

$$3H_0''' + 2H_0 H_0'' - H_0'^2 = 0, \quad (2.142)$$

$$3H_1''' + 2H_0 H_1'' - 5H_0' H_1' + 5H_0'' H_1 = 0. \quad (2.143)$$

As in the case of a boundary layer on a plate, one of the boundary conditions is at infinity. To overcome this suppose $\bar{H}_0$ satisfies (2.142) together with the boundary conditions $\bar{H}_0(0) = \bar{H}_0''(0) = 0$ and $\bar{H}_0'(0) = 1$. If as s becomes large, then $\bar{H}_0$ also becomes large, the asymptotic form of (2.142) is

$$2\bar{H}_0 \bar{H}_0'' - \bar{H}_0'^2 \sim 0, \quad (2.144)$$

so that $\bar{H}_0' \sim c\sqrt{\bar{H}_0}$ where c is some constant of integration. Integrating again,

$$\bar{H}_0 \sim \gamma_0(s + \mu_0)^2 \quad \text{as } s \to \infty, \quad (2.145)$$

where γ_0 and μ_0 are unknown constants. Solving (2.142) numerically for $\bar{H}_0$ gives $\gamma_0 = 0.24454\ldots$ and $\mu_0 = 1.13213\ldots$. Finally, if the required solution is written $H_0(s) = \lambda_0 \bar{H}_0(\lambda_0 s)$ we will have the correct large s asymptotic behaviour if

$$\lambda_0^3 \gamma_0 = \frac{\alpha_0^{-3/2}}{2\sqrt{2}}, \quad (2.146)$$

so that $\lambda_0 = 0.8789\ldots$. It can also be seen that if

$$\bar{H}_0 = \gamma_0(s + \mu_0)^2 + \bar{h}_0(s), \quad (2.147)$$

then keeping only the highest order terms in (2.143) gives

$$3\bar{h}_0''' + 2\gamma_0 s^2 \bar{h}_0'' \sim 0, \quad (2.148)$$

or

$$\bar{h}_0'' \sim e^{-2\gamma_0 s^3/9}, \quad (2.149)$$

so that all terms in $\bar{h}_0$ in the far field of the inner wake are exponentially small and finally

$$H_0(s) \sim \lambda_0 \gamma_0(\lambda_0 s + \mu_0)^2 \quad \text{as } s \to \infty. \quad (2.150)$$

This removed the possibility that $sH_0' - 2H_0 \to 0$ as $s \to \infty$ and established that the inner wake would be singular at the trailing edge.

For small s,

$$\bar{H}_0 \sim s + \frac{1}{3\,3!}s^3 - \frac{2}{9\,5!}s^5 + \cdots, \tag{2.151}$$

so that

$$H_0 \sim \lambda_0^2 s + \frac{\lambda_0^4}{3\,3!}s^3 - \frac{2\lambda_0^6}{9\,5!}s^5 + \cdots, \tag{2.152}$$

and an $O(x^{1/3})$ contribution to the centreline velocity will come from the first term.

The analysis can be extended to (2.143) and subsequent equations. If $p = \lambda_0 s$ and $H_1(s) = \lambda_1 \bar{H}_1(p)$ where λ_1 has to be determined, $\bar{H}_1$ satisfies (2.143) together with boundary conditions $\bar{H}_1(0) = \bar{H}_1''(0) = 0$, $\bar{H}_1'(0) = 1$. For large $s' = p + \mu_0$, $\bar{H}_1$ approximately satisfies

$$3\bar{H}_1''' + 2\gamma_0 s'^2 \bar{H}_1'' - 10\gamma_0 s' \bar{H}_1' + 10\gamma_0 \bar{H}_1 \sim 0, \tag{2.153}$$

and the last three terms (which formally are asymptotically greater than the first) imply either $O(s'^5)$ or $O(s')$ behaviour. However, a term $O(s'^5)$ will give an $O(s'^2)$ term if the last three terms are solved as an inhomogeneous equation against the first. Consequently, in our notation (which is a little different to that in Goldstein's original work),

$$\bar{H}_1 \sim \gamma_1(s'^5 + \frac{30}{\gamma_0}s'^2 + \mu_1 s') + O[\exp(-\frac{2}{9}\gamma_0 s'^3)] \text{ as } s' \to \infty, \tag{2.154}$$

where $\gamma_1 = 0.0041395$ and $\mu_1 = -155.59$ are evaluated from a numerical solution. That having been done, we have for $s \to \infty$,

$$H_1(s) \sim \lambda_1 \gamma_1 \lambda_0^5 s^5 \equiv -\frac{2^{-5/2}\alpha_0^{-3}}{5!}s^5, \tag{2.155}$$

so that λ_1 is determined as $\lambda_1 = -0.1496$.

If we turn to the small s expansion of H_1, first

$$\bar{H}_1 \sim p - \frac{5}{18}p^3 + \cdots, \tag{2.156}$$

and then

$$H_1(s) \sim \lambda_1 \lambda_0(s - \frac{5}{18}\lambda_0^2 s^3 + \cdots) \text{ as } s \to 0. \tag{2.157}$$

Thus we can write the first two terms of the centreline velocity,

$$u(x,0) \sim x^{1/3}(\lambda_0^2 + \lambda_0 \lambda_1 x + \cdots) + o(1) \text{ as } R \to \infty \tag{2.158}$$

It is worth remarking that to leading order for large Reynolds number, the velocity returns to the free stream value on a length scale that depends only on the body length scale and not on the free stream velocity or fluid viscosity. Departures from this will be seen in the far wake and in higher order terms of the expansion of the stream-function against Reynolds number.

Goldstein examined the first three functions H_0, H_1 and H_2 (remember that these are merely terms in a small x expansion of the leading term of a high Reynolds number expansion) and a key result was that for large s, $H_0 \sim s^2$, $H_1 \sim s^5$ and $H_2 \sim s^8$ which of course means that all of these functions would of course fail were they to be carried to the boundary layer edge, since there the stream-function was at most linear in s. Hence it was necessary to introduce another expansion, valid between the inner wake and the edge of the oncoming boundary layer.

2.6.2 Goldstein outer wake

Goldstein argued that if for large s we write $H_0 \sim A_0 s^2 + B_0 s + C_0$, $H_1 \sim A_1 s^5 + B_1 s^4 + \cdots$ and so on, where the constants A_0, B_0, $\cdots$ are all known, for instance $A_0 = \lambda_0^3 \gamma_0$, then the stream-function expansion for large s would be

$$\psi \sim R^{-1/2}\left[x^{2/3}(A_0 s^2 + B_0 s + C_0) + x^{5/3}(A_1 s^5 + B_1 s^4 + ...) + ...\right] + o(R^{-1/2}), \tag{2.159}$$

and provided rearrangement of terms is allowable, the expansion in (x, Y) coordinates is

$$\psi \sim R^{-1/2}\left[(A_0 Y^2 + A_1 Y^5 + ...) + x^{1/3}(B_0 Y + B_1 Y^4 + ...) + ...\right] + o(R^{-1/2}). \tag{2.160}$$

This suggested that the expansion which should be used to match this part of the wake to both the oncoming boundary layer and the inner wake should have the form

$$\psi \sim R^{-1/2}\left[\bar{\psi}_0(Y) + x^{1/3}\bar{\psi}_1(Y) + x^{2/3}\bar{\psi}_2(Y) + ...\right] + o(R^{-1/2}). \tag{2.161}$$

To match to (2.136) as $x \to 0^+$ Goldstein needed

$$\bar{\psi}_0(Y) \equiv \psi_0(0, Y) = \sqrt{2} F_0(Y/\sqrt{2}). \tag{2.162}$$

The whole $O(R^{-1/2})$ expansion still had to satisfy the leading order boundary layer equation (2.23), so that, $\bar{\psi}_1$, $\bar{\psi}_2$ satisfy

$$\bar{\psi}_0' \bar{\psi}_1' - \bar{\psi}_0'' \bar{\psi}_1 = 0, \tag{2.163}$$

$$\bar{\psi}_0' \bar{\psi}_2' - \bar{\psi}_0'' \bar{\psi}_2 = \frac{1}{2}[\bar{\psi}_1 \bar{\psi}_1'' - \bar{\psi}_1'^2]. \tag{2.164}$$

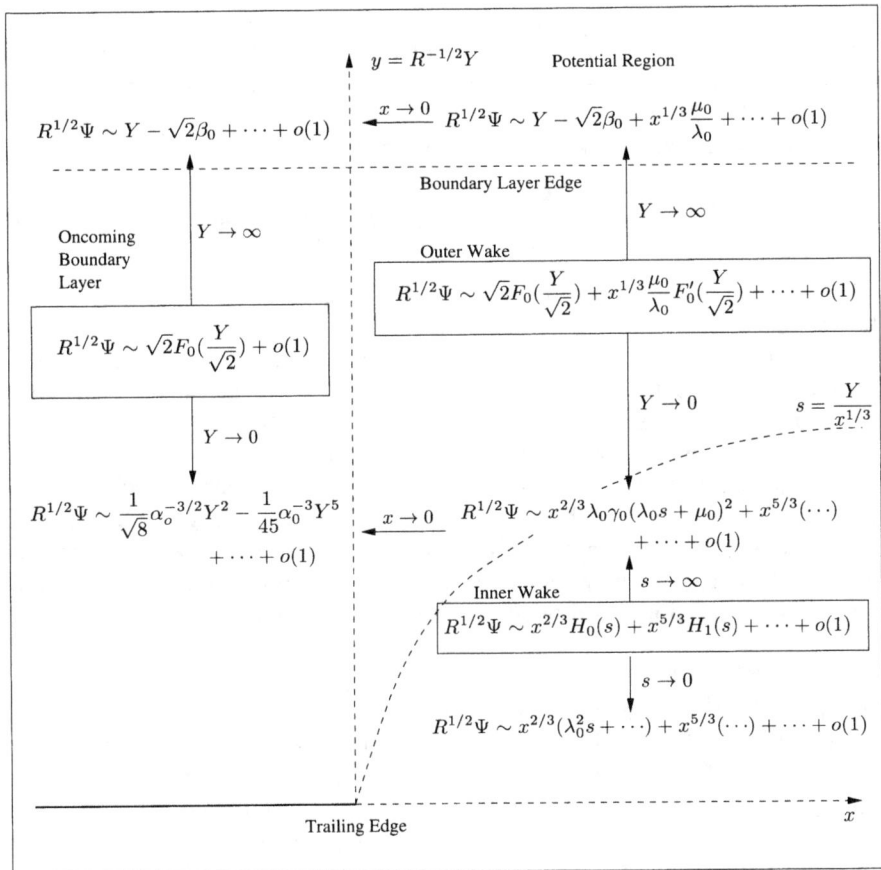

FIG. 2.6. Overview of Goldstein near wake structure showing the oncoming boundary layer, the inner wake region, outer wake region and potential flow region.

It was easy to see that the required solutions were

$$\bar{\psi}_1(Y) = A\bar{\psi}_0'(Y), \qquad (2.165)$$

$$\bar{\psi}_2(Y) = \frac{A^2}{2}\bar{\psi}_0''(Y), \qquad (2.166)$$

where A was an undetermined constant. The outer wake expansion could be matched to the inner wake by having $\bar{\psi}_1 \sim B_0 Y + B_1 Y^4 + \ldots$ and $\bar{\psi}_2 \sim C_0 + C_1 Y^3 + \cdots$ as $Y \to 0$. Using the form of $\bar{\psi}_0$ given by (2.162) in (2.165) and (2.166) then

$$\bar{\psi}_1 \sim \frac{A\alpha_0^{-3/2}}{\sqrt{2}} Y \quad \text{and} \quad \bar{\psi}_2 \sim \frac{A^2 \alpha_0^{-3/2}}{2\sqrt{2}} Y, \quad \text{as } Y \to 0 \qquad (2.167)$$

56   *The Boundary Layer on a Flat Plate*

so that $\frac{A\alpha_0^{-3/2}}{\sqrt{2}} = B_0$ and $\frac{A^2\alpha_0^{-3/2}}{2\sqrt{2}} = C_0$. Substitution of the known values of B_0 and C_0 gave for each expression $A = \mu_0/\lambda_0$, each match having to give the same value of A or the expansion would have failed.

Thus the outer wake expansion could be written

$$R^{1/2}\Psi \sim \sqrt{2}F_0(\frac{Y}{\sqrt{2}}) + x^{1/3}\frac{\mu_0}{\lambda_0}F_0'(\frac{Y}{\sqrt{2}}) + \cdots + o(1). \tag{2.168}$$

As with the expansion for the inner wake, this expansion too gave a singularity in the vertical velocity $v = -\psi_x$ as $x \to 0^+$.

Goldstein (1930) was able to calculate the first nine terms of the expansion in this region with each term being a polynomial in Y combined with derivatives of F_0. Thus the structure of the near wake was described in a way that would remain unchanged until 1969. This relatively complicated set of expansions is illustrated graphically in Figure 2.6. The singularity in the vertical velocity at the trailing edge was recognised but could not be explained away. Tollmien (1931) and Goldstein (1933a) later added a description of the far wake and we shall come to that shortly. Next we shall insert a modern note by applying matched asymptotics to examine a composite near wake solution.

2.6.3 *Composite near wake expansion*

In his original 1930 paper Goldstein calculated a velocity field using the inner wake expansion (2.137) and separately the outer wake expansion (2.161) to obtained a picture of the development of the velocity field by patching the two sets of calculations manually. We can apply the more modern idea of a composite expansion to the near wake so as to graph the development of the velocity perhaps not any more effectively than in Goldstein's original work, but certainly a little more formally. To do this we add the inner wake and the outer wake expansions of the leading high Reynolds number term in each region and subtract the expansion in outer coordinates of the inner wake. Recalling that the inner wake is

$$\psi \sim R^{-1/2}\{x^{2/3}H_0(s) + x^{5/3}H_1(s) + \ldots\} + o(R^{-1/2}), \tag{2.169}$$

and if this is expanded for $s \to \infty$,

$$\psi \sim R^{-1/2}\{\lambda_0\gamma_0 x^{2/3}(\lambda_0 s + \mu_0)^2 \\ + \lambda_1\gamma_1 x^{5/3}[(\lambda_0 s + \mu_0)^5 + \frac{30}{\gamma_0}(\lambda_0 + \mu_0)^2 \\ + \mu_1(\lambda_0 s + \mu_0)]\} + o(R^{-1/2}). \tag{2.170}$$

If we now rewrite this using outer coordinates (x, Y), then the two-term composite expansion is

$$\psi_{2-2} = R^{-1/2}\{x^{2/3}H_0(Yx^{-1/3}) + x^{5/3}H_1(Yx^{-1/3})$$
$$+\sqrt{2}F_0(\frac{Y}{\sqrt{2}}) + \frac{\mu_0 x^{1/3}}{\lambda_0}F_0'(\frac{Y}{\sqrt{2}}) - \lambda_0\gamma_0(\lambda_0 Y + \mu_0 x^{1/3})^2 \qquad (2.171)$$
$$-\lambda_1\gamma_1[(\lambda_0 Y + \mu_0 x^{1/3})^5 + \frac{30}{\gamma_0}(\lambda_0 Y + \mu_0 x^{1/3})^2 + \mu_1(\lambda_0 Y + \mu_0 x^{1/3})]\}.$$

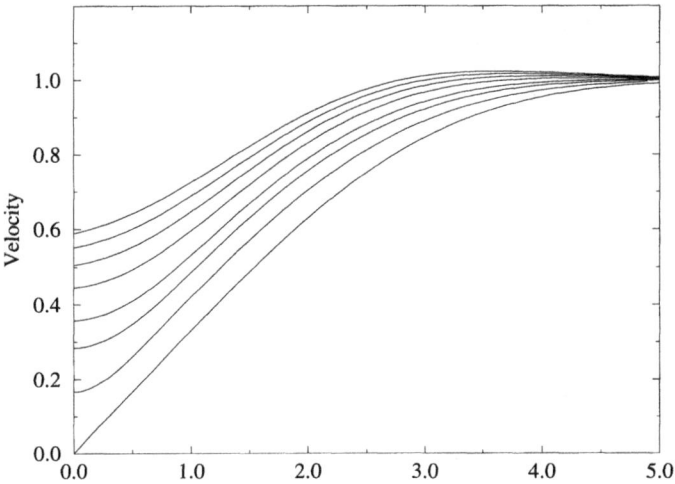

FIG. 2.7. Calculated velocity profiles using a 2-2 term composite expansion. The lowest profile is the velocity leaving the plate, $x = 0$, successive profiles are at $x = 0.01, 0.05, 0.1\ 0.2, 0.3, 0.4, 0.5$.

We have used this expansion to calculate the evolution of the streamwise velocity with distance from the trailing edge, see Figure 2.7. One problem which quickly emerges as x increases is that the two-term outer wake solution leads to velocities which are greater than the free stream value near the outer edge of the wake.

As a contrast we have also calculated the velocity using three terms of the outer wake solution,

$$\psi_{2-3} = R^{-1/2}[\psi_{2-2} + \frac{\mu_0^2 x^{2/3}}{4\lambda_0^2}F_0''(\frac{Y}{\sqrt{2}})], \qquad (2.172)$$

and the streamwise velocity calculated from this expansion is shown in Figure 2.8. It seems clear that the outer wake is much better represented using a composite expansion with three terms of the outer expansion rather than two terms. It is also the case that the expansion is only usable for relatively small values of the streamwise coordinate, roughly half the plate length from the trailing edge.

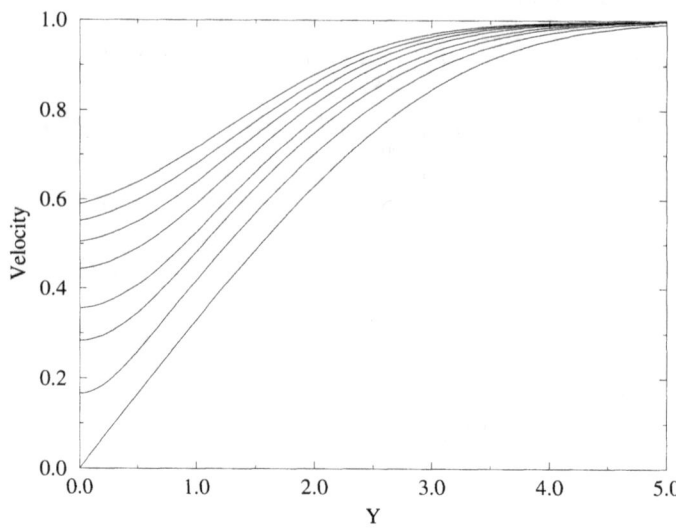

FIG. 2.8. Calculated velocity profiles using a 2-3 term composite expansion. The x locations are the same as in Figure 2.7.

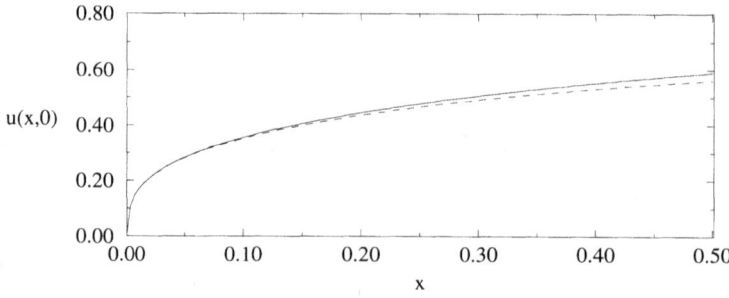

FIG. 2.9. Calculated centreline velocity for Goldstein's inner expansion (- - -) and for a 2-3 composite expansion (———).

The centreline velocity predicted by Goldstein's inner wake and by the composite expansions is illustrated in Figure 2.9. The velocity from the 2-2 and 2-3 composite expansions is identical on the centreline as $F'''(0) = 0$. It should be kept in mind that the plots are only the **leading** order term for the Reynolds number dependence in a double asymptotic expansion for both x and R and consequently show no effect of varying the Reynolds number. Such dependence would require solution of the next order term in the expansion (2.136) and (2.168). We shall consider this again towards the end of the next chapter. An example of the

Reynolds number dependence which has been observed for the wake centreline velocity can be found in Grove et al. (1964a) who saw that Goldstein's inner expansion gave slightly too high a centreline velocity compared to observations so that the composite solution, being larger, is even further from observations.

As a matter of summary, the constants which appear in section 2.6.3 and which have to be found numerically can be tabulated as follows.

$$\begin{array}{rcl} \beta_0 &=& 1.2168\ldots \\ \alpha_0 &=& 1.6552\ldots \\ \gamma_0 &=& 0.2445\ldots \\ \mu_0 &=& 1.1321\ldots \\ \lambda_0 &=& 0.8789\ldots \\ \gamma_1 &=& 0.0041\ldots \\ \mu_1 &=& -155.56\ldots \\ \lambda_1 &=& -0.1496\ldots \end{array}$$

Table 2.1 *Computed values of constants*

The application of formal composite expansions to compute solutions in boundary layer problems seems rather rare.

2.7 Far wake expansion

The near wake expansion, valid possibly as much as half a body length from the trailing edge is based on the boundary layer edge remaining the same transverse distance from the line on which the plate lies. Obviously far from the trailing edge we expect viscosity to have had a spreading effect in the transverse direction. Note that when we deal with wakes of jets into stationary fluid, we are used to thinking of the lateral spreading *outwards* of the jet due to viscous entrainment of fluid. In this case, near the plate the situation is better described as a lateral spreading *inwards* of momentum from the free stream. This is seen clearly in the successive profiles of Figure 2.6 where the position at which the velocity has returned to say 99% of the free stream velocity creeps inwards as the centreline velocity rises (it is also useful to look at diagrams such as Figure 2.6 in terms of a velocity defect, where the maximum defect on the centreline is decreasing faster than viscous effects diffuse laterally so that the velocity defect is both diminishing in magnitude and in width with increasing distance from the trailing edge). However, if we go further from the plate then although the region of disturbance from the wake widens the streamlines are always directed towards the centerline.

The first analysis of the far wake was by Tollmien (1931) and this was improved by Goldstein (1933a) who was not able to extend his approximation beyond a few terms. Later Meksyn (1951) considered the problem but it was the works of Imai (1951) and Stewartson (1957) which showed that the asymptotic sequence for large distance involved logarithmic terms as well as powers of the distance, which provided the correct understanding of the far wake.

The original analysis of Tollmien considered a linear perturbation to a uniform far field but retaining a stretched transverse coordinate, so that variations in the streamwise direction are much smaller than those in the transverse direction. This assumption was necessary to allow a similarity type analysis: without it, even for small perturbations, the far wake would have elliptic terms involving ∇^2.

Consider an expansion for the stream-function in (x, Y) coordinates with $Y = R^{1/2}y$,

$$R^{1/2}\psi \sim Y + \Psi_1(x, Y) + o(1), \text{ as } R \to \infty, \quad (2.173)$$

then the component of the stream-function, Ψ_1, must satisfy

$$(1 + \Psi_{1Y})\Psi_{1xY} - \Psi_{1x}\Psi_{1YY} = \Psi_{1YYY}, \quad (2.174)$$

in other words, a full non-linear boundary layer equation. Now suppose that for x large, the function Ψ_1 has small derivatives: then we can consider a linearisation of (2.174). In formal terms expand

$$\Psi_1 \sim \Psi_{1,0} + o(\Psi_{1,0}), \text{ as } x \to \infty, \quad (2.175)$$

where $\Psi_{1,0} = o(Y)$ for large x. Then the leading term $\Psi_{1,0}$ will satisfy

$$\Psi_{1,0xY} = \Psi_{1,0YYY}, \quad (2.176)$$

with boundary conditions $\Psi_{1,0Y} \to 0$ as $Y \to \infty$ and $\Psi_{1,0}(x, 0) = \dfrac{\partial^2 \Psi_{1,0}}{\partial Y^2}(x, 0) = 0$. This equation is a form of diffusion equation (if distance downstream is associated as a 'time'-like coordinate) so we also need an 'initial condition' that for some value of $x = x_0$ we are given $\Psi_{1,0}(x_0, Y)$. It will not be correct to use the boundary layer profile at the exit of the boundary layer on the plate (i.e. $x_0 = 0$) since we cannot be sure that disturbances at the rear of the plate do occur on transverse and longitudinal length scales which we are using in the far wake. Hence we have to accept an unspecified value where the 'initial' condition will be given. It will shortly be clear that for this term in the approximation, we do not need to specify x_0.

The solution of (2.176) is

$$\Psi_{1,0}(x, Y) = -A \int_0^{Y/\sqrt{2x}} e^{-\frac{1}{2}p^2} \, dp, \quad (2.177)$$

where A is an arbitrary constant. The source of this arbitrary constant is that although we have satisfied the boundary conditions at $Y = 0$ and as $Y \to \infty$, we have not yet used the 'initial' condition. Tollmien's solution to this difficulty was to use Karman's result that assuming the pressure had recovered far downstream to its upstream value and that the flow was essentially parallel, then conservation of momentum would give that an integral of $u(1-u)$ across the wake (and at our order of approximation in the far wake, the linearised version of this) must be constant and equal to the drag on the plate. If the drag on the plate is taken from the boundary layer drag then the constant A can be found. Of course, there are a number of matters which will have to be carefully dealt with. The drag on the plate is not that given by a Blasius solution over a section the same length as the plate since there will be corrections to that both because the plate is of finite length and from higher order solutions of the boundary layer equations which improve on the Blasius solution; however, we might expect that careful asymptotic analysis would resolve those technicalities.

Using (2.177), and returning to the unstretched coordinate y, the velocity in the x direction is

$$u(x,y) \sim 1 - \frac{A}{\sqrt{2x}}\exp(-\frac{y^2 R}{4x}) \text{ as } R \to \infty, \, x \to \infty, \qquad (2.178)$$

while the transverse velocity is

$$v(x,y) \sim -\frac{Ay}{2x^{3/2}}\exp(-\frac{y^2 R}{4x}). \qquad (2.179)$$

It is important to note again that v is negative so that the wake flow is always moving towards the centreline axis.

The errors in this approximation are $o(x^{-1/2})$ as $x \to \infty$ in the first term of an expansion which neglects terms $o(R^{-1/2})$ as $R \to \infty$.

Using Karman's integral for the drag per unit width, $\hat{D}$, on the two sides of the plate,

$$\hat{D} = \rho U^2 L \int_{-\infty}^{\infty} u(1-u)dy. \qquad (2.180)$$

As $u \sim 1$ a common approximation for the momentum flux defect integral is

$$\int_{-\infty}^{\infty} u(1-u)dy \sim \int_{-\infty}^{\infty}(1-u)dy, \qquad (2.181)$$

but care should be applied when considering the development of higher order expansions.

Substituting (2.178) into this integral, we have a drag coefficient c_D

$$c_D = \frac{\hat{D}}{\frac{1}{2}\rho U^2 \times 2L} \sim 2A\sqrt{\frac{\pi}{2R}} + \cdots. \qquad (2.182)$$

The leading order drag coefficient for the plate which generated the wake was given by (2.122)

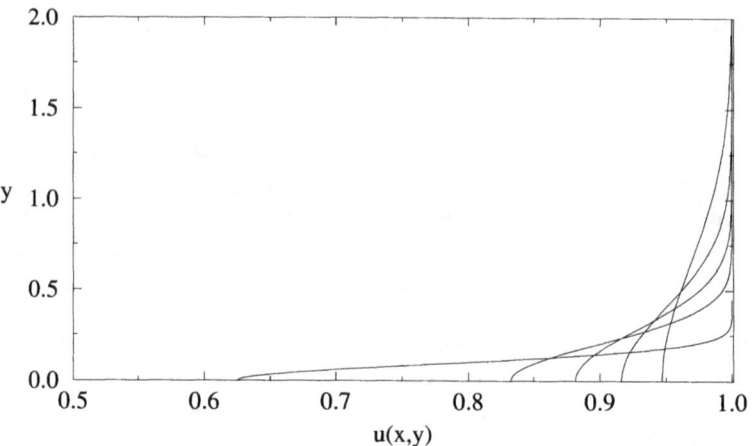

FIG. 2.10. Far wake velocity u from Tollmien's solution for (in increasing centreline velocity) $x = 1, 5, 10, 20, 50$.

$$c_D \sim \frac{1.33...}{\sqrt{R}} + o(R^{-1/2}), \qquad (2.183)$$

and if the leading term in (2.182) and (2.183) are equated, then

$$A = 0.662...\sqrt{2/\pi}$$

and for instance, the centreline velocity is

$$u(x,0) \sim 1 - \frac{0.662...}{\sqrt{\pi x}} + o(x^{-1/2}), \text{ as } x \to \infty, \qquad (2.184)$$

showing that it is possible to determine at least the leading order behaviour without knowing anything about the location of the 'initial' condition for the far wake expansion.

Calculation of the wake velocity $u(x,y)$ is shown in Figure 2.10 where the spreading of the far wake velocity profile is evident even though, as we have seen, the velocity v is negative and towards the $y = 0$ axis.

2.7.1 Higher order expansion for the far wake

The question of how to calculate further terms in the far field expansion and in the expansion for the drag coefficient, (2.183), is more problematical. We shall consider extensions due to Goldstein (1933a,b) Meksyn (1951) and Stewartson (1957) here before deferring further consideration until the end of the next chapter.

Goldstein was the first to attempt to find the higher order asymptotic expansion for the far wake, expanding the velocities in powers of $x^{-1/2}$. In terms of the function Ψ_1 defined in (2.173), and using

$$s = \frac{Y}{\sqrt{2x}}, \tag{2.185}$$

expand

$$\Psi_1 \sim AG_1(s) + \frac{A^2}{\sqrt{2x}}G_2(s) + \frac{A^3}{2x}G_3(s) + \cdots. \tag{2.186}$$

We have already seen that

$$G_1(s) = -\int_0^s e^{-\frac{1}{2}p^2}\,dp,$$

and Goldstein was able to calculate G_2,

$$G_2(s) = -\int_0^s [e^{-p^2} + \sqrt{\frac{\pi}{2}}p\,e^{-\frac{1}{2}p^2}\text{erf}(\frac{p}{\sqrt{2}})]\,dp,$$

so that the three-term centreline velocity was

$$u(x,0) \sim 1 - \frac{0.662...}{\sqrt{\pi x}} - \frac{0.438...}{\pi x} + o(x^{-1}). \tag{2.187}$$

However, Goldstein's attempt to extend the series resulted in the solution for the next term, G_3, which included a term with algebraic decay as $Y \to \infty$ and so the expansion failed since the wake must decay exponentially in Y for large Y (see also section 2.3.3).

It is interesting to note that the second order contribution to the streamfunction is significantly different from the first contribution since although both G_1 and G_2 tend to no-zero constants as $Y \to \infty$, their combination with the appropriate x dependence means that the second order term provides a non-zero transverse velocity which would have to be matched to a suitable outer potential. In this case the transverse velocity is

$$R^{1/2}v \sim \frac{A^2 G_2(\infty)}{2\sqrt{2}x^{3/2}}, \tag{2.188}$$

showing again that while the wake may be thought of as spreading outwards due to viscous action, the outer potential flow 'sees' a body shrinking back to the axis from the parabolic shape established from flow around the leading edge of the plate (as $G(\infty) < 0$). It is an important aspect of boundary layer flows that while exponential decay is demanded in the transverse direction to reflect the diffusive nature of viscous action that should not be applied to variation in the streamwise direction.

Goldstein conjectured that the difficulty in obtaining a third term lay in the form of the expansion (2.186) and as we shall see shortly, the expansion does need to include logarithmic terms, a result shown by Stewartson (1957).

An attempt to correct the difficulty which emerged in Goldstein's expansion was made by Meksyn (1951) who used the same wake variable, (2.185). He expanded the stream-function by

$$R^{1/2}\psi \sim \sqrt{2x}f(x,s), \tag{2.189}$$

so that the equation for the function f could be written in the form

$$\frac{\partial^3 f}{\partial s^3} + f\frac{\partial^2 f}{\partial s^2} = 2x\{\frac{\partial f}{\partial s}\frac{\partial^2 f}{\partial s \partial x} - \frac{\partial^2 f}{\partial s^2}\frac{\partial f}{\partial x}\}. \tag{2.190}$$

The boundary conditions were that

$$f(x,0) = \frac{\partial^2 f}{\partial s^2}(x,0) = 0, \text{ and } \frac{\partial f}{\partial s} \to 1 \text{ as } s \to \infty, \tag{2.191}$$

so that defining

$$F(x,s) = \int_0^s f(x,p)\mathrm{d}p, \tag{2.192}$$

and regarding (2.190) as an ordinary differential equation in s for fixed x, then Meksyn could write down a formal implicit solution,

$$\frac{\partial^2 f}{\partial s^2} = \mathrm{e}^{-F(x,s)}\int_0^s 2x\{\frac{\partial f}{\partial s}\frac{\partial^2 f}{\partial s \partial x} - \frac{\partial^2 f}{\partial s^2}\frac{\partial f}{\partial x}\}\mathrm{e}^{F(x,p)}\mathrm{d}p. \tag{2.193}$$

Meksyn proposed that an iterative solution of (2.193) would both converge and be free of the difficulty encountered by Goldstein.

Letting

$$a(x) = u(x,0) = \frac{\partial f}{\partial s}(x,0),$$

denote the centreline velocity, Meksyn attempted a power series expansion

$$f(x,s) \sim a(x)s + \frac{2xa(x)a'(x)}{6}s^3 + O(s^5), \tag{2.194}$$

and substituting (2.194) into the right-hand side of (2.193)

$$\frac{\partial^2 f}{\partial s^2} = \mathrm{e}^{-\frac{1}{2}as^2+O(s^4)}\int_0^s \{2xaa' + O(p^2)\}\mathrm{e}^{\frac{1}{2}ap^2+O(p^4)}\mathrm{d}p. \tag{2.195}$$

Meksyn then made a rather severe approximation, expanding the exponential inside the integral as a power series and integrating to obtain

$$\frac{\partial^2 f}{\partial s^2} = 2xaa'se^{-\frac{1}{2}as^2} + o(s), \qquad (2.196)$$

and this was integrated again to give

$$\frac{\partial f}{\partial s} \sim a + 2xaa'(1 - e^{-\frac{1}{2}as^2}). \qquad (2.197)$$

Finally he let $s \to \infty$ to obtain a differential equation for $a(x)$,

$$2xaa' + a = 1, \qquad (2.198)$$

with leading order solution

$$a(x) \sim 1 + Ax^{-1/2} \text{ as } x \to \infty, \qquad (2.199)$$

where the constant A was found from a momentum flux balance against the drag on the plate and so provided the same two leading terms as had been found by Tollmien.

The two-term solution for f was then

$$f(x,s) \sim s + \frac{A}{\sqrt{x}} \int_0^s e^{-\frac{1}{2}ap^2} dp. \qquad (2.200)$$

It seems important to note two reservations about Meksyn's expansion. First, it is very much an ad-hoc procedure, which by retaining the function a within the exponentials (and in other terms at higher order) for f introduces an inconsistent expansion. Secondly, basing the expansion around a power series and then integrating over s from zero to infinity may cause difficulty. Of course, this may not be significant since the terms involving f in the argument of the integral on the right-hand side of (2.193) should decay as e^{-F} so that the overall right-hand side of (2.193) should decay exponentially. This is however not the case for the very first term in an expansion

$$f(x,s) \sim a(x)s,$$

since through the term $f_s f_{xs} \sim aa'$ the right-hand side of (2.193) cannot be guaranteed to decay exponentially for large s. Meksyn obtained exponential decay by expanding the exponential term e^F within the integral as a power series whereas the correct form, (2.195) shows that it is dangerous to subsequently take $s \to \infty$ which of course is necessary to match the expansion to the velocity outside the wake.

Meksyn extended his approximation to obtain a third term and obtained a centreline velocity which had a different coefficient to Goldstein's third term. Meksyn obtained

$$u(x,0) \sim 1 - \frac{0.662...}{\sqrt{\pi x}} - (\frac{5}{3} - \frac{\pi}{2})\frac{0.438...}{\pi x} + o(x^{-1}), \qquad (2.201)$$

differing from Goldstein by the leading factor in the third term. He thought that determining higher order terms would not lead to the inconsistency which had been observed by Goldstein and would only be algebraically involved.

The resolution of how the wake expansion should be formulated was provided by Stewartson (1957) who showed that as Goldstein had thought, the simple power series expansion in $x^{-1/2}$ was the cause of Goldstein's technical difficulty and that the expansion would have to involve logarithmic terms as well as powers of $x^{-1/2}$. Stewartson's method was to suppose, as had Goldstein, that the velocity should be expanded

$$u(x,Y) \sim 1 - u_1(x,Y) - u_2(x,Y) - \cdots,$$

and that the terms u_1 and u_2 calculated by Goldstein were correct. Then, as Goldstein had derived, the problem for the term u_3 was

$$\frac{\partial^2 u_3}{\partial Y^2} - \frac{\partial u_3}{\partial x} = \frac{1}{x^{5/2}} F_3\left(\frac{Y}{\sqrt{2x}}\right), \qquad (2.202)$$

where the function $F_3(s)$ was

$$F_3(s) = \frac{1}{4} A^3 \left\{ 3 e^{-\frac{3}{2}s^2} + \sqrt{2\pi} s e^{-s^2} \mathrm{erf}\left(\frac{s}{\sqrt{2}}\right) + \sqrt{\pi} s e^{-\frac{1}{2}s^2} \mathrm{erf}(s) \right\}. \qquad (2.203)$$

Stewartson looked at (2.202) in a wider context, supposing that at some value x_a the velocity field is specified by $u = u_a(Y)$ (an 'initial' condition) and the subsequent development is sought using results from the theory of unsteady heat conduction (with x being the time-like variable). Then the solution of (2.202) could be written down and would involve integrals over the 'initial' solution including an integral

$$\int_{-\infty}^{\infty} d\bar{Y} \int_{x_a}^{x} d\bar{x} \frac{\bar{x}^{-5/2} F_3(\bar{Y}/\sqrt{2\bar{x}})}{\sqrt{2\pi(x-\bar{x})}} \exp -\frac{(Y-\bar{Y})^2}{4(x-\bar{x})}, \qquad (2.204)$$

which is equal to

$$\int_{x_a}^{x} \frac{\bar{x}^{-2}}{\sqrt{\pi(x-\bar{x})}} \int_{-\infty}^{\infty} F_3(p) \exp -\frac{\bar{x}(Y/\sqrt{2\bar{x}} - p)^2}{2(x-\bar{x})} \, dp \, d\bar{x}.$$

It was the behaviour of this integral which Stewartson examined for large x. To determine the asymptotic behaviour as $x \to \infty$ he integrated by parts in $\bar{x}$, and for simplicity he took the case when $Y = 0$ where the integral became

$$\frac{2}{x_a \sqrt{\pi(x-x_a)}} \int_0^{\infty} F_3(p) \exp\left[-\frac{1}{2}\frac{x_a p^2}{x - x_a}\right] dp$$

$$+ \int_{x_a}^{x} \frac{\bar{x}^{-1}}{\sqrt{\pi}(x-\bar{x})^{3/2}} \int_0^{\infty} F_3(p)\left(1 - \frac{xp^2}{x-\bar{x}}\right) \exp\left[-\frac{1}{2}\frac{\bar{x}p^2}{x-\bar{x}}\right] dp \, d\bar{x}.$$

When this was expanded for large x the integral had leading terms

$$\frac{2}{x_a \sqrt{\pi x}} \int_0^{\infty} F_3(p) dp + \frac{1}{\sqrt{\pi} x^{3/2}} \log x \int_0^{\infty} (1-p^2) F_3(p) dp + O(x^{-3/2}). \qquad (2.205)$$

Since there was no other source of a logarithm term which might cancel the term in the integral, Stewartson deduced that the expansion used by Goldstein was indeed incorrect and the correct expansion for the stream-function perturbation should begin

$$\Psi_1 \sim AG_1(s) + \frac{A^2}{\sqrt{2x}}G_2(s) + \frac{A^3}{2x}(G_3(s) + H_3(s)\log x) + \cdots, \quad (2.206)$$

instead of (2.186). In that case the functions G_3 and H_3 satisfied

$$H_3''' + sH_3'' + 3H' = 0, \quad (2.207)$$

and

$$G_3''' + sG_3'' + 3G_3' = F_3 - H_3, \quad (2.208)$$

with boundary conditions at $G_3(0) = G_3''(0) = H_3(0) = H_3''(0) = 0$ and both functions had first derivatives vanishing exponentially as $s \to \infty$. Then the solution for H_3 was

$$H_3(s) = c \int_0^s (1-p^2)e^{-\frac{1}{2}p^2} dp, \quad (2.209)$$

where c was a constant which had to be determined. The solution for G_3 was written in the form

$$G_3(s) = b(1-s^2) + \tilde{G}_3(s), \quad (2.210)$$

where b was a further constant and $\tilde{G}_3$ was a particular integral of (2.208) satisfying the boundary conditions. Goldstein had found that the coefficient of the term with algebraic decay was

$$\int_0^\infty (1-p^2)F_3(p)dp,$$

and of course the whole difficulty he had faced was determined by this integral not vanishing. In Stewartson's expansion, the corresponding coefficient of the algebraically decaying term was

$$\int_0^\infty (1-p^2)[F_3(p) - H_3(p)]dp,$$

and this would indeed vanish if the constant c was chosen from

$$c\int_0^\infty (1-p^2)^2 e^{-\frac{1}{2}p^2} dp = \int_0^\infty (1-p^2)F_3(p)dp.$$

There was, however, no way to determine the constant b by considering the far wake solution only. Stewartson conjectured that the value might be found from matching to the initial velocity profile $u_a(Y)$ (or equivalently, the near wake).

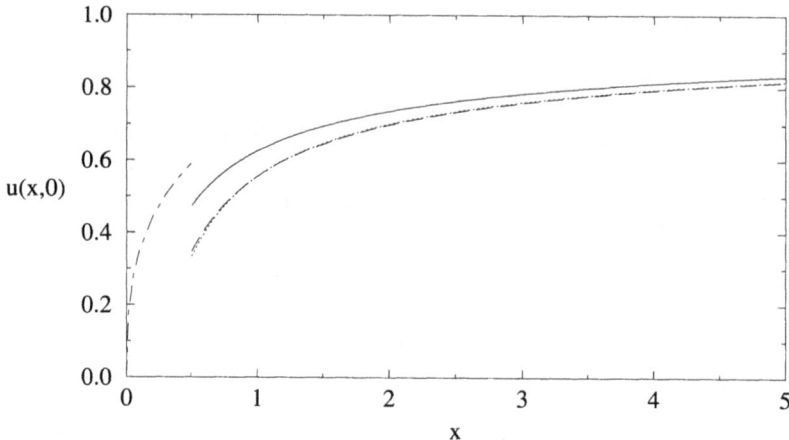

FIG. 2.11. Centreline velocity for the far wake. Two-term expansion, (———). Three-term expansion, (· · ·). Expansion with logarithmic term included, (- - - -). Near wake centreline velocity, (- · - · -).

This did, however, enable Stewartson to express the wake centreline velocity as

$$u(x,0) \sim 1 - \frac{A}{(2x)^{1/2}} - \frac{1}{2}\frac{A^2}{(2x)} - \frac{1}{4\sqrt{3}}\frac{A^3}{(2x)^{3/2}}\log x + O(x^{-3/2}). \quad (2.211)$$

The constant A has again to be found from a momentum flux balance. The expansion developed for the far wake is what might be called a reactive wake, it merely considers the development from a given starting condition without allowing the wake development to interact with the boundary layer flow around the body. At first sight, applying a momentum flux balance at a point x using (2.180) would lead to a complicated series (in A and x) expansion so it is necessary to refine the definition (2.180) to

$$\hat{D} = \lim_{x\to\infty} \rho U^2 L \int_{-\infty}^{\infty} u(1-u)dy = \lim_{x\to\infty} \rho U^2 L \int_{-\infty}^{\infty}(1-u)dy, \quad (2.212)$$

so that to order $R^{-1/2}$ the coefficient $A = 0.662..\sqrt{2/\pi}$ as before.

The predicted centreline velocity from (2.211) is illustrated in Figure 2.11 where it is clear that the major contributions to the velocity are given by a three term expansion (taking the leading $u \sim 1$ as the first term) although if the contribution from a term of order $x^{-3/2}$ were included the distinction between the three term and the full four term expansion might be more evident. Also shown for comparison is the centreline velocity predicted from the near wake solution.

2.8 The drag on a finite plate

We would expect that the leading order drag coefficient from a finite leading section ($x = 1$) of a semi-infinite plate, (2.122),

$$c_D \sim \frac{1.33...}{R^{1/2}},$$

would be all that might be needed for most practical purposes. However, the question of determining the correct asymptotic structure and coefficients for the drag has been extensively studied for this simplest of cases both for its own interest and for its relevance to how drag coefficients could be calculated in more complicated circumstances. The most obvious deficiency in using the leading section of a semi-infinite plate to predict drag is that it allows no feedback from the wake region to the plate. As fluid is accelerated in the wake immediately behind the plate there will be entrainment of fluid leading to modification of the wall shear on the plate and the viscous drag. This extra motion will also affect the outer potential flow field and so modify the overall pressure drag. We have seen that examination of the far wake leads to no new insight about the drag since the far wake solution revolves around knowing the drag in order to determine the wake. This question will to some extent be resolved in the next chapter since the triple deck structure does provide feedback from the wake into the shear stress on the plate wall near the trailing edge. In the remainder of this section we shall discuss two ideas relevant to a finite plate, the Oseen approximation for a finite plate which provides an estimate of the drag coefficient for low Reynolds numbers, and Kuo's theory of a constant width outer wake as a means of allowing the wake to influence the wall shear stress.

2.8.1 *The Oseen limit*

Although our main interest is in high Reynolds number flow, there was a time when it was hoped that expansions for low Reynolds number might provide useful information about moderate Reynolds number flows, so for completeness we shall briefly consider the Oseen limit for flow around a flat plate.

The singular solution determined by Lamb, (1.32), was used by Bairstow et al. (1923) to model flow around a flat plate by a distribution of doublets. This followed an earlier work, Bairstow et al. (1922), which had modelled Stokes flow in limited regions by distributions of singular solutions (since the regions were bounded in some way, Stokes paradox was avoided and that allowed Stokes flow around a cylinder in a channel to be considered). Assuming a distribution of dipoles with strength $\sigma(x)$ on the segment of line $0 \leq x \leq 1$ and that the velocity there was $(-1, 0)$ then the distribution σ had to satisfy

$$\int_0^1 \{e^p K_0(p) - \frac{1}{R}\frac{\partial}{\partial p}[e^p K_0(p) + \frac{1}{2}\log p^2]\}\,d\sigma(\tilde{x}) = \frac{\pi}{R}, \qquad (2.213)$$

where $p = R(x - \tilde{x})/2$.

Bairstow, Cave & Lang sought a numerical solution for σ using quadrature. Later Piercy and Winny (1933) developed an approximate solution for small R by assuming a given functional form for $\sigma(x)$. In the case

$$\frac{d\sigma}{dx} = \frac{A}{\sqrt{x(1-x)}}, \qquad (2.214)$$

they found from substitution into (2.213) and retention of leading order terms for small R,

$$A = \frac{1}{1 - \gamma - \log R/2}, \qquad (2.215)$$

where $\gamma = 0.5772\cdots$ was Euler's constant. The drag coefficient was then

$$c_D \sim \frac{4}{R} \int_0^1 d\sigma(\tilde{x}) = \frac{4\pi}{R(1 - \gamma - \log R/2)} \text{ as } R \to 0. \qquad (2.216)$$

The approximation has been be carried to higher terms and these Oseen solutions provided behaviour for the drag coefficient at low Reynolds numbers to compare (and in some cases, match) with high Reynolds number expansions.

2.8.2 Kuo's constant outer width wake model

The difficulty with wake models being purely reactive, that is determining the spatial evolution a velocity profile taken from the solution on an **infinite plate** without that evolution having any feedback into the boundary layer development, was addressed by Kuo (1953) who used an ad-hoc model to examine the mechanism by which a wake could influence the boundary layer development on the upstream plate by providing an adjustment to the shear on the plate.

We have already seen (from (2.54)) that for a semi-infinite plate the outer potential flow is determined by a transverse velocity

$$v \sim \frac{\beta_0}{\sqrt{2Rx}},$$

which corresponds to sources along the plate. We have also seen that for a finite plate, the far wake determines an outer potential flow with transverse velocity (from (2.188))

$$v \sim -R^{-1/2}(x-1)^{-3/2} \sim -R^{-1/2}x^{-3/2}.$$

Thus to the outer potential flow the body and the wake appears as a distribution of first sources and then sinks. These sinks have the effect of accelerating fluid into the wake centreline region from the boundary layer and that will have an effect on the boundary layer over the plate, increasing wall shear and thus increasing the drag on the plate.

Kuo's idea was to model a finite plate as an object with a special outer potential flow field, one corresponding to a transverse velocity field

$$v \sim \begin{cases} \dfrac{\beta_0}{\sqrt{2Rx}}, & 0 < x \le 1, \\ 0, & x > 1, \end{cases} \qquad (2.217)$$

so that the boundary layer on the object remained of constant thickness for $x > 1$. It turned out that this formulation also led to a singularity in the boundary layer formulation at the trailing edge, $x = 1$ but as Kuo's solution was not integrated past that point in the x-direction, the conditions at $y = 0$ for $x > 1$ were not relevant to Kuo's main result. However, a physically plausible object which might have the same outer potential flow as used by Kuo would be a semi-infinite plate with a variable distribution of suction, $v_W(x)$, $x > 1$ along it where the distribution v_W is unknown but might be expected to behave as

$$v_w \sim -\frac{\beta_0}{\sqrt{2Rx}}, \quad x \gg 1,$$

so as to maintain the boundary layer thickness constant. Note that this still would not provide the outer flow (2.8.2) to match the far wake.

The importance of Kuo's work was that while he did not model a wake-plate interaction directly, he did attempt to understand the effect of terminating the plate boundary layer growth on the boundary layer over the plate and in doing so showed that the feedback from the wake to the boundary layer would provide an important contribution to any second order estimate of the drag. Kuo's view was that because the wake potential decayed to a small value so quickly as x became large, taking $v = 0$ would provide a good estimate of the effect of a wake on the plate boundary layer.

The solution for the outer flow determined by (2.217) was a complex potential

$$w \sim z + \frac{\beta_0}{\pi}\sqrt{\frac{2}{R}} \int_0^1 t^{-1/2} \log(t - z) \, dt, \qquad (2.218)$$

with $z = x + iy$ so that excepting a constant,

$$w = z + \beta_0 \sqrt{\frac{2}{R}} \left[-i\sqrt{z} + \frac{1}{2\pi} \log(1 - z) + \frac{1}{2\pi} \sqrt{z} \log \frac{1 + \sqrt{z}}{1 - \sqrt{z}} \right]. \qquad (2.219)$$

The last term of (2.219) will introduce a factor of $i\pi/\sqrt{z}$ if a minus sign is introduced inside the logarithm, reducing the v velocity to zero for $x > 1$, $y = 0$ as required by the boundary condition (2.217).

The complex velocity for $|z| < 1$ was

$$u - iv \sim 1 - i\beta_0 \sqrt{\frac{1}{2Rz}} + \frac{\beta_0}{\pi\sqrt{2Rz}} \log \frac{1 + \sqrt{z}}{1 - \sqrt{z}}, \qquad (2.220)$$

and on $y = 0$ the velocity along the plate was

$$u \sim 1 + \frac{\beta_0}{\pi\sqrt{2Rx}} \log \frac{1+\sqrt{x}}{1-\sqrt{x}}. \tag{2.221}$$

This velocity was expanded for x on the plate as

$$u \sim 1 + \frac{\sqrt{2}\beta_0}{\pi\sqrt{R}}\left(1 + \frac{x}{3} + \frac{x^2}{5}\cdots\right) + o(R^{-1/2}), \quad 0 < x < 1, \tag{2.222}$$

and that corresponds to a pressure on the plate,

$$p \sim -\frac{\sqrt{2}\beta_0}{\pi\sqrt{R}}\left(1 + \frac{x}{3} + \frac{x^2}{5}\cdots\right) + o(R^{-1/2}), \quad x < 1, \tag{2.223}$$

that is a favourable pressure gradient, indicating as expected, an acceleration for fluid in the boundary layer on the plate.

When we examined the boundary layer over a semi-infinite plate in the absence of a pressure gradient, we saw that the stream-function expansion would not have a term of order R^{-1}. Here, however, the pressure (2.223) implies that there should be a term of this order. Using a boundary layer coordinate Y and similarity variable $s = Y/\sqrt{2x}$ as before, the stream-function can be expanded according to,

$$\psi \sim R^{-1/2}\Psi_0(x, Y) + R^{-1}\Psi_1(x, Y) + o(R^{-1}), \tag{2.224}$$

where the leading order term has already been given by (2.25)

$$\Psi_0 = \sqrt{2x}F_0(s),$$

and satisfies (2.26). If we denote the pressure by $p = R^{-1/2}p_1(x)$ the second term in the stream-function expansion satisfies an inhomogeneous version of (2.58),

$$\frac{\partial^3 \Psi_1}{\partial Y^3} + \Psi_{0x}\frac{\partial^2 \Psi_1}{\partial Y^2} + \Psi_{0YY}\frac{\partial \Psi_1}{\partial x} - \Psi_{0Y}\frac{\partial^2 \Psi_1}{\partial x \partial Y} - \Psi_{0xY}\frac{\partial \Psi_1}{\partial Y} = p_1'(x). \tag{2.225}$$

As p_1' is known as a power series in x, Kuo sought an analogous power series solution of (2.225). The stream-function was expanded

$$\Psi_1 \sim \frac{\beta_0}{\pi}\left(\sqrt{2x}k_1(s) + \frac{1}{2\cdot 3}(2x)^{3/2}k_2(s) + \frac{1}{2^2\cdot 5}(2x)^{5/2}k_3(s) + \cdots\right), \tag{2.226}$$

so that the functions k_r all satisfied the same boundary conditions:

$$k_r(0) = k_r'(0) = 0, \quad k_r'(s) \to 1 \text{ as } s \to \infty. \tag{2.227}$$

Substituting (2.226) into (2.225), the functions k_r are determined by

$$k_r''' + F_0 k_r'' - 2(r-1)F_0'k_r' + (2r-1)F_0''k_r = -(r-1). \tag{2.228}$$

Note that there are minor differences compared with Kuo's original work because the definition of s is slightly different. In general the functions k_r have to be found by integration. That being done, the additional drag on the plate could be found. The shear on the plate would come from

$$\frac{\partial^2 \psi}{\partial Y^2} \sim \frac{1}{\sqrt{2xR}} F_0''(0) + \frac{\beta_0}{\pi R} \left(\frac{1}{\sqrt{x}} k_1''(0) + \frac{1}{3}\sqrt{x} k_2''(0) + \frac{1}{5} x^{3/2} k_2''(0) + \cdots \right), \quad (2.229)$$

and the drag coefficient defined by (2.120) is

$$c_D = 2 \int_0^1 \frac{\partial^2 \psi}{\partial Y^2}(x,0) dx \sim \frac{2\sqrt{2} F_0''(0)}{\sqrt{R}} + \frac{4\beta_0}{\pi R} \left(k_1''(0) + \frac{1}{3^2} k_2''(0) + \cdots \right). \quad (2.230)$$

Using numerical solutions for the functions k_r, Kuo estimated the coefficient of R^{-1} in (2.230) to be 4.12. His method does of course neglect the leading edge drag found by Imai, (2.121) and higher order drag from the plate boundary layer. If Kuo's result is just added to that of Imai, the prediction of drag coefficient is

$$\boxed{c_D \sim \frac{1.33}{(R)^{1/2}} + \frac{6.44}{R} - \frac{2.20}{(R)^{3/2}} \log \sqrt{R} + O(R^{-3/2}).} \quad (2.231)$$

Van Dyke (1962b) thought that Kuo's estimate for the R^{-1} coefficient was too high because the series expansion used by Kuo converged very slowly and that a value of 3 might be better giving a coefficient of 5.3 when combined with the leading edge drag. On the other hand Kuo had neglected the extra 'sink' distribution needed to provide the far wake transverse velocity, (2.8.2), so if anything the coefficient should be higher than that value.

The choice by Kuo of a constant width for the boundary layer downstream of $x = 1$ should not be a serious objection to the idea behind his work. A downstream transverse velocity which is closer to the asymptotic form could be used with the result that a slightly higher drag coefficient would be calculated.

As an example, the outer velocity for $x > 1$ could be specified by the asymptotic form from the far wake solution of Goldstein,

$$v \sim \begin{cases} \dfrac{\beta_0}{\sqrt{2Rx}}, & 0 < x \le 1, \\[2mm] -\dfrac{\gamma_W}{Rx^{3/2}}, & x > 1, \end{cases} \quad (2.232)$$

where $\gamma_W = -A^2 G_2(\infty)/\sqrt{8} > 0$ has been determined from (2.188). This uses the leading term of Goldstein's expansion (so that $(x-1)^{-3/2}$ has been replaced by $x^{-3/2}$) and consequently is likely to underestimate the increase in wall shear due to the wake development.

In that case the outer complex velocity field would be given by

$$u - iv \sim 1 + \frac{\beta_0}{\pi\sqrt{2Rz}} \log \frac{1+\sqrt{z}}{1-\sqrt{z}} - \frac{\gamma_W}{\pi R} \int_1^\infty \frac{1}{t^{3/2}(t-z)} dt. \quad (2.233)$$

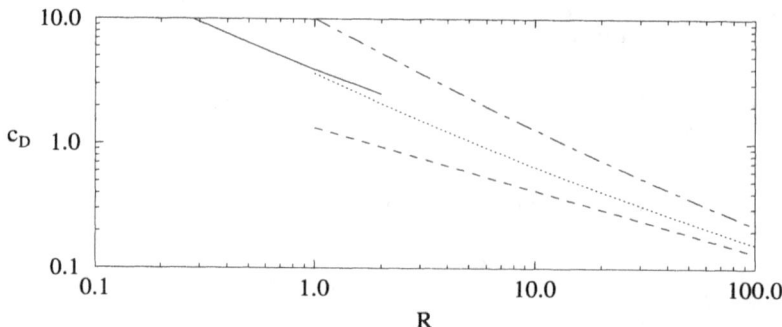

FIG. 2.12. Drag coefficient for a finite plate calculated using (——) Oseen's low Reynolds number approximation, (- - -) leading order Blasius boundary layer, (···) two terms from Blasius boundary layer, (— · — · —) Kuo's two-term approximation.

The additional term can be integrated to give a change to the velocity

$$-\frac{\gamma w}{\pi R}\int_1^\infty \frac{1}{t^{3/2}(t-z)}dt = \frac{\gamma w}{\pi R}\left[z^{-3/2}\log\frac{1+\sqrt{z}}{1-\sqrt{z}} - 2z^{-1}\right], \quad (2.234)$$

so that on the plate the u velocity will be increased by

$$\frac{\gamma w}{\pi R}\left[x^{-3/2}\log\frac{1+\sqrt{x}}{1-\sqrt{x}} - 2x^{-1}\right] \sim \frac{2\gamma w}{\pi}\left[\frac{1}{3} + \frac{1}{5}x + \frac{1}{7}x^2 + \cdots\right], \quad (2.235)$$

which has exactly the same polynomial terms as the expansion in (2.222). Thus the additional contribution would add term by term to the drag coefficient by a small amount (unfortunately the amount would be different for each term). This increase in u velocity is only what would be expected from the additional sink strength associated with (2.232) compared to (2.217), although the details are complicated by the need to retain extra terms in the pressure gradient from the square of the order $R^{-1/2}$ contribution which would be the same size as this extra contribution. Of course, what we shall find in the next chapter is that proper account of the flow in the wake near the trailing edge leads to a trailing edge drag which is asymptotically larger than just order R^{-1} and the 'sink' effect behind the trailing edge perturbs the main boundary layer much more than has been predicted here.

To finish this section the drag coefficient using the various results we have considered is shown in Figure 2.12. Further discussion of the drag on a plate is deferred to the next chapter.

2.9 Summary

This concludes our study of boundary layer theory for a flat plate up to 1969. There were a surprising number of unresolved difficulties and for which triple deck theory will only account for the structure of flow near the trailing edge.

For flow around a semi-infinite plate the general structure of the high Reynolds number asymptotic expansion is understood but how to calculate coefficients for the higher order terms is still not known. It is assumed that having full details of the flow near the leading edge would enable the unknown constants which enter the expansion to be found by matching the boundary layer expansions to the leading edge solutions. One important feature of boundary layer expansions for a plate was the need for logarithmic terms to occur in the expansion in order for the vorticity to decay exponentially far from the plate.

If the plate is of finite extent then conventional boundary layer scaling leads to a singularity at the trailing edge for the transverse velocity. The wake immediately behind the plate was considered in two regions, one near the plate centreline and one further away but still of a boundary layer thickness. Here the complexity of the expansions in distance should not disguise that they are but part of the leading order Reynolds number expansion. Further downstream the wake evolution is in a sense independent of the body, requiring only knowledge of the overall drag for the leading term to be found.

3

THE TRIPLE DECK

3.1 Introduction

The singularity at the trailing edge is very severe. It is not the case that some local expansion near the trailing edge but deep within the boundary layer could remove the singularity. If you study Figure 2.6, page 55, then it is clear that in the inner and outer wake expansions there is a singularity in the vertical velocity at $x = 0^+$. Thus to remove the singularity will require quite a major change in the way the expansions are formulated near the trailing edge. That it took so long to resolve this problem has a parallel in the development of higher order expansions for the boundary layer on a semi-infinite flat plate. It was by examining the nature of the outer potential flow solution that understanding of how to resolve the trailing edge singularity most easily came about (Stewartson (1969), Messiter (1970)).

The history of the 'triple deck' idea is very interesting. It must be remembered that the major challenge was not the singularity at the trailing edge of a plate but rather the singularity at separation in the boundary layer equations. We will discuss that problem in more detail in later chapters. Thus the major thrust of boundary layer research was into whether the boundary layer equations might be adjusted to predict separation and indeed how separation came about. Lighthill (1953b) had considered linearised perturbations to a boundary layer in order to use Fourier transforms in the streamwise direction. He was able to derive two important results. First that the decay of perturbations to the boundary layer would occur on a scale $R^{1/8}$ times the boundary layer thickness and although he did not write it down explicitly, he had reached the longitudinal scale factor $R^{-3/8}L$ which we shall shortly see is a key scale. Secondly he was able to show that the interaction between viscosity and the *perturbation* to the boundary layer would occur in a *sublayer*. As he was dealing with linearised perturbations, he was only able to demonstrate that this sublayer would be $R^{-1/4}$ the thickness of the undisturbed boundary layer, that is $R^{-3/4}L$. In Stewartson (1968) a linearised theory near the trailing edge of a plate was developed, but he too held to a lateral length scale of order $R^{-3/4}$ and so did not bring out the pressure interaction.

The derivation of the suitable length scales to study disturbances to a boundary layer and their important consequence of an interaction between the pressure and the boundary layer emerged in four different papers around 1969. Stewartson and Williams (1969) considered the how separation might come about and formulated their attack on this problem using a longitudinal scale $R^{-3/8}L$ and a sublayer scale $R^{-5/8}L$. Stewartson (1969) considered the trailing edge problem

Introduction

using the same scales and introduced the notation of a 'triple deck'. Messiter (1970) also dealt with the trailing edge problem and derived essentially the same results as Stewartson. Neiland (1969) considered the problem of a separating supersonic boundary layer and derived the triple deck scalings although his work was not translated into English until 1972 and so there was a delay in recognising the important contribution he too had made to this area. The extent to which all of these authors should be associated with the 'triple deck' idea is shown by the fact that the submission dates of the various manuscripts for publication were only between December 1968 and March 1969. In this chapter we focus on the resolution of the trailing edge singularity and so on the work of Stewartson and Messiter; in later chapters we shall look at separation and the consequences of the work of Stewartson & Williams and Neiland, most notably for incompressible flow by Sychev (1972), Messiter and Enlow (1973) and Smith (1977a).

Retain coordinates centred on the trailing edge with x, y scaled by the plate length and $Y = R^{1/2}y$ a boundary layer scale coordinate. In that case the stream-function upstream of the trailing edge is given to leading order in the boundary layer by

$$R^{1/2}\psi \sim \sqrt{2(1+x)} F_0\left(\frac{Y}{\sqrt{2(1+x)}}\right), \tag{3.1}$$

and for x small and negative,

$$R^{1/2}\psi \sim \sqrt{2} F_0\left(\frac{Y}{\sqrt{2}}\right) + \frac{1}{\sqrt{2}} x [F_0\left(\frac{Y}{\sqrt{2}}\right) - \frac{Y}{\sqrt{2}} F_0'\left(\frac{Y}{\sqrt{2}}\right)], \tag{3.2}$$

where all higher order terms in R will give terms $O(x)$ and hence considerable care will be needed to extend ideas to higher order. If the properties of F_0 already derived are used as $Y \to \infty$ they show that in the potential flow region the stream-function must satisfy

$$\boxed{\psi \sim y - R^{-1/2}\sqrt{2}\beta_0 + O(x), \text{ as } y \to 0, \ x < 0.} \tag{3.3}$$

On the other hand, downstream of the trailing edge and in the outer wake, the stream-function has to satisfy

$$R^{1/2}\psi \sim \sqrt{2} F_0\left(\frac{Y}{\sqrt{2}}\right) + x^{1/3}\frac{\mu_0}{\lambda_0} F_0'\left(\frac{Y}{\sqrt{2}}\right), \tag{3.4}$$

so that in the potential flow region,

$$\boxed{\psi \sim y + R^{-1/2}[x^{1/3}\frac{\mu_0}{\lambda_0} - \sqrt{2}\beta_0], \text{ as } y \to 0, \ x > 0.} \tag{3.5}$$

If these two expansions, (3.3) and (3.5), are to be correct, the leading terms of the stream-function in the potential flow should be given by a complex potential

$$w(z) \sim z - i\beta_0\sqrt{2}R^{-1/2} + (a+ib)z^{1/3}R^{-1/2}, \text{ as } z \to 0, \qquad (3.6)$$

where a and b are real numbers which must to be chosen so that the stream-function satisfies (3.3) and (3.5) on $y = 0$.

If $z \to |x|e^{0i}$, then

$$\Im w \to y - \beta_0\sqrt{2}R^{-1/2} + b|x|^{1/3}R^{-1/2},$$

so it must be that $b = \mu_0/\lambda_0$ to match (3.5). While if $z \to |x|e^{\pi i}$ it is necessary that

$$\Im[(a+ib)e^{\pi i/3}] = 0$$

to remove the dependence on $|x|^{1/3}$ and to match (3.3): hence

$$a = -b/\sqrt{3}.$$

It is interesting to see just what shape 'body' is given by a potential flow (3.6). In Figure 3.1 we show the body shape near the trailing edge for a Reynolds number $R = 500$. It is clear that in the potential region there is some readjustment upstream of the trailing edge and that as expected, the 'body' shrinks to the axis as you go further from the trailing edge. It is important to stress that there is a singularity in the vertical velocity at $x = 0$, $y = 0$ but not elsewhere in the potential flow region.

The key idea of Stewartson and Messiter was that the potential flow (3.6) must have associated with it a pressure perturbation which will affect flow near the trailing edge and in particular the boundary layer as it approached the trailing edge. It was clear from the shape that the boundary layer takes in Figure 3.1 that there needed to be an acceleration of fluid near the trailing edge and that such acceleration must occur both upstream and downstream of the trailing edge. If we use Bernouilli's equation in the potential flow region then the pressure perturbation can be found from the velocity with

$$\frac{dw}{dz} = u - iv = 1 - \frac{1}{3}\frac{\mu_0}{\lambda_0}(\frac{1}{\sqrt{3}} - i)R^{-1/2}z^{-2/3}, \qquad (3.7)$$

and $p \sim 1 - u$ so that if we go to the boundary layer, this would imply a pressure perturbation as $y \to 0$ of

$$p \sim \begin{cases} -\dfrac{2}{3\sqrt{3}}\dfrac{\mu_0}{\lambda_0}R^{-1/2}|x|^{-2/3}, & x < 0, \\ \dfrac{1}{3\sqrt{3}}\dfrac{\mu_0}{\lambda_0}R^{-1/2}x^{-2/3}, & x > 0. \end{cases} \qquad (3.8)$$

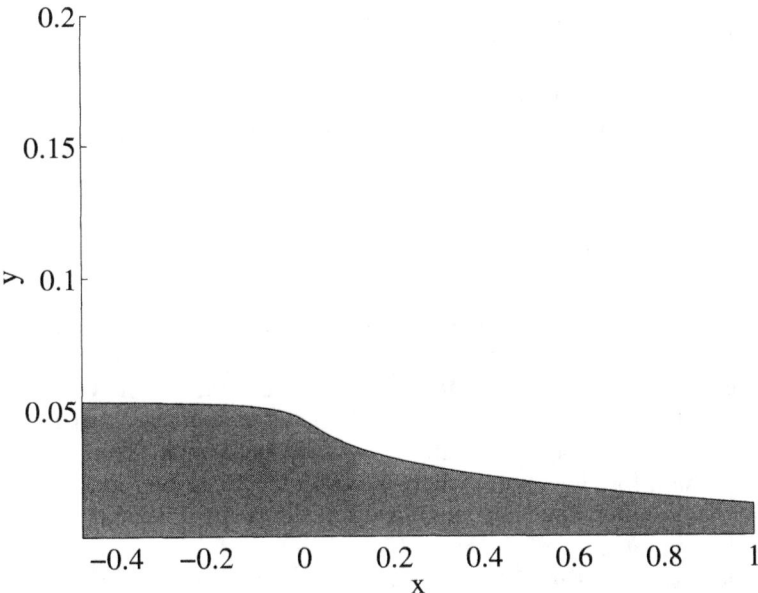

FIG. 3.1. Body shape 'seen' by potential flow near a trailing edge at $x = 0$ for $R = 500$.

It is clear that this does indeed represent a pressure acceleration (since the gradient is negative throughout) but it is also clear that this provides another difficulty since although the pressure starts at zero far upstream and recovers back to zero far downstream of the trailing edge there will be a singularity in the form of a discontinuous jump from a negative infinite value to a positive infinite value in the pressure at the trailing edge. To make the pressure continuous there will have to be a region of adverse pressure gradient very near the trailing edge and a pressure profile which only falls and rises to finite values.

However, we can begin to see how the Goldstein wake expansions may be adjusted near the trailing edge to try to remove the singularity there. Retaining at present as an assumption that the pressure is uniform across the boundary layer, we can ask what flow perturbation will be induced in the various upstream and wake regions. The pressure gradient is of order $O(R^{-1/2}x^{-5/3})$ so in the inner wake the significant balance can most easily be seen from the viscous terms, where the leading term for the stream-function in the Navier–Stokes equations is $R^{-1}\psi_{yyy}$, which using the variable $s = Y/x^{1/3}$, is of order $R^{-1}R^{3/2}x^{-1}O(\psi')$ where ψ' is a shorthand for the perturbation to the stream-function induced by the pressure gradient. Balancing this with the pressure gradient will induce a perturbation in the stream function of order $O(R^{-1}x^{-2/3})$ so that the expansion (2.137) should be adjusted to

$$R^{1/2}\psi \sim x^{2/3}H_0(s) + \cdots + R^{-1/2}[x^{-2/3}H_0^1(s) + \cdots] + \cdots, \qquad (3.9)$$

where $H_0^1, \cdots$, are further unknown functions.

This expansion shows clearly that the inner wake expansion will fail to be asymptotic when

$$x^{2/3} \sim R^{-1/2}x^{-2/3}, \text{ or } x = O(R^{-3/8}). \qquad (3.10)$$

This is an important result since although it shows x to be small compared with the plate length, it is long compared with the boundary layer thickness at the trailing edge and on a boundary layer scale, the start of the Goldstein inner wake will go off to infinity as the Reynolds number increases and the pressure singularity may be relegated to being a far field behaviour. Thus providing the correct expansion for the region where $x = o(R^{-3/8})$ can be found and matched to the Goldstein inner wake any singularity in the vertical velocity at the trailing edge may be removed.

If we now turn to the Goldstein outer wake region, the expansion for the stream-function, (2.161), will be also affected by the pressure gradient. To see the most significant balance it is now easiest to look at the inertia terms, in particular the term vu_y in the Navier–Stokes x-momentum equation. Again, using a generalised perturbation ψ' to (2.161), the vu_y inertia term is order $O(R^{-1/2}x^{-2/3}R\psi'_{YY})$ so to balance a pressure gradient of order $O(R^{-1/2}x^{-5/3})$ we need $\psi' = O(R^{-1}x^{-1})$ and hence the stream-function expansion should be

$$R^{1/2}\psi \sim \sqrt{2}F_0(\frac{Y}{\sqrt{2}}) + x^{1/3}\frac{\mu_0}{\lambda_0}F_0'(\frac{Y}{\sqrt{2}}) + \cdots + R^{-1/2}[x^{-1}F_0^1(Y) + \cdots] + \cdots,$$
$$(3.11)$$

where $F_0^1, \cdots$, are further unknown functions.

This expansion will fail to be asymptotic when

$$x^{1/3} \sim R^{-1/2}x^{-1}, \text{ or } x = O(R^{-3/8}), \qquad (3.12)$$

exactly the same result as obtained for breakdown of the inner wake expansion. We can also note that if the inner wake expansion is defined for $s = Y/x^{1/3} = O(1)$ then the lateral extent that must be considered for the inner wake breaking down is $Y = O(x^{1/3}) = O(R^{-1/8})$ corresponding to $y = O(R^{-5/8})$.

Whatever expansion is used for $x = O(R^{-3/8})$, in the potential flow region a perturbation is needed which has the same scale in x and y, that is $R^{-1/2} << y << R^{-3/8}$.

Upstream of the trailing edge the pressure gradient will cause a sublayer near the wall which has to match the oncoming boundary layer. This has a linear velocity gradient near the wall (and so a quadratic stream-function). Writing

Introduction

$s = Y/(-x)^{1/3}$ for $x < 0$, this means upstream near the wall the expansion should provide a match with the oncoming boundary layer by

$$R^{1/2}\psi \sim \frac{1}{\sqrt{8}}\alpha_0^{-3/2}(-x)^{2/3}s^2 + \cdots + R^{-1/2}[(-x)^{-2/3}g_0^1(s) + \cdots] + \cdots, \quad (3.13)$$

where $g_0^1, \cdots$, are yet more unknown functions. Using the same argument about balances in the x momentum Navier–Stokes equation as for the downstream inner layer. This gives the same limit as before for the series to fail to be asymptotic: $-x = O(R^{-3/8})$ and $Y = O(R^{-5/8})$.

In the main part of the oncoming boundary layer, if an inertial term, for instance uu_x, is balanced against the pressure gradient then a term of order $O(R^{-1}(-x)^{-2/3})$ is required, but to match with the sublayer near the wall a term of order $(R^{-1}(-x)^{-1})$ is needed (this is not obvious within a heuristic outline such as this, later we shall see that the function g_0 in (3.13) satisfies $g_0 \sim s$ as $s \to \infty$). Thus in the main part of the oncoming boundary layer we expect an expansion

$$R^{1/2}\psi \sim \sqrt{2}F_0(\frac{Y}{\sqrt{2}}) + \frac{x}{\sqrt{2}}[F_0(\frac{Y}{\sqrt{2}}) - \frac{Y}{\sqrt{2}}F_0'(\frac{Y}{\sqrt{2}})] + \cdots + O(R^{-1/2}x^{-1}),$$
$$(3.14)$$

so this expansion will not be asymptotic when $-x = O(R^{-1/2})$. Nevertheless, to match with the sublayer expansion, a leading order perturbation on a scale $-x = O(R^{-3/8})$ has to be considered in this region too although it is likely that at a higher level in the expansion, the length scale $|x| \sim R^{-1/2}$ will become important.

This sets out heuristically how the Goldstein near wake can be modified downstream of the trailing edge. Stewartson described the structure as a series of 'decks', in this case a triple deck structure and the three decks are characterised by

(i) a potential perturbation in the outer deck,

(ii) an inviscid rotational disturbance in the middle deck, the outer part of the boundary layer, as seen in the discussion where we have balanced inertial terms against the pressure gradient,

(iii) a boundary layer type disturbance in the lower deck where near the wall there is effectively a perturbation to the linear shear in the oncoming flow leading to an inhomogeneous boundary layer equation.

Connecting the three decks is a pressure perturbation (and as we shall see later, a displacement function) which has to be determined by allowing the decks to interact with each other via matched asymptotic expansions. The general form of the interaction regions is shown in Figure 3.2. Note that the picture is a little more complicated because very near the trailing edge it is necessary to have a region where the boundary layer assumption of y derivatives being much greater than x derivatives in the viscous terms is abandoned. There, assuming

82   The Triple Deck

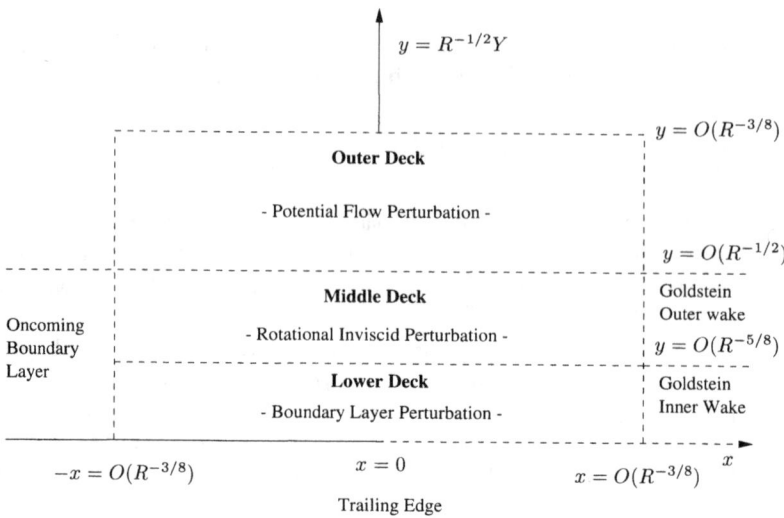

FIG. 3.2. Schematic triple deck structure based on heuristic examination of asymptotic expansions in the near wake and upstream of the trailing edge.

the same scale for x and y variation, an oncoming velocity of order $O(R^{1/2}y)$ and a perturbation balance between uu_x and $R^{-1}(u_{xx} + u_{yy})$ leads to consideration of length scales $O(R^{-3/4})$ since if $x, y \sim R^{-a}$ and u' is a generalised perturbation to a flow similar to the oncoming linear shear, then $R^{1/2}yu'_x \sim R^{-1}(u'_{xx} + u'_{yy})$ gives $a = 3/4$.

3.2 Formulation

One difficulty in understanding the interaction near the trailing edge is that of developing a clear and easily visualised notation for such a complex problem. The physical coordinates have been non-dimensionalised by the plate length to give coordinates (x, y) with the plate between $x = -1$ and $x = 0$ on $y = 0$. Hence at the trailing edge the boundary layer on the plate will have thickness of order $R^{-1/2}$. There is already a scaled transverse coordinate

$$\text{Middle deck} \quad \boxed{Y = R^{1/2}y,} \quad (3.15)$$

so that Y is $O(1)$ at the edge of the boundary layer. Now define two further transverse coordinates,

$$\text{Inner deck} \quad \boxed{Z = R^{5/8}y = R^{1/8}Y,} \quad (3.16)$$

and

$$\text{Outer deck} \quad \boxed{W = R^{3/8}y = R^{-1/8}Y,} \quad (3.17)$$

so that there will be two matching scales:

(i) $Z \to \infty$ with $Y \to 0$,
(ii) $Y \to \infty$ with $W \to 0$.

In addition there needs to be a scaled longitudinal variable,

$$\text{Interaction region} \qquad \boxed{X = R^{3/8}x,} \qquad (3.18)$$

and have matching (i) $X \to \infty$ with $x \to 0^+$ and (ii) $X \to -\infty$ with $x \to 0^-$.

The analysis of the three decks follows the sequence of: middle deck where an inviscid rotational disturbance introduces two functions of X, a pressure gradient and a displacement function; the outer deck where a potential flow solution provides a link between the pressure gradient and displacement function; and finally the inner deck where the solution of a non-linear boundary layer type equation (which matches to the middle deck) finally generates the unknown pressure and displacement functions.

3.3 The middle deck

Expand for large R,

$$R^{1/2}\psi \sim \sqrt{2}F_0(\frac{Y}{\sqrt{2}}) + R^{-1/8}\psi_1^m(X,Y) + R^{-1/4}\psi_2^m(X,Y) + o(R^{-1/4}), \quad (3.19)$$

so that if we denote $U_0(Y) = F_0'(Y/\sqrt{2})$ for the oncoming boundary layer velocity, the physical quantities u, v, p are formally expanded as

$$u \sim U_0(Y) + R^{-1/8}\frac{\partial \psi_1^m}{\partial Y}(X,Y) + R^{-1/4}\frac{\partial \psi_2^m}{\partial Y}(X,Y) + o(R^{-1/4}),$$

$$v \sim \quad - R^{-1/4}\frac{\partial \psi_1^m}{\partial X}(X,Y) - R^{-3/8}\frac{\partial \psi_2^m}{\partial X}(X,Y) + o(R^{-3/8}),$$

$$p \sim \quad R^{-1/8}p_1^m(X,Y) + R^{-1/4}p_2^m(X,Y) + o(R^{-1/4})$$

Although the expansion for the pressure is consistent with the velocity expansion, if we recall that the pressure perturbation is expected to behave according to

$$p \sim R^{-1/2}x^{-2/3} \sim R^{-1/4}X^{-2/3}, \qquad (3.20)$$

so that $p_1^m \equiv 0$ to agree with the pressure perturbation coming from the interaction at the trailing edge.

If the expansion (3.19) is substituted into the Navier–Stokes equations, then the y momentum equation shows that the pressure perturbations p_1 and p_2 are independent of Y (note that this is not the case if the expansion is continued to higher order terms) and using $p_1^m = 0$, the leading order terms in the x momentum equation give

$$U_0(Y)\psi_{1\ XY}^m - U_0'(Y)\psi_{1\ X}^m = 0. \tag{3.21}$$

The solution of this equation involves an unknown function, denoted $A_1(X)$,

$$\boxed{\psi_1^m(X, Y) = A_1(X)U_0(Y)} \tag{3.22}$$

where any arbitrary constant in the stream-function is determined to be zero so that ψ_1^m can vanish far upstream. The function $A_1(X)$ is called a displacement function and $-A_1$ represents a lateral displacement of the oncoming flow in the middle deck.

Next the equation for ψ_2^m is

$$U_0(Y)\psi_{2\ XY}^m - U_0'(Y)\psi_{2\ X}^m = -p_2'(X) - \psi_{1\ Y}^m\psi_{1\ XY}^m + \psi_{1\ X}^m\psi_{1\ YY}^m, \tag{3.23}$$

so substituting the solution for ψ_1^m,

$$U_0(Y)\psi_{2\ XY}^m - U_0'(Y)\psi_{2\ X}^m = -p_2'(X) - A_1(X)A_1'(X)[U_0'^2 - U_0U_0''], \tag{3.24}$$

where there need be no confusion in using a dash to represent derivatives of functions of either X or Y with respect to their dependent variable.

The simplest way to solve this is to introduce another function $A_2(X)$ in the solution of the homogeneous equation, determine a particular integral for the pressure gradient of the form $p_2(X)G(Y)$ where $U_0 G' - U_0' G = -1$ and note that $\frac{1}{2}A_1^2(X)U_0'(Y)$ is a particular integral for the remaining terms. Again a final constant of integration is set to zero so the stream-function perturbation can vanish far upstream.

The solution for G is found by noting that as $Y \to \infty$, $U_0 \to 1$ so that $G' \to -1$, and as the equation for G can be rewritten

$$U_0^2 \frac{d}{dY}\left(\frac{G}{U_0}\right) = -1, \tag{3.25}$$

by trying a solution of the form $G = -YU_0(Y) + H(Y)$ one obtains

$$G(Y) = -YU_0(Y) + U_0(Y)\int_Y^\infty \left(1 - \frac{1}{U_0^2(Y)}\right)dY + c_2, \tag{3.26}$$

where c_2 is an arbitrary constant which cannot be discarded at this point. Thus the solution for ψ_2^m is

$$\psi_2^m(X, Y) = p_2(X)\{U_0(Y)\int_Y^\infty \left(1 - \frac{1}{U_0^2(Y)}\right)dY - YU_0(Y) + c_2\}$$
$$+ \frac{1}{2}A_1^2(X)U_0'(Y) + A_2(X)U_0(Y). \tag{3.27}$$

This expansion has to be matched to the upstream and downstream expansions. Upstream, (3.5) gives

$$R^{1/2}\psi \sim \sqrt{2}F_0(\frac{Y}{\sqrt{2}}) + \frac{1}{\sqrt{2}}x[F_0(\frac{Y}{\sqrt{2}}) - \frac{Y}{\sqrt{2}}F_0'(\frac{Y}{\sqrt{2}})],$$
$$\sim \sqrt{2}F_0(\frac{Y}{\sqrt{2}}) + \frac{1}{\sqrt{2}}R^{-3/8}X[F_0(\frac{Y}{\sqrt{2}}) - \frac{Y}{\sqrt{2}}F_0'(\frac{Y}{\sqrt{2}})], \quad (3.28)$$

whereas downstream, (2.161) gives

$$R^{1/2}\psi \sim \sqrt{2}F_0(\frac{Y}{\sqrt{2}}) + x^{1/3}\frac{\mu_0}{\lambda_0}F_0'(\frac{Y}{\sqrt{2}}),$$
$$\sim \sqrt{2}F_0(\frac{Y}{\sqrt{2}}) + R^{-1/8}X^{1/3}\frac{\mu_0}{\lambda_0}F_0'(\frac{Y}{\sqrt{2}}). \quad (3.29)$$

Consequently in the middle deck

Middle deck $\boxed{R^{1/2}\psi \sim \sqrt{2}F_0(\frac{Y}{\sqrt{2}}) + R^{-1/8}A_1(X)F_0'(\frac{Y}{\sqrt{2}}).}$ (3.30)

The first two terms of this expansion will match upstream and downstream provided the displacement function satisfies

$$A_1(X) \to 0, \text{ as } X \to -\infty, \text{ and } A_1(X) \to \frac{\mu_0}{\lambda_0}X^{1/3}, \text{ as } X \to \infty. \quad (3.31)$$

3.4 The outer deck

In considering the behaviour of the expansion for the stream-function from the middle deck in the outer deck, $Y \to \infty$, $U_0 \to 1$ is used and then

$$R^{1/2}\psi \sim Y - \sqrt{2}\beta_0 + R^{-1/8}A_1(X) + R^{-2/8}[p_2(X)(-Y + c_2) + A_2(X)]. \quad (3.32)$$

Since $Y = R^{1/8}W$ in the outer deck scaling and retaining only terms to $O(R^{-1/8})$,

$$R^{1/2}\psi \sim R^{1/8}W - \sqrt{2}\beta_0 + R^{-1/8}[A_1(X) - p_2(X)W] \text{ as } W \to 0. \quad (3.33)$$

Thus in the outer region the stream-function should be expanded

Outer deck $\boxed{R^{1/2}\psi \sim R^{1/8}W - \sqrt{2}\beta_0 + R^{-1/8}\psi_1^o(X,W),}$ (3.34)

where $\psi_1^o \to 0$ as $X, W \to \infty$ and matches (3.33), as $W \to 0$. Since the perturbation is potential it is only necessary to find a harmonic function which satisfies the boundary conditions, most importantly that it should vanish at infinity so that the expansion (3.34) can match to a potential expansion outside the outer deck. In this case

$$\psi_1^o(X, W) = \frac{1}{\pi} \int_{-\infty}^{\infty} \frac{A_1(\zeta)W}{(X-\zeta)^2 + W^2} d\zeta. \tag{3.35}$$

As $W \to 0$ part of the argument within the integral behaves like a delta-function reducing the integral to $A_1(X)$ when $W = 0$. However, this is not the only form needed, since the velocities are derivatives of the stream-function. Thus an alternative form is required where derivatives will have the form of a convolution with a 'delta-function' like argument. Integrating by parts and using the properties of A_1 for large X to rewrite

$$\psi_1^o(X, W) = -\frac{1}{\pi} \int_{-\infty}^{\infty} A_1'(X)[\tan^{-1}(\frac{X-\zeta}{W}) + \frac{\pi}{2}]d\zeta, \tag{3.36}$$

then a form suitable for calculating velocities is obtained. Now if the matching condition with (3.33) on ψ_{1W}^o at $W = 0$ is imposed the result is a link between the displacement function A_1 and the unknown pressure perturbation p_2,

$$\boxed{p_2(X) = \frac{1}{\pi} \int_{-\infty}^{\infty} \frac{A_1'(\zeta)}{X-\zeta} d\zeta,} \tag{3.37}$$

(where the integral is interpreted as a Cauchy principal value). Thus the derivative of the displacement function and the pressure perturbation are linked by a form of Hilbert transform.

This completes setting up the expansion in the outer deck, the form (3.34), satisfies the conditions needed as $X \to \pm\infty$ through properties of the function $A_1(X)$ and matches to both the middle deck as $W \to 0$ and to the potential far field as $W \to \infty$.

3.5 The inner deck

In order to determine A_1 and p_2 the analysis and solution of the lower deck has to be completed. This has to start with an examination of how the expansion from the middle deck behaves as $Y \to 0$.

Recall from (2.30), that $F_0(s) \sim \alpha_0^{-3/2} s^2/2$ for small s, then using $Y = R^{-1/8}Z$ in the middle deck expansion gives as $Y \to 0$,

$$R^{1/2}\psi \sim \alpha_0^{-3/2} R^{-1/4}[\frac{1}{2\sqrt{2}}Z^2 + \frac{1}{\sqrt{2}}A_1(X)Z + \frac{1}{2}A_1^2(X) + c_2 p_2(X)] + o(R^{-1/4}). \tag{3.38}$$

This indicates that in the inner region, the stream-function expansion should be

Inner deck $\boxed{R^{1/2}\psi \sim R^{-1/4}\psi_1^i(X,Z) + o(R^{-1/4}),}$ (3.39)

and for the pressure,

$$p \sim R^{-1/4}p_1^i(X,Z) + o(R^{-1/4}). \tag{3.40}$$

If these are substituted into the Navier–Stokes equations, firstly from x momentum

$$\psi_{1Z}^i \psi_{1XZ}^i - \psi_{1X}^i \psi_{1ZZ}^i = -p_{1X}^i + \psi_{1ZZZ}^i, \tag{3.41}$$

and from y momentum,

$$p_{1Z}^i = 0, \tag{3.42}$$

so that the pressure can still be taken to be uniform across the inner deck and $P_1^i(X,Z) = p_2(X)$.

This brings us to the fundamental interaction problem in the triple deck formulation. This can be restated in more familiar velocity component formulation. Let $u = \psi_{1Z}^i$ and $v = -\psi_{1X}^i$ where no subscripts are used for clarity and we keep in mind that u and v are only the leading order velocity terms in a complex expansion. The triple deck procedure will only work if it is possible determine a function $A_1(X)$ such that

$$u_X + v_Z = 0, \tag{3.43}$$

$$uu_X + vu_Z = -\frac{1}{\pi}\int_{-\infty}^{\infty} \frac{A_1''(\zeta)}{X-\zeta}d\zeta + u_{ZZ}, \tag{3.44}$$

together with

$$u \to \frac{\alpha_0^{-3/2}}{\sqrt{2}}[Z + A_1(X)], \quad \text{as } Z \to \infty. \tag{3.45}$$

Further, the solution has to satisfy boundary conditions $u = v = 0$ on $Z = 0$, $x \leq 0$, $u_z = v = 0$ on $Z = 0$, $x > 0$ and A_1 has to have the asymptotic properties

$$A_1(X) \to 0, \text{ as } X \to -\infty, \quad A_1(X) \to \frac{\mu_0}{\lambda_0}X^{1/3}, \text{ as } X \to \infty, \tag{3.46}$$

together with u, v having appropriate asymptotic properties as $X \to \pm\infty$.

In his original paper, Stewartson posed but did not solve this problem and Messiter attempted a solution by assuming a form for A_1 and carrying out one iteration of an approximation scheme. It was not until Jobe and Burggraf (1974) that a numerical solution of this problem showed (although obviously without mathematical 'proof') that the function A_1 could be determined. At the same time Daniels (1974) and Smith (1974) provided numerical solutions for the analogous compressible flow problem although Smith (1972) could claim to provide the first numerical solution of an interactive problem. We will defer to the next chapter a detailed examination of different numerical techniques to solve interactive boundary layer equations.

3.6 Computed results

The computed solution of the interactive equations, (3.43)-(3.46), shows an interaction region which extends a considerable distance upstream with the 'displacement function' A_1 growing slowly from zero and the pressure falling, not to a singular value but to a finite negative value. Then there is a region of pressure recovery (and hence adverse pressure gradient) immediately after the trailing edge before a pressure maximum is reached after which the pressure decreases and eventually reaches its correct asymptotic value. In this region the pressure gradient is once more negative. The displacement function rises rapidly and while it approaches its asymptotic values upstream and downstream, the difference between the computed values and the asymptotic values is much greater for the displacement function than the pressure. These are illustrated in figures 3.3 and 3.4. It is interesting however to examine the streamlines, Figure 3.5, since when we started we were faced with a singularity in the transverse velocity so that while the streamlines were continuous, there was a vertical tangent in the streamline slope as $x \to 0^+$. The leading order triple deck expansion has overcome that problem so that the vertical velocity is no longer singular; however, it is discontinuous so that now it is v_X which is singular. Similarly the pressure is no longer singular near $X = 0$ but the pressure gradient is discontinuous and so the second derivative of the pressure is singular. In order to move these 'singularities' to even higher derivatives it would be necessary to continue the expansion to higher terms. The relation between the streamlines and $-A_1$ should be evident.

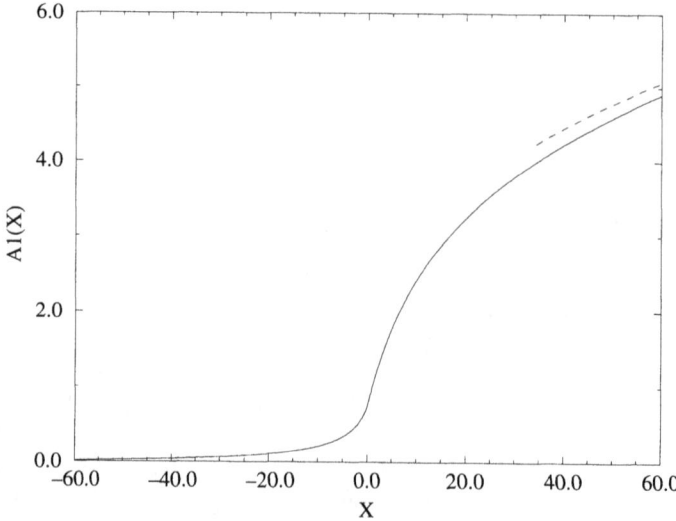

FIG. 3.3. Computed 'displacement' function $A_1(X)$ versus distance (X) from the trailing edge. The asymptotic behaviour for $X \to \infty$ is shown by $----$.

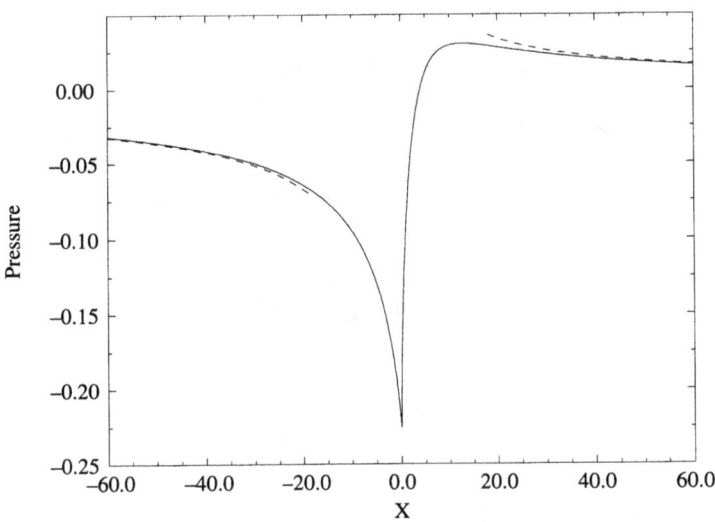

FIG. 3.4. Computed pressure $p_1(X)$ versus distance (X) from the trailing edge. The asymptotic behaviour for large X is shown by – – – –.

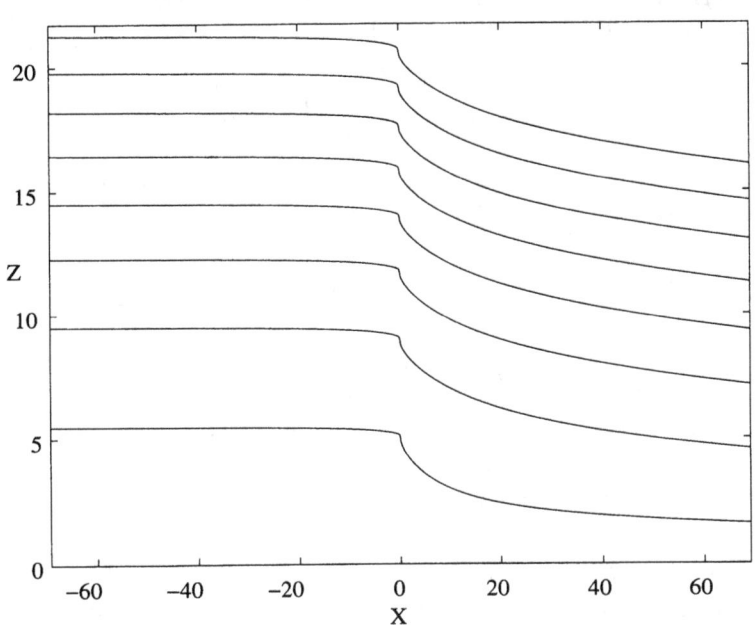

FIG. 3.5. Computed inner deck streamlines near the trailing edge. Note the discontinuity in v_X at the trailing edge.

3.7 Drag

Having determined the expansion near the trailing edge, the viscous drag on the plate would still be given by an integral of the shear stress over the plate, only now the stress near the trailing edge will be altered from the Blasius or leading order value according to the solution for ψ_1^i. Formally, with the plate between $x = 0$ and $x = 1$, the contribution from viscous drag to the drag coefficient can be written

$$c_D = 2R^{-1} \int_{-1}^{0} \frac{\partial u}{\partial y}(x,0) dx, \qquad (3.47)$$

and if we denote

$$u_B(x,Y) = U_0\left(\frac{Y}{\sqrt{1+x}}\right) = F_0'\left(\frac{Y}{\sqrt{2(1+x)}}\right),$$

as the Blasius velocity profile then

$$c_D = 2R^{-1/2} \int_{-1}^{0} \frac{\partial u_B}{\partial Y}(x,0) dx + 2R^{-1/2} \int_{-1}^{0} [\frac{\partial u}{\partial y}(x,0) - \frac{\partial u_B}{\partial Y}(x,0)] dx \qquad (3.48)$$

The first term is of course the leading order Blasius drag which we have already seen in (2.122) gives for a unit length plate,

$$2R^{-1/2} \int_{-1}^{0} \frac{\partial u_B}{\partial Y}(x,0) dx = \frac{1.33...}{R^{1/2}}.$$

The second term in (3.48) can only be expanded asymptotically. To do this change to the triple deck coordinate X so that

$$2R^{-1/2} \int_{-1}^{0} [\frac{\partial u}{\partial y}(x,0) - \frac{\partial u_B}{\partial Y}(x,0)] dx =$$

$$2R^{-7/8} \int_{-R^{3/8}}^{0} [\frac{\partial^2 \psi_1^i}{\partial Z^2}(X,0) - \frac{\partial u_B}{\partial Y}(R^{3/8}X,0)] dX. \qquad (3.49)$$

Now the expansion of the wall shear far upstream has

$$\frac{\partial^2 \psi_1^i}{\partial Z^2}(X,0) \sim \frac{\partial u_B}{\partial Y}(R^{3/8}X,0) + O((-X)^{-4/3}) \text{ as } -X \to \infty,$$

so that there will be an error of order $R^{-7/8}(R^{3/8})^{-1/3} = R^{-1}$ if the integral is extended to $X = -\infty$, and an error of order $R^{-7/8}R^{-3/8} = R^{-10/8}$ if the Blasius shear is replaced with its value at $x = 0$, leading to an expansion

$$c_D \sim \frac{1.33...}{R^{1/2}} + \frac{2}{R^{7/8}} \int_{-\infty}^{0} [\frac{\partial^2 \psi_1^i}{\partial Z^2}(X,0) - \frac{1}{\sqrt{2}} F_0'''(0)] dX + O(R^{-1}), \qquad (3.50)$$

or alternately

$$c_D \sim \frac{1.33...}{R^{1/2}} + \frac{2}{R^{7/8}} \int_{-\infty}^{0} [\frac{\partial^2 \psi_1^i}{\partial Z^2}(X,0) - \frac{\partial^2 \psi_1^i}{\partial Z^2}(-\infty,0)] dX + O(R^{-1}). \quad (3.51)$$

The second term gives the increase in viscous drag from acceleration of fluid into and through the triple deck region which is needed to provide the extra fluid associated with the wake centreline velocity increasing immediately downstream of the trailing edge. The expansion (3.51) also has to be augmented by the term found by Imai and by second order effects on the Blasius solution of the type considered by Kuo each of order R^{-1}. As we have seen, Kuo's work was only a pragmatic attempt to consider changes to the leading order Blasius solution. However, any further augmentation is difficult because first the triple deck contribution provides an unknown component of order R^{-1} and secondly, there will be a viscous region of size $O(R^{-1})$ around the leading edge of the plate and that contribution too cannot be calculated. Thus the two-term expansion, (3.51) may provide the best asymptotic estimate which can be made analytically for the drag.

The integral in (3.51) was calculated by Jobe and Burggraf (1974) to give the coefficient of $R^{-7/8}$ as 2.694.... Contemporaneously the coefficient was estimated by A.I. van de Vooren to be 2.66 ± 0.05 and by R.E. Melnik to be 2.645...(see Jobe and Burggraf (1974), page 111, and Veldman and van de Vooren (1974) who give a slightly different value 2.651 ± 0.003).

An accepted value for the coefficient is 2.66... so that the first two terms for the drag coefficient are

$$\boxed{c_D \sim \frac{1.33...}{R^{1/2}} + \frac{2.66...}{R^{7/8}} + O(R^{-1}).} \quad (3.52)$$

3.8 Numerical solution of the Navier–Stokes equations

The very simple geometry involved for flow about a flat plate might appear to lend itself to straightforward numerical treatment: the boundaries are aligned with rectangular meshes, there is no separation and the far field is known on the incoming boundary and the wake boundary (at least the asymptotic form of the flow is known on the wake boundary, the details of course depend on the drag of the plate which is not known). In this coordinate system the leading edge might be considered a source of difficulty but that can be largely removed by using parabolic coordinates. Thus it might be considered that the computational problem is straightforward; this of course is far from true since there are regions near the leading edge and the trailing edge where there are integrable singularities which contribute significantly to quantities such as drag, and while we know the size of these regions their small size as the Reynolds number becomes large does mean that carefully graded meshes are required at both the leading and trailing edges.

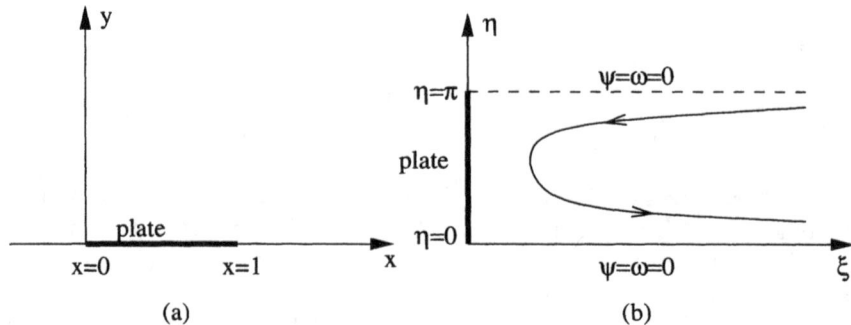

FIG. 3.6. Coordinate transformation of flow region to a semi-infinite strip used by Dennis and Dunwoody (1966) (a) Plate coordinates (b) Transformed coordinates.

Dennis and Dunwoody (1966) attempted to obtain a numerical solution by solving for a Fourier series representation in a particular transformed coordinate system, what might be considered almost a spectral method. They used not the conventional parabolic transformation but rather a transformation of the flow region to a semi-infinite strip. In their original work they located the (x,y) coordinates at the centre of the plate, here with the plate occupying $0 \le x \le 1, y = 0$, and let

$$x = \frac{1}{2}(1 + \cosh\xi \cos\eta), \quad y = \frac{1}{2}\sinh\xi \sin\eta, \qquad (3.53)$$

so the flow region was transformed to $0 \le \eta \le \pi, \xi \ge 0$ and the plate was on $\xi = 0$; see Figure 3.6. Note that the leading edge was at $\eta = \pi$ and the trailing edge at $\eta = 0$, the wake was located along $\eta = 0, \xi > 0$

Using this transformation the vorticity equation was

$$\frac{\partial \psi}{\partial \eta}\frac{\partial \omega}{\partial \xi} - \frac{\partial \psi}{\partial \xi}\frac{\partial \omega}{\partial \eta} = \frac{1}{R}\left(\frac{\partial^2 \omega}{\partial \eta^2} + \frac{\partial^2 \omega}{\partial \xi^2}\right). \qquad (3.54)$$

where the vorticity was related to the stream-function by

$$\frac{\partial^2 \psi}{\partial \eta^2} + \frac{\partial^2 \psi}{\partial \xi^2} = \frac{1}{2}(\cos 2\eta - \cosh 2\xi)\omega. \qquad (3.55)$$

The stream-function and vorticity both vanished on the lines $\eta = 0$, $\eta = \pi$, $\xi > 0$ while $\psi = \frac{\partial \psi}{\partial \xi} = 0$ on $\xi = 0, 0 \le \eta \le \pi$. The velocity far field $\psi \to y$ was satisfied provided

$$\frac{1}{\cosh\xi \sin\eta}\frac{\partial \psi}{\partial \xi} \to \frac{1}{2}, \quad \frac{1}{\sinh\xi \cos\eta}\frac{\partial \psi}{\partial \eta} \to \frac{1}{2} \text{ as } \xi \to \infty. \qquad (3.56)$$

By considering an Oseen approximation for ξ large, Dennis and Dunwoody determined the far field condition for the stream-function as

$$\psi \sim \frac{1}{4}e^{\xi}\sin\eta - \frac{1}{2}C_D(1 - \frac{\eta}{\pi}),\ 0 < \eta \leq \pi, \tag{3.57}$$

the equivalent result to that derived by Imai (1951), see section 2.4 above. The far field stream-function has a jump across the wake.

Dennis and Dunwoody derived a numerical solution in terms of a Fourier expansion,

$$\psi = \sum_n f_n(\xi)\sin n\eta, \tag{3.58}$$

although computing power available to them limited their solution to a small number of terms, typically five. Their calculation of the drag was in broad agreement with theoretical prediction but did not have the resolution to examine the fine detail of the asymptotic structure of the drag coefficient. Their transformation does not seem to have been used for a full numerical solution although it has some very attractive properties, see appendix A.

Plotkin and Flügge-Lotz (1968) undertook a direct numerical solution near the trailing edge of a plate using rectangular coordinates. Rather than try to solve the whole flow field they focused only on a small region around the trailing edge using a finite difference representation in primitive variables. In doing this they found a solution which can only be described as a remarkable anticipation of the triple deck.

Their method was in two steps: first, to solve for velocities in the absence of a pressure gradient so as to obtain a continuous transverse velocity at the outer edge of their calculation and then to use that result as a new boundary condition to solve the full Navier–Stokes equations for the velocities and pressure. Using this pragmatic approach they observed that the region about the trailing edge where the flow was different to the Blasius solution was greater than the order $R^{-3/4} \times R^{-3/4}$ region where is was previously assumed the Blasius boundary layer would fail. An illustration of how close they were to revealing one of the triple deck scales comes from their derivation of an empirical correlation for the pressure from their numerical results,

$$p \sim R^{-0.2} f(R^{0.45} x), \tag{3.59}$$

(f is an unspecified function) whereas the triple deck scaling for this problem, (3.40), discovered later by Stewartson and Messiter gave

$$p \sim R^{-0.25} f(R^{0.375} x). \tag{3.60}$$

Another computation of the local behaviour of flow near a trailing edge was by Schneider and Denny (1971) who set out to use a rectangular mesh which would resolve a region of order $R^{-3/4} \times R^{-3/4}$ around the trailing edge by using stretched coordinates. Their transformation

$$s = \sqrt{R}(x + \frac{4}{1+x^{-\sqrt{R}}}),\ t = 1 - e^{-\sqrt{R}y}, \tag{3.61}$$

was used with a two-step method which incorporated three flow regions. In the first step a boundary layer equation and a potential flow equation were solved in sequential iteration, the boundary layer solution giving a displacement thickness with which a potential flow solver gave a pressure and outer flow velocity for feedback into the boundary layer equation solver. Once a boundary layer solution was determined it provided Dirichlet boundary conditions for a full Navier–Stokes solver around the trailing edge.

Their results were limited by available computer time to only one Reynolds number but that case did indicate a sharp rise in wall stress very near the trailing edge indicative of a region which agreed with the prediction of Carrier and Lin (1948) and Weinbaum (1968) (see also the discussion of Weinbaum's work in section 8.7 and in particular Figure 8.21 for a corner half angle of π) that the stream-function should behave like $r^{3/2}$ near the trailing edge. What they did not do was to look at the solution from their boundary layer – potential flow interaction in the light of Stewartson's and Messiter's triple deck theory.

Chen and Patel (1987) developed a numerical solution to the full Navier–Stokes equations for a region which started part way along the plate, so as to have a Blasius boundary layer profile as an inlet condition. A rectangular mesh was used with longitudinal stretching near the trailing edge as well as lateral stretching away from the plate. Finite difference equations for velocity and pressure were solved using a form of pressure correction procedure. Their results supported the general triple deck structure and showed that the computational domain size (and by implication the boundary conditions there) was a significant factor in the solution leading them to conclude that a more accurate solution would have to include the flow at and upstream of the leading edge.

The most comprehensive numerical solution of the full Navier–Stokes equations for flow about a plate we have to date is due to McLachlan (1991a) who used a transformation to parabolic coordinates

$$x = \xi^2 - \eta^2, \quad y = 2\xi\eta, \tag{3.62}$$

and symmetry about the line $y = 0$ to give a computational region in the first quadrant of the (ξ, η) plane.

The upper surface of the plate was transformed to $0 \leq \xi \leq 1$, $\eta = 0$ and with a mesh of $M + 1$ points in the ξ direction and $N + 1$ in the η direction, McLachlan solved the transformed equations

$$\frac{\partial \psi}{\partial \eta}\frac{\partial \omega}{\partial \xi} - \frac{\partial \psi}{\partial \xi}\frac{\partial \omega}{\partial \eta} = \frac{1}{R}\left(\frac{\partial^2 \omega}{\partial \eta^2} + \frac{\partial^2 \omega}{\partial \xi^2}\right). \tag{3.63}$$

and

$$\frac{\partial^2 \psi}{\partial \eta^2} + \frac{\partial^2 \psi}{\partial \xi^2} = -4(\xi^2 + \eta^2)\omega, \tag{3.64}$$

in a region $0 \leq \xi \leq \xi_M$, $0 \leq \eta \leq \eta_N$ but it is the boundary conditions for this region which are most noteworthy. The computational problem is illustrated in Figure 3.7.

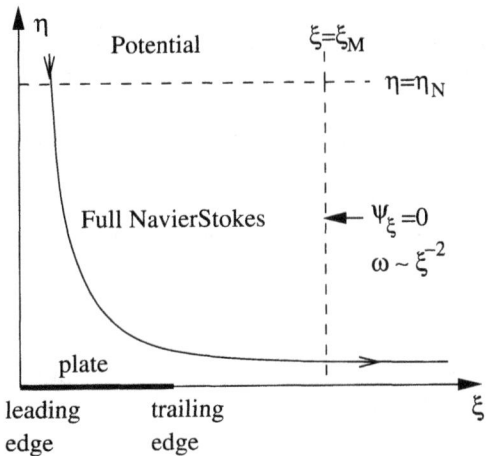

FIG. 3.7. Transformed coordinates and computation region used by McLachlan (1991a).

There were of course the usual boundary conditions on $\xi = 0$ and $\eta = 0$: $\psi = 0$ and $\omega = 0$ on the streamlines and $\psi = \psi_\eta = 0$ on the plate. As we have seen, in parabolic coordinates the leading order boundary layer remained of constant thickness as ξ became large, so McLachlan applied $\psi_\xi = 0$ and $\omega \sim \xi^{-2}$ on the computational boundary $\xi = \xi_M$. On the remaining boundary the formal condition was only $\psi \sim 2\xi\eta$ as $\eta \to \infty$ but that was very difficult to implement directly since the decay of irrotational disturbances was relatively slow and getting that wrong fed back into the solution near the plate. The correction of Imai which was proportional to the drag was also difficult to apply directly for the same reason, the slow decay of irrotational terms: that is, if at $\eta = \eta_N$ the stream-function is set equal to the far field value $2\xi\eta_N$ plus Imai's correction, the error between this value and the true solution at η_N was still sufficiently large to affect the computed drag *unless* η_N were to be very large.

Fornberg (1980) has proposed an ingenious practical method to overcome this problem (see also discussion later of Fornberg's numerical solutions for flow about a cylinder). He suggested that since the vorticity decays exponentially for large η (and with it any rotational contributions to the stream-function) the values of the stream-function at η_{N-1} could be used to solve for potential flow in the region $\eta \geq \eta_{N-1}$, that is outside the last but one row of the computational region and then that solution could be used to determine a Dirichlet boundary condition on ψ at the last row, $\eta = \eta_N$, for a further iteration of the Navier–Stokes solver. Since the potential flow problem could in principle be solved analytically, this led to a relation between the stream-function at each boundary grid point and the values of the stream-function at the grid points along $\eta = \eta_{N-1}$. The vorticity was of course set equal to zero on that boundary. This great care with boundary conditions, together with very carefully mesh refinement enabled McLachlan to resolve the flow with sufficient accuracy to demonstrate that the triple deck contribution to the drag was indeed the next term after the Blasius term. He also showed that at Reynolds numbers below a few thousand, the order R^{-1} terms were sufficient to mask the correct asymptotic behaviour of the solution although in the prediction of drag the different order R^{-1} effects (for instance, leading edge drag, far wake–boundary layer interaction and higher order triple deck) offered some cancellation so that the overall coefficient of R^{-1} may be quite small.

3.9 Summary

This chapter completes the description of our understanding of the theory for flow around the simplest of objects, a flat plate aligned with the flow. We have seen that the trailing edge singularity described by Goldstein in the 1930s has been resolved by using a modification to the Blasius boundary layer which included an interaction between the pressure and flow displacement near the trailing edge. The correctness of this complicated application of matched asymptotic expansions is now accepted and is supported by computational evidence. The development of the triple deck description, our understanding of the important physical balances in each deck and how the decks interact is one of the outstanding achievements in theoretical fluid mechanics. Two general surveys about the triple deck are Meyer (1983) and Rothmayer and Smith (1998a).

The flat plate problem remains one of active interest in fluid mechanics, see for instance Berger and Scalise (1995) for a recent computation of flow in the wake. There remain a number of unresolved problems but triple deck theory provides important ideas about how other, more complicated flow problems should be tackled. An example is the flow in a boundary layer past a small wall perturbation where provided the wall perturbation is sufficiently small it will provide a perturbation to a boundary layer which can be described by a triple deck structure, see Smith et al. (1981)

4

NUMERICAL SOLUTION OF TRIPLE DECK EQUATIONS

4.1 Introduction

Neither Stewartson (1969) nor Messiter (1970) computed a solution to the triple deck problem. Stewartson offered no assurance that a solution even existed, Messiter used a guessed displacement function $A_1(X)$ to calculate an improved function and this way indicated how a numerical solution might be determined without directly solving the problem.

As we shall see later, there are a number of variants of the fundamental problem for an interactive boundary layer depending on the context of the flow. The common thread for all of the problems is that solutions are needed for a nonlinear boundary layer problem with a specified upstream shear and an unknown pressure distribution. In all cases the longitudinal velocity has to match some perturbation far from the wall (after all it is the perturbation of the velocity far from the wall which is forcing the modification of the boundary layer) although in some cases that modification is explicitly known and in other cases it has to emerge from the solution through a coupling with the pressure solution.

The equations we have already seen (3.43–3.45), have the upstream wall shear intimately bound into the problem. It is possible to rescale the equations to remove any dependence on the upstream wall shear (or the outer local velocity, see Rothmayer and Smith (1998a)). If the upstream wall shear is denoted λ (so that in (3.45) $\lambda = \alpha_0^{-3/2}/\sqrt{2}$) then the transformation

$$X = x_s \hat{X}, \ Z = z_s \hat{Z}, u = u_s \hat{u}, \ v = \frac{z_s u_s}{x_s}\hat{v}, \ A_1 = z_s \hat{A}_1,$$

gives in (3.44)

$$\frac{u_s^2}{x_s} = \frac{z_s}{x_s^2} = \frac{u_s}{z_s^2},$$

and in (3.45)

$$u_s = \lambda z_s,$$

with solution

$$x_s = \lambda^{-5/4}, \ z_s = \lambda^{-3/4}, \ u_s = \lambda^{1/4}, \tag{4.1}$$

and a corresponding pressure gradient scale $\lambda^{1/2}$. Using this transformation gives what might be called a fundamental problem for interactive boundary layer theory. In the discussion which follows we shall retain the shear upstream as having value λ but the significance of the transformation (4.1) in simplifying the computational problem needs to be appreciated. In the rest of this chapter we will use $A(X)$ for the displacement function rather than A_1.

The numerical solution of interactive boundary layer problems is well understood and in addition to initial solutions such as Smith (1974), Jobe and Burggraf (1974) there have been a number of further studies and excellent summaries are available in Rothmayer and Smith (1998c) and Sychev et al. (1998). Some methods to deal with marching in parabolic boundary layer type problems are reviewed in Rubin (1982) and numerical methods are discussed in Davis and Werle (1982)

Our plan here is to describe fully what might be thought of as the simplest method, that originally used by Jobe & Burggraf and a program to implement that method is available through material in appendix A. We shall also look at the method used by Smith, a method which mimics the two layer wake structure started by Goldstein and we will look briefly at a spectral method developed by Burggraf and Duck (1982) which can deal efficiently with the complicated pressure–displacement function interaction.

4.2 Numerical solution in rectangular coordinates

As we have seen, the archetypal triple deck equations which need solving for the velocities u, v and the pressure p are a momentum equation

$$u\frac{\partial u}{\partial X} + v\frac{\partial u}{\partial Z} = -p'(X) + \frac{\partial^2 u}{\partial Z^2}, \qquad (4.2)$$

a continuity equation

$$\frac{\partial u}{\partial X} + \frac{\partial v}{\partial Z} = 0, \qquad (4.3)$$

where there are very specific conditions on the velocities upstream and as $Z \to \infty$ and on the pressure gradient. Our plan is to present a simple but general scheme for solving this type of problem, initially specifically in the context of the trailing edge problem but then with the modifications which are needed later for separation and channel flows.

In the trailing edge problem this set of coupled non-linear equations has to have a solution satisfying the boundary conditions that

$$u = v = 0, \text{ on } Z = 0, \ X \leq 0,$$

$$\frac{\partial u}{\partial Z} = v = 0, \text{ on } Z = 0, \ X > 0,$$

and that

$$u \to \lambda(Z + A(X)), \text{ as } Z \to \infty,$$

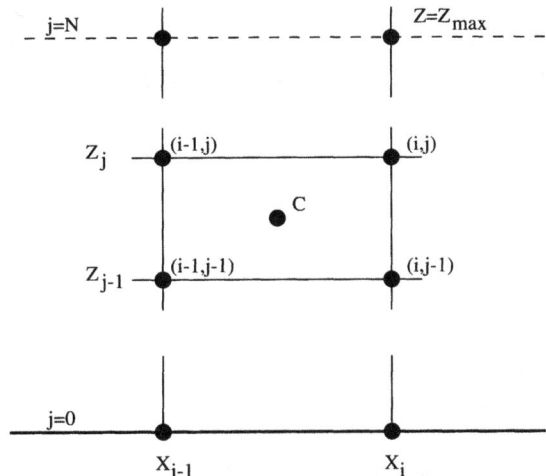

FIG. 4.1. Computational grid for triple deck solution.

where the 'displacement' function $A(X)$ satisfies (3.37) linking it with the pressure,

$$p(X) = \frac{1}{\pi} \int_{-\infty}^{\infty} \frac{A'(\zeta)}{X - \zeta} d\zeta, \qquad (4.4)$$

or equivalently

$$A'(X) = -\frac{1}{\pi} \int_{-\infty}^{\infty} \frac{p(\zeta)}{X - \zeta} d\zeta,$$

We suppose that these equations are to be discretised at points (X_i, Z_j) where $-M_u \leq i \leq M_d$ and $0 \leq j \leq N$. There the mesh can be non-uniform and the maximum extent in the Z direction is for $Z_{max} = Z_N$. Use the notation $h_i = X_i - X_{i-1}$ and $k_j = Z_j - Z_{j-1}$.

The general strategy is to suppose we are given an approximate value for $A(X)$ then by marching in X we can obtain an approximation for $p(X)$ and that being done, an improved guess for A can be found from the Hilbert integral relating p and A.

The solution from marching in the X direction has to be solved iteratively at each X value as the discretisation produces a set of discrete non-linear equations. Usually the solution at each X step has to be found using Newton's method. Then sweeps in the X direction are repeated until the solution converges to some predetermined criterion. Thus at the heart of the solution is a simple step: given a value of A and the solution at X_{i-1}, what is the solution at X_i? What we shall do is answer this specific question before showing how that step is incorporated into the overall solution.

Consider a cell vertex representation of the solution using a finite volume approximation over the cell $(i-1, j-1)$, $(i-1, j)$, $(i, j-1)$ (i, j), see Figure 4.1. Before doing so the equations are replaced by the set

$$u = \frac{\partial f}{\partial Z},$$
$$w = \frac{\partial u}{\partial Z}, \qquad (4.5)$$
$$u\frac{\partial u}{\partial X} - w\frac{\partial f}{\partial X} - \frac{\partial w}{\partial Z} + p' = 0,$$

where f is used for the stream-function and $-w$ will be the vorticity on $Z = 0$ but not elsewhere. Use capital letters to denote the discrete values of the variables f, u, w at the cell vertices and $P'_{i+\frac{1}{2}}$ for the pressure gradient at the cell centre. If the equations are then thought of as being evaluated at the centre of the cell, C in Figure 4.1, the discrete forms of the first two equations in (4.6) are

$$U_C \equiv \frac{1}{4}(U_{i,j} + U_{i-1,j} + U_{i,j-1} + U_{i,j} + U_{i-1,j-1}) =$$
$$\frac{1}{2k_j}(F_{i,j} + F_{i-1,j} - F_{i,j-1} - F_{i-1,j-1}) \equiv F_Z, \qquad (4.6)$$

$$W_C \equiv \frac{1}{4}(W_{i,j} + W_{i-1,j} + W_{i,j-1} + W_{i,j} + W_{i-1,j-1}) =$$
$$\frac{1}{2k_j}(U_{i,j} + U_{i-1,j} - U_{i,j-1} - U_{i-1,j-1}) \equiv U_Z, \qquad (4.7)$$

where U_C, W_C, F_Z, U_Z represent estimates for $u, w, \frac{\partial f}{\partial Z}, \frac{\partial u}{\partial Z}$ at C. If we denote estimated for $\frac{\partial u}{\partial X}, \frac{\partial f}{\partial X}, \frac{\partial w}{\partial Z}$ at C by U_X, F_X and W_Z,

$$U_X = \frac{1}{2h_i}(U_{i,j} + U_{i,j-1} - U_{i-1,j} - U_{i,j}),$$

$$F_X = \frac{1}{2h_i}(F_{i,j} + F_{i,j-1} - F_{i-1,j} - F_{i,j})$$

$$W_Z = \frac{1}{2k_j}(W_{i,j} + W_{i-1,j} - W_{i,j-1} - W_{i-1,j-1})$$

then the boundary layer equation in (4.6) has discrete form

$$U_C U_X - W_C F_X - W_Z + P'_{i+\frac{1}{2}} = 0. \qquad (4.8)$$

Non-linearity enters the equations in (4.8) through the products in the first two terms.

Applying (4.6)-(4.7) at N cell centres (that is $j = 1, \cdots, N$) gives $3N$ equations for $3N+4$ unknowns, $\{F_j, U_j, W_j\}, j = 0, \ldots, N$ and $P'_{i+\frac{1}{2}}$. These equations are completed by the boundary conditions

$$F_{i,0} = U_{i,0} = 0, \qquad (4.9)$$

on the plate or

$$F_{i,0} = W_{i,0} = 0, \qquad (4.10)$$

in the wake, and

$$U_{i,N} = \lambda(Z_N + A_i), \ W_N = \lambda, \qquad (4.11)$$

at the outer edge of the computation.

A solution for the non-linear discrete equations can be obtained using Newton's method. Let

$$\begin{aligned} F_{i,j} &\to F_{i,j} + \delta F_j, & U_{i,j} &\to U_{i,j} + \delta U_j, \\ W_{i,j} &\to W_{i,j} + \delta W_j, & P'_{i+\frac{1}{2}} &\to P'_{i+\frac{1}{2}} + \delta P', \end{aligned} \qquad (4.12)$$

and linearise by neglecting second order quantities. If the vector $\delta \mathbf{W}$ is defined by

$$\delta \mathbf{W}^{\mathrm{T}} = (\delta F_0, \delta U_0, \delta W_0, \cdots, \delta F_N, \delta U_N, \delta W_N, \delta P'), \qquad (4.13)$$

then the result is a matrix equation for $\delta \mathbf{W}$. Substituting (4.12) into (4.6)-(4.8) gives

$$\begin{aligned} -\delta F_{j-1} + a_j \, \delta U_{j-1} \qquad\qquad\qquad + \delta F_j + a_j \, \delta U_j \qquad\qquad\qquad &= 2k_j b_j, \\ -\delta U_{j-1} + a_j \, \delta W_{j-1} \qquad\qquad + \delta U_j + a_j \, \delta W_j \qquad\qquad &= 2k_j c_j, \\ e_j \, \delta F_{j-1} + g_j \, \delta U_{j-1} + q_j \, \delta W_{j-1} + e_j \, \delta F_j + g_j \, \delta U_j + s_j \, \delta W_j + 2k_j \delta P'_j &= 2k_j d_j, \end{aligned}$$

where the constants on the left-hand side are

$$a_j = -\frac{1}{2}k_j, \ e_j = \frac{k_j}{h_i}W_C, \ g_j = -\frac{1}{2}k_j U_X - \frac{k_j}{h_i}U_C,$$
$$q_j = -1 + \frac{1}{2}k_j F_X, \ s_j = 1 + \frac{1}{2}k_j F_X,$$

while the residuals on the right-hand side are

$$b_j = F_Z - U_C, \ c_j = U_Z - W_C, \ d_j = W_Z - U_C U_X + W_C F_X - P'_{i+\frac{1}{2}}.$$

Hence the matrix equation for the Newton update is

$$\mathbf{M}\delta\mathbf{W} = \mathbf{G}, \qquad (4.14)$$

where **G** will come from the terms on the right-hand side of (4.2) together with zero entries in the first two and last two rows. On the plate, the matrix **M** is

$$\begin{pmatrix}
1 & 0 & 0 & 0 & 0 & 0 & \cdots & 0 & 0 & 0 & 0 & 0 & 0 & \cdots & 0 & 0 & 0 & 0 & 0 & 0 & 0 \\
0 & 1 & 0 & 0 & 0 & 0 & \cdots & 0 & 0 & 0 & 0 & 0 & 0 & \cdots & 0 & 0 & 0 & 0 & 0 & 0 & 0 \\
-1 & a_1 & 0 & 1 & a_1 & 0 & \cdots & 0 & 0 & 0 & 0 & 0 & 0 & \cdots & 0 & 0 & 0 & 0 & 0 & 0 & 0 \\
0 & -1 & a_1 & 0 & 1 & a_1 & \cdots & 0 & 0 & 0 & 0 & 0 & 0 & \cdots & 0 & 0 & 0 & 0 & 0 & 0 & 0 \\
e_1 & g_1 & q_1 & e_1 & g_1 & s_1 & \cdots & 0 & 0 & 0 & 0 & 0 & 0 & \cdots & 0 & 0 & 0 & 0 & 0 & 0 & 2k_1 \\
\cdot & \cdot & \cdot & \cdot & \cdot & \cdot & \cdots & \cdot & \cdot & \cdot & \cdot & \cdot & \cdot & \cdots & \cdot & \cdot & \cdot & \cdot & \cdot & \cdot & \cdot \\
0 & 0 & 0 & 0 & 0 & 0 & \cdots & -1 & a_j & 0 & 1 & a_j & 0 & \cdots & 0 & 0 & 0 & 0 & 0 & 0 & 0 \\
0 & 0 & 0 & 0 & 0 & 0 & \cdots & 0 & -1 & a_j & 0 & 1 & a_j & \cdots & 0 & 0 & 0 & 0 & 0 & 0 & 0 \\
0 & 0 & 0 & 0 & 0 & 0 & \cdots & e_j & g_j & q_j & e_j & g_j & s_j & \cdots & 0 & 0 & 0 & 0 & 0 & 0 & 2k_j \\
\cdot & \cdot & \cdot & \cdot & \cdot & \cdot & \cdots & \cdot & \cdot & \cdot & \cdot & \cdot & \cdot & \cdots & \cdot & \cdot & \cdot & \cdot & \cdot & \cdot & \cdot \\
0 & 0 & 0 & 0 & 0 & 0 & \cdots & 0 & 0 & 0 & 0 & 0 & 0 & \cdots & -1 & a_N & 0 & 1 & a_N & 0 & 0 \\
0 & 0 & 0 & 0 & 0 & 0 & \cdots & 0 & 0 & 0 & 0 & 0 & 0 & \cdots & 0 & -1 & a_N & 0 & 1 & a_N & 0 \\
0 & 0 & 0 & 0 & 0 & 0 & \cdots & 0 & 0 & 0 & 0 & 0 & 0 & \cdots & e_N & g_N & q_N & e_N & g_N & s_N & 2k_N \\
0 & 0 & 0 & 0 & 0 & 0 & \cdots & 0 & 0 & 0 & 0 & 0 & 0 & \cdots & 0 & 0 & 0 & 0 & 1 & 0 & 0 \\
0 & 0 & 0 & 0 & 0 & 0 & \cdots & 0 & 0 & 0 & 0 & 0 & 0 & \cdots & 0 & 0 & 0 & 0 & 0 & 1 & 0
\end{pmatrix}$$

In the wake the second row is changed to reflect the boundary condition $W_0 = 0$ instead of $U_0 = 0$.

The Newton iteration can be started from the $i-1$ values or from the values at the previous sweep in the X-direction and usually converges in only a small number of iterations.

The pressure gradient being determined at the end of a sweep in the X direction, the pressure at the points $\{X_i\}$ can be found by integration. Once the pressure has been found a Hilbert transform has to be applied to calculate the displacement function $A(X)$. In practice this proves an unstable scheme and severe under-relaxation has to be applied to any update of the displacement function. If the value determined from a Hilbert transform of the pressure was denoted $\tilde{A}$ then the update which needed to be applied was

$$A \leftarrow (1-\varepsilon)A + \varepsilon\tilde{A}. \tag{4.15}$$

A typical value for the under-relaxation parameter was $\varepsilon = 0.05$.

The evaluation of the Hilbert transform was itself a critical element. The pressure was given at the nodal values X_i and the transform applied to calculate values of A' at the mid-points. The integral could be discretised using simple quadrature

$$A'_{i+\frac{1}{2}} = -\frac{1}{\pi} \sum_{j=-\infty}^{\infty} \frac{P_j}{X_{i+\frac{1}{2}} - X_j} \frac{h_{j-1} + h_j}{2},$$

but since values in the tail of the sum decay very slowly a simple truncation of the sum would not give a satisfactory transform. However, since the asymptotic form for the pressure was known,

$$p(X) \sim \begin{cases} -0.3433\lambda^{-1/3}(-X)^{-2/3}, & X \to -\infty, \\ 0.1717\lambda^{-1/3}X^{-2/3}, & X \to \infty, \end{cases}$$

a numerical estimate could be made for the value of the tails and that resulted in a suitable numerical transform. Once the values $A'_{j+\frac{1}{2}}$ were found, the displacement function came from a further quadrature of A' to provide $\tilde{A}$ in (4.15). For a discussion on numerical evaluation of the Hilbert integral in this problem see also the appendix in Ragab and Nayfeh (1982).

4.3 Solution using sublayer coordinates

A numerical method implemented by Smith (1972, 1974) was based on an analogy with Goldstein's inner and outer wake solution to solve a triple deck problem at a discontinuity in the wall conditions in compressible flow. The problem was slightly different to that considered here since in compressible flow the pressure–displacement function relationship is not as difficult to deal with, the pressure just being minus the gradient of the displacement function. Whereas Smith (1974) was motivated by injection problems, Daniels (1974) considered trailing edge flow.

We have seen that the development downstream of the trailing edge gave different expansions, near the centreline an appropriate variable was (using the variables X and Z here)

$$s = \frac{Z}{X^{1/3}}, \tag{4.16}$$

and the leading order contribution to the stream-function (2.137) had the form

$$\psi \sim X^{2/3} G(X, s), \tag{4.17}$$

where G was developed as a power series in X. Smith developed a numerical solution by using new variables

$$\xi = X^{1/3}, \quad s = \frac{Z}{X^{1/3}},$$

near the centreline and variables

$$\xi = X^{1/3}, \quad Z,$$

further from the centreline. The use of the same streamwise variable in both regions made patching between the two regions straightforward. The analogy with Goldstein's expansions was continued by supposing that in the inner region the stream-function was given by (4.17) where G satisfied

$$\frac{\partial^3 G}{\partial s^3} + \frac{2}{3}G\frac{\partial^2 G}{\partial s^2} - \frac{1}{3}[\frac{\partial G}{\partial s}]^2 - \frac{1}{3}\xi(\frac{\partial G}{\partial s}\frac{\partial^2 G}{\partial \xi \partial s} - \frac{\partial G}{\partial \xi}\frac{\partial^2 G}{\partial s^2}) = \frac{1}{3\xi}\frac{dp}{d\xi}. \tag{4.18}$$

In the same way as (4.6) by writing

$$H = \frac{\partial G}{\partial s}, \quad W = \frac{\partial H}{\partial s},$$

then (4.18) written in the form

$$\begin{aligned} G_s &= H, \\ H_s &= W, \\ W_s &= -\frac{2}{3}GW + \frac{1}{3}H^2 + \frac{1}{3}\xi\left(H\frac{\partial H}{\partial \xi} - W\frac{\partial G}{\partial \xi}\right) + \frac{1}{3\xi}\frac{dp}{d\xi}, \end{aligned} \quad (4.19)$$

could be discretised using a compact box method for the dependent variables G, H, W and p on a mesh ξ_i, s_j.

In the outer layer Smith used

$$\psi = F(\xi, Z), \quad (4.20)$$

where F satisfied

$$\frac{\partial^3 F}{\partial Z^3} - \frac{1}{3\xi^2}\left(\frac{\partial F}{\partial Z}\frac{\partial^2 F}{\partial \xi \partial Z} - \frac{\partial F}{\partial \xi}\frac{\partial^2 F}{\partial Z^2}\right) = \frac{1}{3\xi^2}\frac{dp}{d\xi}, \quad (4.21)$$

and again writing F_Z and F_{ZZ} as dependent variables then a box scheme can be used to obtain non-linear discretise equations for F and its first two derivatives with respect to Z. Since both regions share the same ξ values the two regions can be easily patched together provided the discrete Z values were chosen with care. Let the discretisation of ξ be at points ξ_i. If the inner region extended $0 < s < s_m$ then the outer layer equation (4.21) would be applied at points defined by $Z_i = s_m \xi_i$ so that patching between the two meshes occurred at nodal values and no interpolation was necessary. Given the values of the dependent variables at ξ_i the resulting non-linear equations for values of G, H, W, F, F_Z, F_{ZZ} at ξ_{i+1} were solved by a Newton iteration, giving a similar matrix problem for each Newton iteration to that already described.

4.4 A spectral method

In the case of an incompressible triple deck the pressure–displacement function relationship occurring in the form of a convolution integral adds a level of complexity and difficulty to any numerical solution. Burggraf and Duck (1982) used the simplification of the convolution form of the pressure–displacement function to a product when Fourier transforms were taken in the X direction. Of course such simplification would be at the expense of the non-linear inertial terms in the boundary layer equation, products in physical space, becoming convolution integrals in transform space.

A spectral method

If the pressure is eliminated from the boundary layer equation the differential system for the triple deck can be written

$$w = \frac{\partial u}{\partial Z},$$
$$u\frac{\partial w}{\partial X} + v\frac{\partial w}{\partial Z} = \frac{\partial^2 w}{\partial Z^2}, \qquad (4.22)$$
$$\frac{\partial u}{\partial X} + \frac{\partial v}{\partial Z} = 0. \qquad (4.23)$$

The boundary conditions are still those of no slip at the wall $Z = 0$, $u = v = 0$ and far field conditions $w \to \lambda$, $u \to \lambda(Z + A(X))$ as $Z \to \infty$. The pressure and displacement are still connected by the Hilbert integral (4.4).

Using a Fourier transform,

$$\overline{f}(k) = \frac{1}{\sqrt{2\pi}}\int_{-\infty}^{\infty} u(X)e^{-ikX}dX, \qquad (4.24)$$

and transformed variables

$$\overline{U}(k, Z) = \overline{u - \lambda Z},$$
$$\overline{W}(k, Z) = \overline{w - \lambda}, \qquad (4.25)$$
$$\qquad (4.26)$$

then the equations (4.23) become differential equations

$$\frac{d\overline{U}}{dZ} = \overline{W},$$
$$\frac{d^2\overline{W}}{dZ^2} - ik\lambda Z\overline{W} = \overline{(u - \lambda Z)\frac{\partial w}{\partial X} + v\frac{\partial w}{\partial Z}}, \qquad (4.27)$$
$$\frac{d\overline{v}}{dZ} = ik\overline{U}.$$
$$\qquad (4.28)$$

The two first order differential equations for $\overline{U}$ and $\overline{v}$ have boundary condition $\overline{U}(k, 0) = \overline{v}(k, 0) = 0$ while the outer boundary condition for $\overline{W}$ is just $\overline{W} \to 0$ as $Z \to \infty$.

The ingenious solution developed by Burggraf and Duck resulted in the pressure being eliminated from the numerical solution and at the same time providing the second boundary condition needed for the second order differential equation for $\overline{W}$.

This was achieved by using, first, from the boundary condition on the velocity as $Z \to \infty$,

$$A(X) = \lambda^{-1} \int_0^\infty [w(X,Z) - \lambda]\, \mathrm{d}Z, \qquad (4.29)$$

and secondly from the boundary layer equation at $Z = 0$,

$$p'(X) = \frac{\partial w}{\partial Z}(X,0), \qquad (4.30)$$

so that in terms of transforms

$$\frac{\mathrm{d}\overline{W}}{\mathrm{d}Z}(k,0) = \mathrm{i}k|k|\lambda^{-1} \int_0^\infty \overline{W}(k,Z)\mathrm{d}Z. \qquad (4.31)$$

Far upstream the dependent variables all vanish, so excepting the right-hand side of the boundary layer equation in (4.28), the problem has been reduced to solving ordinary differential equations in Z. Burggraf and Duck approached the non-linearity in the boundary layer equation by an iterative solution, regarding the right-hand side of (4.28) as being known at each iteration. They found that the fastest method to evaluate the expression

$$\overline{(u - \lambda Z)\frac{\partial w}{\partial X} + v\frac{\partial w}{\partial Z}} \qquad (4.32)$$

which has a convolution representation was to use fast Fourier transforms to calculate the physical variables u, v, w from their transformed values, evaluate (4.32) in the X domain and then apply a further fast Fourier transform to calculate values in the k domain. They found their method gave a substantial reduction in total computational time. This spectral method should have even greater advantages when applied to periodic channel problems of the type we consider later.

4.5 Channel flow

As we shall see later, problems involving channel flow lead to an identical interactive boundary layer equation but the nature of the middle deck changes and the absence of an outer deck removes the pressure–displacement function relation. Instead, depending on the size and length of any wall perturbation, the displacement function may be known in terms of the wall shapes or it may only be known implicitly and have to emerge from a numerical solution.

When the wall variation is sufficiently small then the displacement function is explicitly known in terms of the two wall shapes and the methods we have described can be applied directly by removing the displacement function from the matrix **M**. When the wall variation is large enough then the two wall layers provide a coupled problem which has to be solved for each wall with the local wall pressure being different but related by the displacement function. In particular there is a lower and an upper layer with a interactive boundary layer equation to be solved and the pressure in the lower layer, P_L, and the upper layer, P_U, are related by

$$P_U(X) = P_L(X) + qA''(X) \qquad (4.33)$$

where q is a known constant.

The boundary conditions far from each plate have a similar form but involve the wall shape functions, so for instance if the lower wall is determined by a function $f(X)$, the stream-function in the lower layer (with normal coordinate η) will have to satisfy

$$\psi_L(X,\eta) \sim \frac{1}{4}\eta^2 + \frac{1}{2}[A(X) + f(X)]\eta \text{ as } \eta \to \infty, \tag{4.34}$$

with a very similar condition for the upper layer. This condition is not suitable for direct use in a scheme since there is also a constant which can appear in the expansion for ψ_L as $\eta \to \infty$ but that constant does not appear in derivatives with respect to η so conditions at the outer edge of the computation on the first two derivatives can be applied using (4.34).

The method of solution in each layer is the same as for a single triple deck deck. If each layer is discretised over $N + 1$ points normal to the wall then there will be $3(N + 1)$ unknowns in each layer corresponding to $F = \psi_L$, $U = F_\eta$ and $W = U_\eta$. In addition there will be a pressure in each layer and the displacement function giving $6N + 9$ unknowns. There are two conditions on each wall, $3N$ equations at each point in each boundary layer, two conditions (on U and W) at the edge of each boundary layer and (4.33) relating the pressures and displacement function giving $6N + 9$ equations. The set of equations at each X station is again non-linear and has to be solved using a Newton iteration. Since it is the pressure gradient which enters the boundary layer equation the pressure relation (4.33) has to be differentiated so that A''' enters the problem. If the X step size is h and we are considering the equations at the X-station $X = ih$ then we have used both $A''' \approx (A_{i+1} - 3A_i + 3A_{i-1} - A_{i-2})/h^3$ and $A''' \approx (A_i - 3A_{i-1} + 3A_{i-2} - A_{i-3})/h^3$ successfully. The latter may have some stability advantage. It is also possible to use a system of equations for A and its derivatives, $B = A'$, $C = B'$ and $D = C' = A'''$ by extending the coefficient matrix $\mathbf{M}$ still further, see for instance Smith (1977b). Further detail of implementation issues can be found via appendix A. Experience is that when there is a significant cross channel pressure variation then convergence is remarkably poor and there are a number of stability difficulties which still need to be resolved.

Part II

Separation

5
INTRODUCTION TO SEPARATION

Steady separated flow is in many ways an idealised or academic part of fluid dynamics. It is the case that virtually all separated flows are either unsteady (implying that the underlying steady separated flow is unstable) or that separation occurs as part of a transient flow process. In the latter case, if the time variation is sufficiently slow, it might be hoped that through a quasi-steady approximation, the time varying flow will be given by a steady solution at a time-local Reynolds number. *The range of flow conditions for which it is possible to observe steady separated flow is, for most geometries, very small.* Since it is the **onset** of separation which is a significant fluid mechanical event, heralding possible changes in lift, drag, heat or mass transfer, the fact that separation occurs at all is an important engineering design parameter, regardless that it is followed by more complicated fluid mechanical events than those described by a steady flow model. Thus the study of idealised two-dimensional steady separation has relevance to practical flow situations by defining regions of design or parameter space where important fluid mechanical events can arise.

Of course, there is no question that high Reynolds number *real* flows are vastly different to any high Reynolds number solution of the steady Navier–Stokes equations. At the centre of what we are trying to understand is whether **moderate** Reynolds number separated flows which are real and are steady, can be usefully modelled by applying high Reynolds number expansions of the steady Navier–Stokes equations at those moderate Reynolds numbers. Clearly this cannot be done without having decided on the correct high Reynolds number limit of the Navier–Stokes equations, since it is asymptotic expansions for that flow which will be used at moderate Reynolds numbers. It is also the case that while a flow may not be globally steady, parts of the flow may be to all intents considered as steady and laminar while only further downstream is the flow significantly unsteady or turbulent. Hence there is also a hope that asymptotic theories, even when not applicable to a whole flow field, may nevertheless be reasonably applied to parts of a flow field. In problems involving separation of a viscous fluid, the question as to which flow is the correct high Reynolds number limit is far from trivial.

In flow of a viscous fluid it is possible for fluid to be moving backwards (relative to some far field flow direction) near a wall without there being any suggestion of separation. This is particularly the case in unsteady low Reynolds number flows where the flow can locally resemble a developing Rayleigh layer and even though backflow occurs the fluid motion is more or less parallel to the wall. In such flow, solutions of the Navier–Stokes equations are more or less identical to solutions of the unsteady Stokes equations.

The situation we commonly denote as separation is where the fluid moves rapidly away from a surface and where the interaction between inertia, represented by the non-linear $\mathbf{u} \cdot \nabla \mathbf{u}$ term), the pressure and viscous stress is so intertwined that all three elements are bound up with the flow development.

A region where the pressure gradient is positive is called a region of adverse pressure gradient and viscous separation is usually associated with an adverse pressure gradient. It is most often explained as an effect whereby the adverse pressure gradient has a greater effect on fluid moving slowly near a wall than further away (an incorrect appeal to Bernouilli's equation within the boundary layer would give $u_x \sim -p_x/u$ so u varies more rapidly with distance near the wall and in particular u_x negative when p_x positive) and so at some point the flow direction near the wall must reverse. Continuity would then give that the fluid coming from upstream and downstream would have to move sharply away from the body surface to accommodate the two approaching streams of fluid.

Unfortunately when applied to an inviscid flow the very general description 'steady separation' does not even bring to mind an unambiguous physical situation and so a corresponding ambiguity extends to which flow might be the correct high Reynolds number limit for separated viscous flow. There are at least three model inviscid flow situations which can be described as separated flow and which have thus coloured ideas about the high Reynolds number limit for viscous separation.

The first model situation is that of a cavitation bubble (a bubble of the vapour phase of the liquid in which the body is moving) attached to the downstream side of a body in steady flow. The underlying idea is that if the fluid motion about a body is sufficiently fast, by appealing to Bernouilli's equation, the pressure will fall to such a low value that cavitation occurs. Cavitation is observed in many situations, for instance near propeller blades in water. There are many pictures of steady, attached cavitation regions (see for instance photographs in Wu (1972)) so it is clear that at a practical level, cavitation regions can develop and become attached quite quickly. In this situation viscosity is not of prime importance to the flow structure and so it is reasonable to believe that the study of steady inviscid flow may directly model the fluid behaviour.

The second model is that of a second but much lower density fluid entrained in an attached bubble behind a body moving into or through a fluid, for instance air entrained behind a body fired from air into water. Again in this case, intuition would suggest that an inviscid flow model might describe many important features of such a flow and be reasonably taken as the small viscosity limit.

In both of these cases, the limiting inviscid flow is taken to have constant pressure within the separated region. This can be expressed quantitatively by a pressure coefficient, if $\hat{p}_c$ is the pressure within the separated 'cavity' and U, $\hat{p}_\infty$ the speed upstream and pressure upstream, then the pressure coefficient is

$$C_p = (\hat{p}_c - \hat{p}_\infty)/\frac{1}{2}\rho U^2. \tag{5.1}$$

The third model is that there is a steady region of the same fluid recirculating (or possibly stationary) on the downstream side of a body. In this case we might still naively expect an inviscid flow model to describe the major features of the flow if the viscosity is small. Unfortunately there is not a unique inviscid flow which might be used and depending on which inviscid flow is taken, consequences can emerge which are difficult to accept. We shall investigate those difficulties in detail shortly. In each of the first two situations it is fairly straightforward to imagine how the flow is set up and viscosity plays little part in the process of establishing the separated flow. In this third case the development can only come about through viscous action so although there may be a number of different inviscid flow configurations, not all may be achievable from letting viscosity vanish in the Navier–Stokes equations.

Thus in attempting to fix ideas about what type of flow might emerge when separation occurs and **then** the viscosity vanishes (neglecting real effects, such as transient disturbances and turbulence, just focusing on steady, laminar solutions of the Navier–Stokes equations) there is agreement that the first two situations are essentially mathematically identical and are modelled adequately by inviscid flow theory, called either free streamline flow or Helmholtz motions. What is not clear or universally accepted is whether we are able to properly specify the inviscid flow which results from letting the viscosity vanish in the Navier–Stokes equations after separation has occurred and in particular whether in the limit of infinite Reynolds number, the flow outside of a separation region is the same inviscid flow as occurs in the case of cavitation or entrainment.

Ideas about the correct inviscid limit are also often coloured by preconceptions about the drag at high Reynolds number. It is difficult to do other than echo Batchelor's comment that *the limit of infinite Reynolds number is so far outside the range of 'experience' that the prediction of a finite value for the drag coefficient ... is not more welcome physically than the prediction of a zero value* (Batchelor (1956b), page 393). There is even confusion as to the meaning of a zero drag **coefficient**. Nothing illustrates the singular nature of the infinite Reynolds number limit better than the distinction between a drag coefficient and the drag. Even if the drag coefficient tends to zero that does not mean that the physical force on a body also vanishes. As an example, coefficient $c_D = cR^{-1/2}$ with appropriate constant c could be taken as a suitable viscous drag coefficient for a plate of length L. This coefficient vanishes as $R \to \infty$. However, the physical drag would be given by

$$\text{Drag} = c\sqrt{\rho\mu}U^{3/2}L^{1/2}.$$

and even though the drag coefficient vanishes as $R \to \infty$, the physical drag becomes infinite if the the Reynolds number is large through U being large at constant viscosity. On the other hand, if the Reynolds number is becoming large because the viscosity is vanishing then the physical drag would go to zero too. The physical drag would only vanish under all circumstances of the limit $R \to \infty$ if $c_D = o(R^{-2})$. This confusion about the asymptotic behaviour of a drag coefficient has influenced views about the suitability of some models.

Thus there have been two sometimes parallel theoretical themes which have been examined to try to explain viscous separation.

One theme is what might be called a classical boundary layer approach. In flow over a curved boundary the free stream velocity, u_e will vary along the boundary and so the pressure will vary according to $-u_e^2/2$ and if u_e is decreasing then the pressure will be rising. From the inception of the boundary layer idea through to the mid 1960s an important area of study in boundary layer theory has been the development of a boundary layer in an adverse pressure gradient. Nearly all of these studies point to the appearance of a singularity in the boundary layer solution at the point of separation when the adverse pressure gradient is specified.

The second theme is based on the ideas for inviscid separation just outlined. These use free streamline theory where in addition to having to be concerned about a singularity at the point of separation (although a very different singularity from that in boundary layer calculations), the position of separation from a curved body is not unique without a further condition and the far field of the solution, specifically the wake, has counter intuitive aspects.

We shall also look at attempts to predict separation by combining these two methods: boundary layer theory and free streamline theory.

Rather than attempt to study these themes in full generality, we shall mostly restrict our study to the case of separation behind a two-dimensional circular cylinder. Thus one objective of the next chapter is to imagine we are at a time before interactive boundary layer theory had been developed and we have been asked to determine this simple flow field when separation occurs. What solution would we arrive at? As indicated above, whether we use boundary layer theory, free streamline theory or a combination of these two, we shall not be able to arrive at a completely satisfactory conclusion and it is the nature of the unsatisfactory conclusion in each case which is one objective of that chapter. In each case, we follow some of the historical development of ideas in some detail but we shall begin with a brief description of the experimental results which have been available to test model predictions. See also Williams (1977) for a review of many aspects of boundary layer separation.

6

SEPARATED FLOW ABOUT A CYLINDER

6.1 Observation at moderate Reynolds number

There are many qualitative observations of flow about a circular cylinder but the ones we shall concentrate on here are those which give quantitative information about the surface pressure distribution, the surface velocity and steady separation. The difficulty in taking reliable measurements should not be underestimated. The range of flow rates for which steady separation occurs is quite small and it is difficult to completely eliminate effects of the boundary of the flow tunnel. Pressures can be very small and in the early part of this century, with manometer type gauges, readings took a long time to stabilise and it was very difficult to construct small pressure port openings on the surface of the cylinder. Temperature variations had to be as small as possible, particularly when using oil. Despite these difficulties there are a number of convincing sets of measurements of steady laminar flow around a cylinder. This section is not intended to be a comprehensive survey of experimental results for either flow separation or flow around a cylinder and only some results which have had an impact on theoretical studies are surveyed. A very comprehensive description of papers about flow around a cylinder can be found in Zdravkovich (1997).

Assume that the flow far from the cylinder has velocity U and the cylinder has radius a. In non-dimensionalising the flow problem, we choose to use the cylinder **radius** as a characteristic length and Reynolds number based on the radius, $R = Ua/\nu$; it is common to base the Reynolds number on cylinder diameter and where necessary we will denote that Reynolds number $R_d = 2R$. Since the factor of two for conversion between radius and diameter based Reynolds numbers is so simple this should not cause any confusion.

The picture we have of steady separated flow is illustrated in Figure 6.1 for a unit radius cylinder. At some angle α_s from the front stagnation point the flow separates from the wall forming two counter rotating vortices. The vortices have centres C and C' behind the cylinder and the separated region is of length L_s and of maximum width W_s. Some authors have used the distance from the rear of the cylinder as the separation length, here we shall measure L_s from the cylinder centre. The stream-function is zero on the lines of separation, a measure of the strength of the vortices is the value of the stream-function at C, ψ_c. Each of the characteristics, α_s, L_s, W_s and ψ_c should be regarded as functions of the Reynolds number and much of what fluid mechanics seeks to do is just to provide the functional dependence of each of these characteristics on the Reynolds number.

The sequence of events associated with separation can be very complicated. The best way to look at these events is through bifurcation theory and we shall consider a particular channel flow in detail in Chapter 10 where the notation for bifurcation theory is also briefly set out. The most general sequence for a flow as the Reynolds number is increased is first one of loss of symmetry in one plane with the flow remaining steady, so that the flow is two-dimensional but asymmetric. Typically this would be via a pitchfork bifurcation if asymmetry occurs with two different 'directions' about the plane of symmetry which has been lost. Then symmetry might be lost in a second plane so that the flow becomes three-dimensional but remains steady. This too may be via a pitchfork bifurcation. Alternately there may be a Hopf bifurcation giving a time periodic flow. Then there can be a progression via further Hopf bifurcations to a complex time dependent flow. The precise sequence of events varies for different geometries. What seems to be observed for flow around a cylinder is transition from steady two-dimensional symmetric flow to periodic shedding of two-dimensional vortices and then more complex time dependent flow (see also Coutanceau and Bouard (1977)). Our concern is with the steady two-dimensional symmetric flow. As shall be stressed again later, that a flow is observed to be asymmetric or time periodic or more complicated does not mean that the Navier–Stokes equations do not have steady two-dimensional symmetric (about the centre plane) solutions for that geometry; it does however mean that the steady symmetric flow is unstable in some way and possibly unobservable. It is also important to stress again how small is the range of Reynolds numbers for which steady symmetric two-dimensional flow can be observed around a cylinder, Reynolds numbers less than 25 although with careful experimentation the flow can be kept steady for somewhat larger Reynolds numbers.

Thus if the global flow field around a cylinder is being considered, including its wake, the range of Reynolds numbers for steady flow is relatively small. However, when the wake is unsteady or turbulent, it is observed for much of the front part of the cylinder that the flow is effectively steady and indeed laminar over a much wider range of Reynolds numbers. A general criterion for transition from a laminar to a turbulent boundary layer was observed to be $U\delta/\nu \approx 10^3$ (with δ being the boundary layer thickness) and if the boundary layer thickness satisfied $\delta/a \sim R^{-1/2}$ then there should be substantial laminar regions on the front side of the cylinder up to Reynolds numbers of order 10^6. The implication is that downstream transient disturbances do not propagate far enough upstream to affect the boundary layer developing over the front part of the cylinder where the flow can be described according to laminar theory. Thus, there is a sense that even though the whole flow field can not be described as steady, the theory for laminar flow can still be applied to part of the surface of the cylinder and that any prediction of separation would be useful at Reynolds numbers far in excess of those for which the whole flow field is laminar and symmetric.

As already indicated, when presenting pressure data it is convenient to use a pressure coefficient, c_p defined in terms of the dimensional pressure $\hat{p}$ and

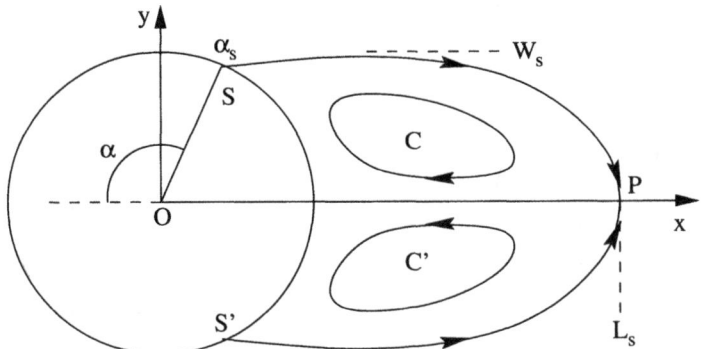

FIG. 6.1. Idealised steady separation behind a cylinder. Separation occurs at angles $\pm\alpha_s$ from the forward stagnation point at points S, S'. The pair of vortices in the separated region have centres C, C', length L_s and width W_s.

upstream conditions for the pressure and velocity, $\hat{p}_\infty$ and U by

$$c_p = \frac{\hat{p} - \hat{p}_\infty}{\frac{1}{2}\rho U^2}. \qquad (6.1)$$

In potential flow with the dimensional surface speed denoted Uq and using Bernouilli's equation to relate the pressure and surface speed, the pressure coefficient can be written

$$c_p = 1 - q^2. \qquad (6.2)$$

This gives $c_p = 1$ at the leading stagnation point, in a viscous flow the pressure coefficient is higher than this value at low Reynolds numbers. In practical situations the pressure is measured at the surface and a surface speed deduced from the pressure assuming that the pressure is constant across the boundary layer on a cylinder and Bernouilli's equation is really being applied at the outer edge of the boundary layer to obtain a surface speed or outer tangential velocity for the boundary layer.

6.1.1 Heimenz (1911)

Early quantitative measurements of pressure around a cylinder were carried out by Heimenz (1911) whose results were later used by Pohlhausen in applying Karman's integral method and we shall discuss that work shortly. Heimenz observations were for a cylinder of diameter 9.75 cm and upstream speed 19 cm/sec resulting in a Reynolds number, $R = 9262$. Heimenz observed that the pressure fell from the stagnation pressure at the front of the cylinder until around an angle $\alpha = 70°$ before rising, indicating an adverse pressure gradient after $\alpha = 70°$. Shortly after the start of the adverse pressure gradient the flow separated, around

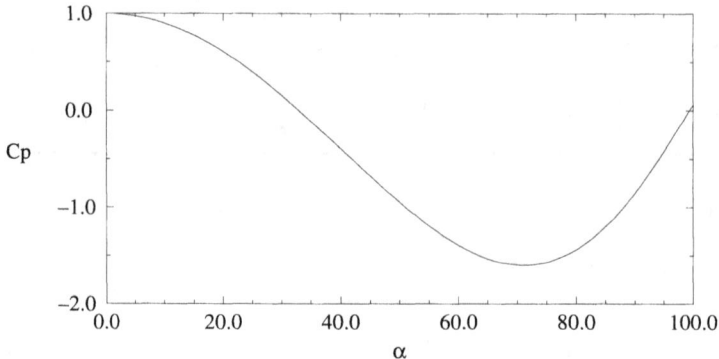

FIG. 6.2. Pressure coefficient observed by Heimenz for flow around a cylinder at $R = 9262$.

an angle $\alpha = 80°$. Heimenz used a ninth order polynomial to represent the pressure gradient, obtaining in our notation

$$\frac{\partial p}{\partial \alpha} \approx -3.37\alpha + 2.01\alpha^3 + 9.13\alpha^5 - 0.10\alpha^7 - 0.01\alpha^9, \tag{6.3}$$

and the pressure coefficient corresponding to Heimenz's observations is illustrated in Figure 6.2. The pressure observations allowed Heimenz to infer a quintic approximation for the speed at the edge of the boundary layer, see Figure 6.3.

6.1.2 Fage (1928), Fage & Falkner (1929)

In a series of papers, Fage (1928) republished as Fage (1929) and Fage and Falkner (1931) described measurements of flow around a circular cylinder. One set of experiments, using a 8.9-inch diameter cylinder in a 4-foot wind tunnel were intended to examine the boundary layer thickness for transition on the cylinder (previous experiments by Burgers and van der Hegge Zijnen examined the boundary layer on a flat plate aligned with the flow) by measuring pressures around the cylinder and across the boundary layer using a traversing fine pressure tube. These observations at Reynolds numbers between 51 300 and 166 500 gave results for the pressure coefficient which were similar to those of Heimenz. In a second set of experiments the stress at the surface was measured using pressure observations close to the surface of two cylinders of diameter 2.93 inches and 5.89 inches, again in a 4-foot wind tunnel. The near wall pressure measurements allowed the velocity to be estimated and the stress found from $\mu \frac{\partial u}{\partial r}$. These measurements also explored the transition region, at Reynolds numbers between 30 000 and 106 000, and at the lower end of Reynolds numbers observed, the wall shear rose from zero at the front stagnation point, reaching a maximum near

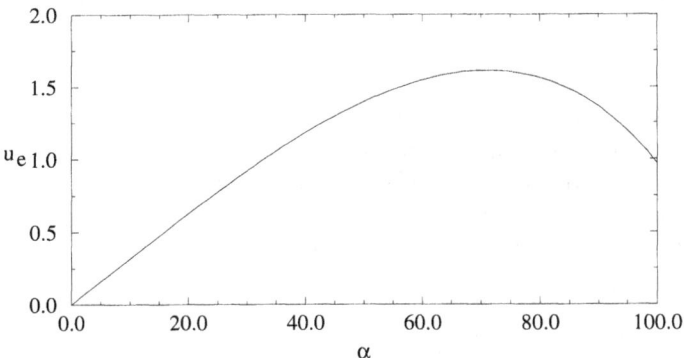

FIG. 6.3. Surface velocity deduced by Heimenz from pressure coefficient for flow around a cylinder at $R = 9262$

$\alpha = 55°$ and then plummeted to zero at an angle just above $\alpha = 80°$ depending on the flow conditions.

6.1.3 Thom (1929), Linke (1931), Thom (1933)

Further experiments by Thom (1928a), Linke (1931) and Thom (1933) included flow visualisation, pressure measurements and estimation of wall stress around cylinders with diameters between 0.3 and 4 cm. These experiments were at smaller Reynolds numbers than Heimenz or Fage, generally of order 1 000-10 000 although Thom's observations went down to very low Reynolds numbers of order 1 and included the range where the global flow field was steady and separated. Thom was able to achieve low Reynolds numbers by using oils with viscosity around .4 cm^2/sec (about 40 times the viscosity of water) as working fluids. Generally all these experiments reproduced the features already described by Heimenz, and Fage & Falkner, for the pressure coefficient and for the wall stress. In the observations the pressure coefficient was approximately constant and negative across the rear of the cylinder inside the separated region.

6.1.4 Homann (1936)

In experiments which produced some very beautiful flow visualisation, Homann (1936b) also used oil as a working fluid to observe flow around cylinders (Homman's flow visualisation was later reproduced in Batchelor (1967)). The oils used by Homann were one of viscosity 1.13 cm^2/sec and one of viscosity 0.174 cm^2/sec. The first of these was of course of very high viscosity (over a hundred times the viscosity of water) so he was able to observe flow at Reynolds numbers between $R = 1.95$ and $R = 140.5$. In addition to the pressure around the cylinder he observed the separation angle and found that it varied from around 145° at $R = 10$ to just under 100° when $R = 140$. Of course, most of his measurements were for flows with an unsteady wake but they did include examples of steady separated flow.

6.1.5 Kovásznay (1949)

An important observation about wake-body interaction came from Kovásznay (1949) who used measurements of heat transfer from a hot wire to study the wake behind a circular cylinder. His main conclusion was that vortex shedding was not directly initiated in the separation regions behind a cylinder, rather an instability in the wake fed back to the vortices and resulted in vortex shedding.

6.1.6 Roshko (1955)

The stabilisation of the wake by a splitter plate was also observed by Roshko (1955) who used two different splitter plates, one long plate going from the rear edge of the cylinder to around five diameters downstream and one shorter plate a little longer than one diameter which could be moved up to seven diameters downstream. Roshko's main concern was oriented towards unsteady flow but his measurements showed a marked dependence of the drag and of vortex shedding on the type and location of the splitter plate. His work was followed by a further study, Roshko (1961), of much higher Reynolds number flows (up to 10^7) where the effect of a splitter plate was much less dramatic.

6.1.7 Grove, Shair, Petersen & Acrivos (1964)

Carrying on the idea of using a splitter plate to stabilise the flow, Grove et al. (1964b) were able to observe steady separated flow behind a cylinder for Reynolds numbers up to $R_d = 400$. Their experiments showed that over the range of Reynolds numbers in their experiments, the length of the separation region increased linearly with Reynolds number, the pressure coefficient in the separated region was negative and close to -0.45 for a wide range of Reynolds numbers and the separation angle changed from around $145°$ at $2R = 20$ to less than $110°$ at $2R = 400$ with the rate of change decreased substantially as the Reynolds number increased. The first observation was entirely consistent with Homann's observation but their second observation (which is a larger separation angle than observed by Homann) indicated that for larger Reynolds numbers, Homann's observation of separation angle may have been influenced by unsteadiness in the wake.

6.1.8 Acrivos et al (1965 & 1968)

In Acrivos et al. (1965) and Acrivos et al. (1968) experiments were carried out to compare with a model for large Reynolds number flow. They argued that the closed vortex behind a cylinder should be long and thin so that the drag on a wake of length L might, using an analogy with the drag of a boundary layer, be given coarsely by

$$c_D \sim c\sqrt{\frac{L}{R}}, \qquad (6.4)$$

where c is some order one constant and c_D tends to some constant for large R. In that case the vortex length would have to satisfy

$$L \sim (\frac{c_D}{c})^2 R. \tag{6.5}$$

On the whole their experiments did not contradict their hypothesis that the length of the vortex might increase linearly with Reynolds number while remaining only of order one width although later Smith (1979b) estimated that the walls of the experimental tunnel may have had a much greater confining effect on the vortex width than was realised when the experiments were carried out.

An important feature raised in these works which has not yet been fully resolved is the question of what flow one might expect on the scale $R(x,y)$ when as we have seen for the flat plate, for instance (2.62), the equations of motion are independent of the Reynolds number which in this problem enters through the boundary conditions on the cylinder. In the stretched coordinates the cylinder has a radius R^{-1} and shrinks to a point as $R \to \infty$. What disturbance remains from the point singularity which the cylinder is becoming?

6.1.9 Shear stress measurement: Dimopoulos & Hanratty (1968)

One quantity which is easy to compute is the shear stress on the wall. This is however a very difficult quantity to measure directly. Dimpopoulos and Hanratty (1968) used an indirect technique to measure the wall shear stress. The rate of an electro-chemical reaction of ferricyanide ions to ferrocyanide ions was dependent on flow rate past a cathode and it was possible by a form of Reynolds analogy to relate the mass transfer rate (that is the current through the cathode) to a averaged wall shear stress over the cathode. Their results for the separation angle were for a variation between $125°$ at $R = 60$ to around $115°$ at $R = 180$ while their wall shear stress measurements showed reasonable agreement with a boundary layer calculation ahead of separation. Later Son and Hanratty (1969b) measured wall shear at much higher Reynolds numbers where the flow was unsteady.

6.1.10 Coutanceau & Bouard (1977)

The final set of observations we shall discuss are due to Coutanceau and Bouard (1977) where comparison to a number of other sets of observations can be found. They used photographs of the movement of magnesium cuttings to visualise flow around a cylinder for Reynolds numbers below $2R = 40$. Their method was to propel a cylinder through a tank of stationary liquid and to take photographs using a camera moving with the cylinder. They were able to vary the ratio of channel width to cylinder diameter so as to extrapolate their results to a cylinder in an unbounded fluid. As they did not use a splitter plate their results were limited to low Reynolds numbers but nevertheless they too found that the separation region grew essentially linearly with Reynolds number and that the angle on the cylinder at which separation occurred was in agreement with previous observations.

6.1.11 Summary

In one sense the preceding sections which describe observations of flow around a cylinder during more than half a century of research are all repeating the same results. Indeed there are available remarkably complete observations in the region where stable separation occurs behind a cylinder. From our viewpoint one important observation was that the length of the separated region appeared to grow linearly in the range of Reynolds numbers for which there was a steady solution, being given approximately in an unbounded fluid by

$$L_s \approx 1 + 0.23(R - 2.2). \tag{6.6}$$

This was a very positive prediction against which any theory or computation might be tested. The prediction of separation angle was less amenable to a simple formula but there are good observations of the separation angle and the separation region shape for $2R = 20$, 30, 40 as well as the drag coefficient.

Unfortunately what the observations do not do is give any form to the high Reynolds number limiting flow about a cylinder since the range of Reynolds numbers for these observations was so low. Even with the added stabilisation of a downstream splitter plate the range of Reynolds numbers for which we have observations was still far too low to have much inviscid character. The best which might be hoped is that any predictions from theory or computation which agree with observations at these Reynolds numbers might give confidence in results for much higher Reynolds numbers, although even that hope may be overstated.

6.2 Free streamline theory

Free streamline theory is a classical branch of fluid mechanics and as we have indicated, of importance in many areas outside of viscous flow. There are a number of texts which describe the theory, although as subject in its own right and not in the context of boundary layer theory. Milne-Thompson (1938) has an extensive and very readable description of theoretical ideas. Thwaites (1960) has sections devoted to these motions. Comprehensive studies of what might be called modern understanding is in Woods (1961), Birkoff and Zarantonello (1957), Gurevich (1965) and Gilbarg (1960). All of these works are important reading material for anyone interested in further details of free streamline theory.

Our plan in this section is to describe some of the detail of the historical development of this subject, drawing out important results as we progress, as well as computing various examples, mostly to illustrate free streamline theory for flow around a cylinder. There are many aspects of free streamline theory which we shall not touch upon, and indeed some models for flow around a cylinder which have been proposed but which are not particularly relevant here. In this group are Riabouchinsky (1920) flow where a second body is introduced to obtain downstream reattachment and re-entrant jet models where fluid 'flows' back through the rear of a body into what is a different Riemann sheet but in doing so, allows a downstream stagnation point. Another example of an ad-hoc model which might of interest is Kiya and Arie (1977). An unusual example of a calculation of unsteady separation from a cylinder using free streamline theory has been considered by Il'ichev and Postolovskii (1972) whose comparisons with observation strengthen the belief that even though the global flow field may be significantly more complex than we can describe at present, many aspects of the local flow near a cylinder may be described using free streamline theory.

It is also worth noting that if the boundaries of a body are polygonal then many problems can be simplified to ones involving Schwarz–Christoffel mapping, see for instance Dias et al. (1987) and Kythe (1998), section 3.4. It is when the boundaries are curved that greater difficulties have to be overcome.

The earliest example of free streamline theory was given by Helmholtz (1868). The development of potential theory had enabled both fluid dynamic and electrostatic problems to be solved but Helmholtz realised that unlike electric fields, fluids would not, near corners, follow streamlines which required an infinite acceleration. As he described, "*every perfect geometrical sharp edge by which a fluid flows must tear it asunder and establish a surface of separation, however slowly the rest of the liquid may move*". He was drawn to this conclusion by the belief that if the velocity "*exceeds a certain amount, the pressure must in fact become negative, and the fluid torn apart*". We know now that under very controlled conditions fluids can withstand quite large tension (negative pressure, see for instance Bull (1956)) but Helmholtz's reasoning continues to influence the study of separation. In addition to understanding that separation might occur, Helmholtz deduced that the pressure would be constant on a free streamline and hence also the velocity in an inviscid irrotational flow. This enabled him to determine the solution for Borda's mouthpiece or a re-entrant jet.

An additional attraction of free streamline theory was the prediction of non-zero drag. It must have been very difficult to be a fluid dynamicist in the early part of the nineteenth century because of d'Alembert's paradox: that known potential flows lead to zero drag whereas common experience was that all bodies experienced a drag of some magnitude. The attraction of a theory which predicted a non-zero drag for a potential flow should be obvious. Present day fluid mechanics is built upon the rescue of theoretical ideas from d'Alembert's paradox by free streamline theory perhaps as much as by Prandtl's boundary layer theory.

6.2.1 *Helmholtz (1868)*

Helmholtz observed that the map from coordinates $z = x + iy$ to the complex potential $w = \phi + i\psi$,

$$z = w + e^w, \tag{6.7}$$

represented the electrostatic field for two plates at $y = \pm\pi$, $-\infty < x \leq -1$ but that this would be unsatisfactory were it used to describe fluid flow near the edges of the plates. The field lines for this map are shown in Figure 6.4 and it is

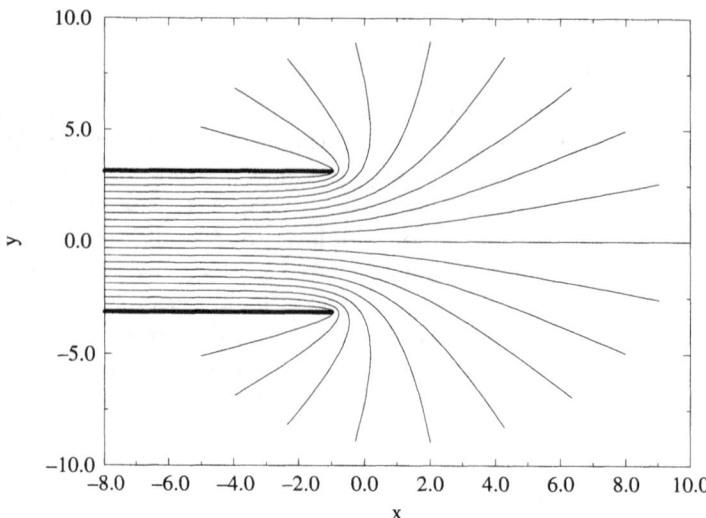

FIG. 6.4. Streamlines for potential flow into a channel.

clear that there would need to be a tremendous acceleration of fluid as it moves around the edges at the entrance to the channel and instantaneously changes direction.

To solve this problem he proposed that the map should be

$$z = w + e^w + f(w), \tag{6.8}$$

where $f = \sigma + i\tau$ had to be determined so that the fluid velocity would be constant on free streamlines and there would be no flow across the plates. On the streamlines $\psi = \pm\pi$, the condition for constant velocity is

$$\left|\frac{dz}{dw}\right|^2 = (1 - e^\phi + \frac{d\sigma}{d\phi})^2 + (\frac{d\tau}{d\phi})^2 = C, \tag{6.9}$$

where C is a constant. Helmholtz observed that one solution of these equations would be given by $\frac{d\sigma}{d\phi} = 0$ and choosing the constant $C = 1$ so that $\frac{d\tau}{d\phi} \to 0$ as $\phi \to -\infty$ (as the streamlines should be parallel there),

$$(1 - e^\phi)^2 + (\frac{d\tau}{d\phi})^2 = 1. \tag{6.10}$$

The solution needed is thus an analytic function f whose imaginary part, τ satisfies

$$\frac{d\tau}{d\phi} = \pm\sqrt{2e^\phi - e^{2\phi}}, \tag{6.11}$$

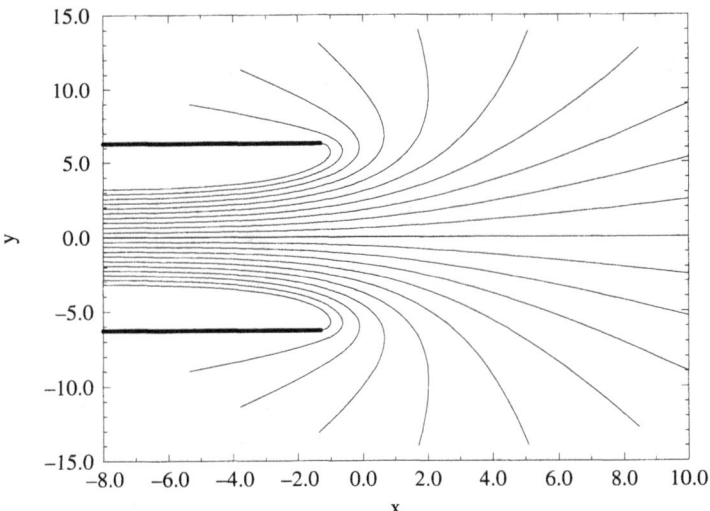

FIG. 6.5. Free surface solution for a two-dimensional Borda's mouthpiece.

for $-\infty < \phi \leq \phi_0$, whose real part vanishes on that interval, where y is constant for $\phi > \phi_0$ and where ϕ_0 has yet to be determined. Helmholtz was able to integrate (6.11) to obtain

$$z = w + e^w + i\{\sqrt{-2e^w - e^{2w}} + 2\sin^{-1}[\frac{i}{\sqrt{2}}e^{w/2}]\}. \tag{6.12}$$

The constant ϕ_0 is determined to be $\phi_0 = \log 2$ and the plates are at $y = \pm 2\pi$ for $-\infty < x \leq 2 - \log 2$. The streamlines for the solution are shown in Figure 6.5. Now the fluid leaves the plates smoothly and the free streamline curves around before tending asymptotically to $y = \pm \pi$ so there is a contraction ratio for the jet of 1/2.

Thus Helmholtz had developed an inverse method: given the form of the map (6.8) it was possible to examine which free streamline flow it represented.

6.2.2 Kirchoff (1869)

The ideas proposed by Helmholtz were very quickly developed further by Kirchoff (1869) who observed that on a free streamline, ψ constant, Helmholtz's condition (6.9) non-dimensionalised for unit free stream velocity, would be just

$$(\frac{\partial x}{\partial \phi})^2 + (\frac{\partial y}{\partial \phi})^2 = 1, \tag{6.13}$$

and that this condition would be automatically satisfied for a map $g(w)$ with

$$\frac{dz}{dw} = g(w) + \sqrt{g(w)^2 - 1}, \tag{6.14}$$

provided the free streamlines could be mapped by g to the real interval $(-1, 1)$. Kirchoff gave two examples,

$$g(w) = k + e^{-w}, \tag{6.15}$$

for flow between two semi-infinite parallel, but otherwise arbitrarily located, plates. In the case $k = 1$ this reduced to Helmholtz's example (provided a rotation of π connects their solutions) and a second example,

$$g(w) = k + w^{-1/2}, \tag{6.16}$$

which described a jet impinging on a plate.

Kirchoff too was dealing with an inverse method, there was still no theory to determine separated flow about an arbitrary curved body.

6.2.3 Rayleigh (1876)

Kirchoff's solution using (6.16) was taken up by Rayleigh (1876b) to calculate the drag on a plate which is oblique to a stream and which has free streamlines. Rayleigh was first to observe that the free streamlines behind a plate at right angles to a stream would continue to separate without bound as they went further downstream. We shall see shortly that this is a very general property of a wide class of free streamline flows and some ways lies at the heart of whether free streamline flows are a correct limit of the Navier–Stokes equations as the Reynolds number becomes infinite.

Rayleigh considered a general oblique flow against a plate; in the case of flow at right angles to a plate then $g(w) = w^{-1/2}$ so that

$$\frac{dz}{dw} = w^{-1/2} + \sqrt{w^{-1} - 1}. \tag{6.17}$$

This is a flow satisfying $\frac{dz}{dw} \to i$ as $w \to \infty$, or $w \sim -iz$ in the far field, so that the flow is parallel to the y-axis and in the $+y$ direction. The free streamlines will occur when $\psi = 0$ and $\phi \geq 1$, and although the streamlines become parallel to the y-axis, on the free streamlines

$$\frac{dz}{d\phi} = \pm\phi^{-1/2} + i\sqrt{1 - \phi^{-1}}, \tag{6.18}$$

and

$$\frac{dx}{d\phi} \sim \pm\phi^{-1/2}. \tag{6.19}$$

Integrating this equation, the streamline trajectory has

$$x \sim \pm 2\phi^{1/2}, \tag{6.20}$$

and the free streamlines diverge without bound far downstream.

6.2.4 General Formulation

Before proceeding further we need to define some notation and to examine the general mathematical problem which lies at the centre of free streamline calculations.

We have already defined the coordinate $z = x + iy$ and the complex potential $w = \phi + i\psi$. If the velocity is (u, v) the speed q and the angle of the flow relative to the $+x$-axis denoted θ then define a complex functions ζ and Ω by

$$\zeta = \frac{dz}{dw} = \frac{1}{q}e^{i\theta}, \qquad (6.21)$$

$$\Omega = \log \zeta = -\log q + i\theta. \qquad (6.22)$$

The essence of the theory is that if either $\zeta(w)$ or $\Omega(w)$ can be determined, then the flow field is known from integration, either $z = \int \zeta(w) dw$ or

$$z = \int e^{\Omega(w)} dw. \qquad (6.23)$$

There are some important features of these maps which should be noted. If the pressure is constant on the free streamlines then the velocity will also be constant so that the free streamlines will always be vertical line segments in the Ω plane. If in addition to being constant, the free streamline pressure is the same as the upstream pressure, then using velocities scaled by the upstream velocity, the free stream velocity will be $q = 1$ and so Ω will be imaginary on a free streamline. On any body in the flow, the angle of the flow must match the tangent to the body; here, however, we do not have the map between body coordinates and θ except if the body is composed of straight line segments (in which case we may still not explicitly know the coordinates of the vertices in w or Ω-planes). This is illustrated in Figure 6.6 for a flat plate at right angles to a stream, where separation occurs at the ends of the plate (S,S'), there is a stagnation point at O and the free streamlines map to J,J' as $x \to \infty$. In this case the angle θ is $+\pi$ in the upper half of the plate and $-\pi$ on the lower part of the plate whilst the free streamlines are on the imaginary axis for $-\pi/2 \leq \theta \leq \pi/2$. In the w plane, the plate is mapped to the line segment $\psi = \pm 0$, $0 \leq \phi \leq \phi_0$ where ϕ_0 has to be determined (having ϕ_0 unknown is not a contradiction with the example considered by Rayleigh since with the upstream velocity given, either the length of the plate or the potential at the trailing edge can be specified and the other determined from the solution).

The general strategy which evolved to solve these problems relied on an auxiliary plane, denoted here the t-plane, such that the map from w-plane to t-plane and the map from the Ω-plane to the t-plane can be determined and then the t-plane eliminated to give the relation between Ω and w. Later Woods

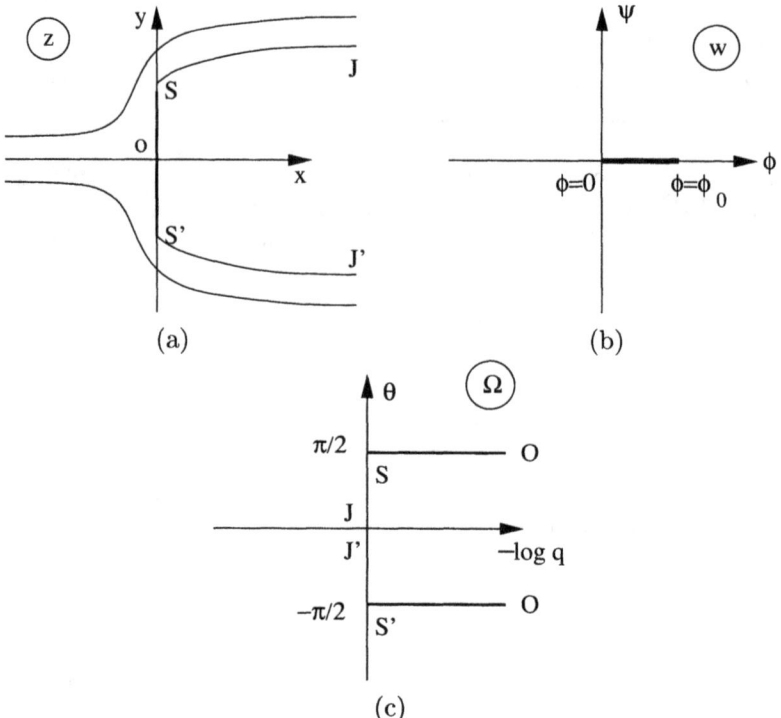

FIG. 6.6. Schematic diagrams of (a) the coordinate plane z, (b) the velocity potential plane w and (c) the transformed plane $\Omega = \log \mathrm{d}z/\mathrm{d}w$.

(1954a) introduced a more direct technique and we shall examine that shortly. In either approach, flow about bodies composed of straight sides can be solved using Schwarz–Christoffel transformations but flow about bodies with curved surfaces presents more difficulty.

As we have commented, free streamline models lead to non-zero drag. We shall consider only the simple case where the free streamline speed is the same as the speed far upstream so that $q=1$ on the free streamline. If we suppose the x and y components of the force are $\rho U^2 a F_D$ and $\rho U^2 a F_L$ respectively then a complex form of the non-dimensional force is

$$F = F_D + iF_L. \tag{6.24}$$

If an element $\mathrm{d}z$ has pressure difference $p-p_\infty$ acting across it then the elemental force acting on $\mathrm{d}z$ is

$$\mathrm{d}F = -i(p - p_\infty)\mathrm{d}z. \tag{6.25}$$

Using Bernouilli's equation

$$\mathrm{d}F = -\frac{i}{2}(1-q^2)\mathrm{d}z = -\frac{i}{2}\left(1 - \frac{\mathrm{d}w}{\mathrm{d}z}\overline{\frac{\mathrm{d}w}{\mathrm{d}z}}\right)\mathrm{d}z. \tag{6.26}$$

so that the force on a body is

$$F = -\frac{i}{2}\int_S (1 - \frac{dw}{dz}\overline{\frac{dw}{dz}})dz, \qquad (6.27)$$

where the contour S is around the wetted part of the body and over arbitrary parts of the free streamline since there are no pressure forces contributing to the drag over any part of the free streamline. The contour S is taken in a clockwise direction. Using (6.22),

$$F = -\frac{i}{2}\int_S (1 - e^{-\Omega-\overline{\Omega}})dz, \qquad (6.28)$$

and since $dz = e^\Omega dw$,

$$F = -\frac{i}{2}\int_S (e^\Omega - e^{-\overline{\Omega}})dw. \qquad (6.29)$$

The precise details of how to proceed further depend on the transformations being used. In the case considered by Levi-Civita, see Figure 6.7, the function Ω is defined in the upper half circle and can be extended to the lower half circle by $\overline{\Omega(t)} = -\Omega(\bar{t})$ when the force can be written

$$F = \frac{i}{2}\int_C e^{\Omega(t)}\frac{dw}{dt}(t)dt, \qquad (6.30)$$

where C is a contour around the unit circle in the anticlockwise direction. The argument of the integral has a pole at the origin and the force can be found from the residue there.

6.2.5 Levi-Civita (1907), Brillouin (1911)

Levi-Civita (1907) proposed a general method to calculate free streamline flows. Firstly the w-plane was mapped to the upper half of a Z-plane using a transformation $w = a(Z - b)^2$ such that in the Z-plane, the body was on the interval $(-1, 1)$ of the real axis and the free streamlines on the rest of the real axis. Then using an auxiliary plane defined by

$$Z = -\frac{1}{2}(t + \frac{1}{t}), \qquad (6.31)$$

the upper half of the Z-plane could be mapped to the upper part of the unit disc in the t-plane. A similar method had been proposed in 1890 by Zhukovskii; see Gurevich (1965), which used the upper half of an auxiliary plane in the same way rather than a unit disc but Levi-Civita seemed not to know of that work.

In this case the body surface was mapped to the upper part of the unit circle and the free streamlines to the interval $(-1, 1)$ on the real axis. Levi-Civita's contribution was to realise that the function θ had a step singularity at the

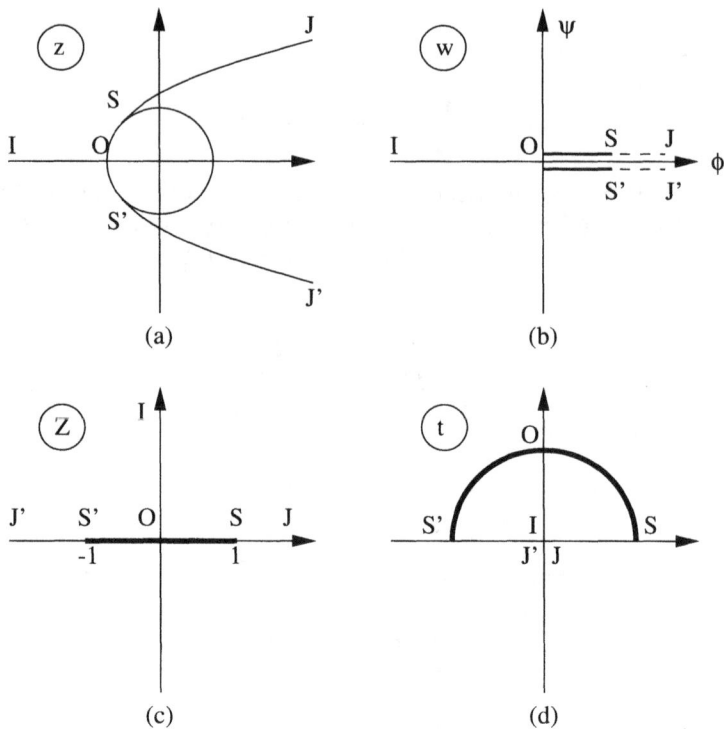

FIG. 6.7. Maps introduced by Levi-Civita (a) coordinate plane, (b) complex potential, (c) Z-plane and (d) auxiliary t-plane.

stagnation point O, but if that were dealt with, the remainder of the function Ω would be analytic within the upper half of the unit disc so that it could be expressed as a power series in t. If the power series were to be specified, then the flow could be computed. Thus his method remained an indirect method and there was still no technique to determine the power series in t for a given body. Levi-Civita's work was extended by Brillouin (1911) who seems to be the first to have realised that specifying an arbitrary power series for the expansion of part of Ω in the upper part of the unit disc could result in free streamlines crossing each other or intersecting the body and by Villat (1914).

6.2.6 Brodetsky (1923)

The first solution for free streamline flow around a curved body was calculated by Brodetsky (1923). He considered symmetric free streamline flow around a cylinder. As we noted above, two of the upstream velocity, body radius and potential at separation can be specified. Assume that the upstream velocity has $q = 1$ and that the potential at S is $\phi = 1$. Brodetsky modified Levi-Civita's transformation between w and the auxiliary variable t slightly so that the w-plane was mapped onto a unit half disc with positive real part,

$$w = -\frac{1}{4}(t - \frac{1}{t})^2. \tag{6.32}$$

Then the variable Ω is expressed in the form

$$\Omega(t) = \log\frac{1+t}{1-t} + A(t), \tag{6.33}$$

where the logarithm deals with the step singularity at the stagnation point O, and the function A will be analytic on the unit disc. Because of the symmetry in the problem, the function A will have a Taylor expansion in odd powers of t, so Brodetsky expanded A as

$$A(t) = A_1 t + \frac{1}{3}A_3 t^3 + \frac{1}{5}A_5 t^5 + \cdots. \tag{6.34}$$

It is convenient to write the auxiliary variable t in polar coordinates, $t = \rho e^{i\sigma}$. On the upper free streamline SJ where $\sigma = \pi/2$, $t = i\rho$, $0 < \rho \le 1$, Ω is imaginary (as required, since $q = 1$) and

$$\theta = 2\tan^{-1}\rho + \sum_{j=0}^{\infty}\frac{(-1)^j}{2j+1}A_{2j+1}\rho^{2j+1}. \tag{6.35}$$

On the upper half of the body where $\rho = 1$, $t = e^{i\sigma}$, $0 \le \sigma \le \pi/2$, the fluid speed is given by

$$q = \sqrt{\frac{1-\cos\sigma}{1+\cos\sigma}}\exp[-\sum_{j=0}^{\infty}\frac{1}{2j+1}A_{2j+1}\cos(2j+1)\sigma]. \tag{6.36}$$

and the flow direction by

$$\theta = \frac{\pi}{2} + \sum_{j=0}^{\infty}\frac{1}{2j+1}A_{2j+1}\sin(2j+1)\sigma. \tag{6.37}$$

Next the streamline curvature can be determined. On the free streamline, since

$$\phi = \frac{1}{4}(\rho + \frac{1}{\rho})^2, \tag{6.38}$$

and as $d\phi/ds = q = 1$ where s is distance measured along the streamline, then

$$\frac{ds}{d\theta} = \frac{ds}{d\phi}\frac{d\phi}{d\rho}/\frac{d\theta}{d\rho}, \tag{6.39}$$

or

$$\frac{ds}{d\theta} = \frac{(\rho^4-1)/2\rho^3}{2/(1+\rho^2) + A_1\rho - A_3\rho^3 + \cdots}. \tag{6.40}$$

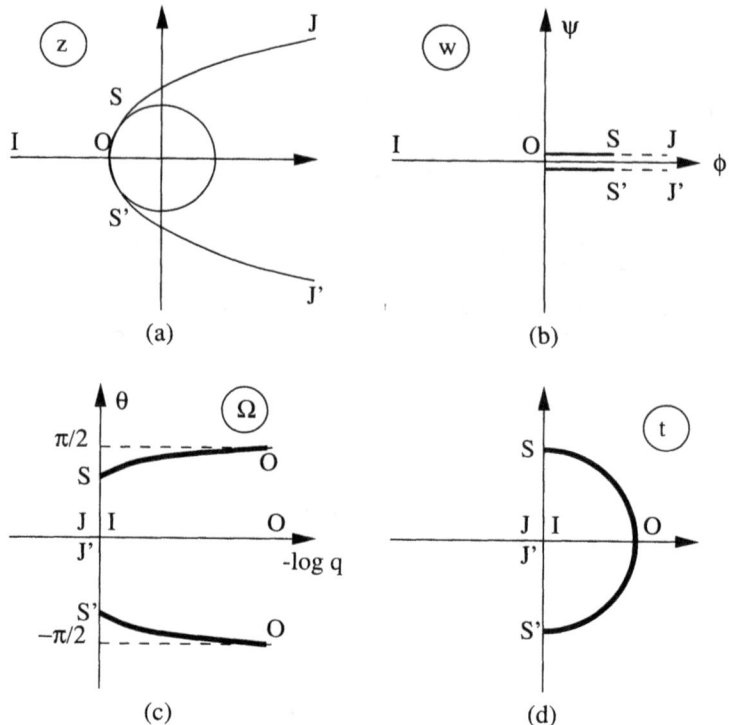

FIG. 6.8. Maps used by Brodetsky (a) coordinate plane, (b) complex potential, (c) Ω-plane and (d) auxiliary t-plane.

However, on the body,

$$\frac{ds}{d\theta} = \frac{2q^{-1}\sin\sigma\cos\sigma}{A_1\cos\sigma + A_3\cos 3\sigma + A_5\cos 5\sigma + \cdots}, \qquad (6.41)$$

so that as the separation point is approached on the body, $\sigma \to \pi/2$, then $q \to 1$ and at separation

$$\left(\frac{ds}{d\theta}\right)_s = \frac{2}{A_1 - 3A_3 + 5A_5 + \cdots}. \qquad (6.42)$$

Brodetsky reasoned that the only sensible limit for the free streamline curvature as the separation point was approached would be if (6.40) were non-zero and equal to the curvature of the body at the separation point. In order for this to be possible the denominator of (6.40) should vanish as $\rho \to 1$, or

$$\Gamma = 1 + A_1 - A_3 + A_5 \cdots = 0. \qquad (6.43)$$

The condition that the streamline curvature is continuous is usually called **smooth separation**.

Brodetsky also proposed that if in addition to (6.43), the curvature $-\mathrm{d}s/\mathrm{d}\theta$ in (6.41) were to be equated with the curvature of the body at a finite number of points ahead of separation, then the coefficients $A_1, A_3, \cdots$ could be determined from the resulting equations. He suggested that if the number of terms were increased, then the process would converge quickly to the smooth separation solution. This was the first direct solution method where a body could be specified and the solution determined for a shape which closely resembled the particular body.

In the case of a circular cylinder, Brodetsky showed that the free streamline which satisfies (6.43) would leave the cylinder at an angle close to 55° from the front stagnation point and streamlines for this case were calculated by Schmieden (1929).

Although Brodetsky regarded the case where the curvature was not continuous across the separation point as physically unacceptable, his method is easily adapted to deal with an arbitrary specified separation angle. The way that this might be done was set out by Schmieden (1932, 1934) who showed that the range of physically acceptable separation angles would lie between 55° and 125°: if the angle was less than 55° the free streamline intersected the body surface, if the angle was greater than or equal to 125° then the free streamline would cross the $y = 0$ axis.

It is particularly straightforward to extend Brodetsky's solution to arbitrary separation angle for the case of a circular cylinder where the radius of curvature is constant, and given by

$$a = \frac{-2}{A_1 - 3A_3 + 5A_5 + \cdots}, \qquad (6.44)$$

so there is no need to calculate the angle from the stagnation point which is associated with any value of σ. For more general bodies it would be necessary to include a calculation of $\theta(\sigma)$ to solve for the flow. For further reading, see Birkoff et al. (1951), Zarantonello (1952) and Birkoff et al. (1953-54).

In the case of a cylinder, for some integer N let $\sigma_j = j\pi/(2N)$ for $j = 0, 1, \cdots, N$ and specify a separation angle from the front stagnation point α_s, then we shall have the set of simultaneous equations:

$$A_1 - \frac{A_3}{3} + \cdots + (-1)^N \frac{A_{2N+1}}{2N+1} = -\alpha_s, \qquad (6.45)$$

and for $j = 0, \cdots, N-1$: using $c_{i,j} = \cos(2i+1)\sigma_j/\cos\sigma_j$ for $i = 1, \cdots, N$,

$$\frac{2\sin\sigma_j \sqrt{\frac{1+\cos\sigma_j}{1-\cos\sigma_j}} e^{\mathcal{A}_\mathcal{R}(\sigma_j)}}{A_1 + c_{1,j}A_3 + \cdots + c_{N,j}A_{2N+1}} = \frac{2}{A_1 - 3A_3 + \cdots + (-1)^N(2N+1)A_{2N+1}}, \qquad (6.46)$$

where $\mathcal{A}_\mathcal{R}(\sigma)$ is the real part of the first $N+1$ terms of $A(e^{i\sigma})$ and noting that the factor $\sin\sigma_j \sqrt{\frac{1+\cos\sigma_j}{1-\cos\sigma_j}} = 2$ for $j = 0$.

	$\theta_s = 55°$	$\theta_s = 75°$	$\theta_s = 90°$	$\theta_s = 110°$	$\theta_s = 125°$
A_1	-.94209	-1.26110	-1.49301	-1.79297	-2.01141
A_3	0.04917	0.12082	0.18814	0.29265	0.38025
A_5	-0.00613	-0.02715	-0.05123	-0.09455	-0.13521
A_7	0.00120	0.00908	0.01939	0.04001	0.06107
A_9	-0.00032	-0.00391	-0.00901	-0.01996	-0.03184
A_{11}	0.00011	0.00200	0.00481	0.01115	0.01834
A_{13}	-0.00004	-0.00115	-0.00284	-0.00678	-0.01138
A_{15}	0.00002	0.00072	0.00181	0.00439	0.00747
A_{17}	-0.00002	-0.00048	-0.00121	-0.00298	-0.00512
A_{19}	0.00002	0.00033	0.00084	0.00210	0.00363

Table 6.1 The first ten Brodetsky coefficients for separation angles between 55° and 125°.

This set of equations can be solved iteratively and the process converges quickly. Having solved these equations then the trajectory for the free streamline can be found using (6.23), putting the stagnation point at the origin, the trajectory which represents the body is given by

$$z_B(\sigma) = \int_0^\sigma \frac{i(1+e^{i\sigma})^2(1+e^{2i\sigma})}{2e^{2i\sigma}} \exp[\mathcal{A}_N(e^{i\sigma})] d\sigma, \qquad (6.47)$$

and the free streamline by

$$z_{fs}(\rho) = z_B(\pi/2) - \int_\rho^1 \frac{(1+i\rho)^2(1-\rho^2)}{2\rho^3} \exp[\mathcal{A}_N(i\rho)] d\rho, \qquad (6.48)$$

with $\mathcal{A}_N(t)$ the sum of the first $N+1$ terms of $A(t)$. The solution can easily be rescaled using the radius a to obtain flow about a unit radius cylinder.

The result of computing trajectories this way can be seen in Figure 6.9. As has been indicated, the solution is only meaningful for a range of angles, if α_s is too small the trajectories intersect with the cylinder, if α_s is too large, the trajectories from the two separation points on either side of the body cross each other.

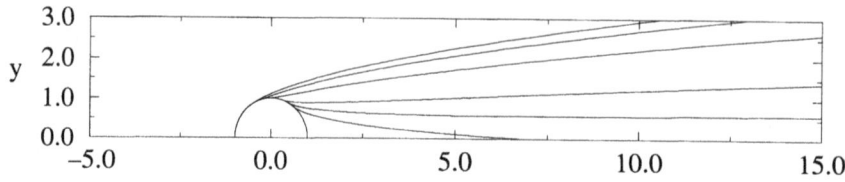

FIG. 6.9. Free streamlines for Brodetsky's model, $\alpha_s = 55, 75, 90, 110, 120, 130$. The last case crosses the axis and is an unphysical solution.

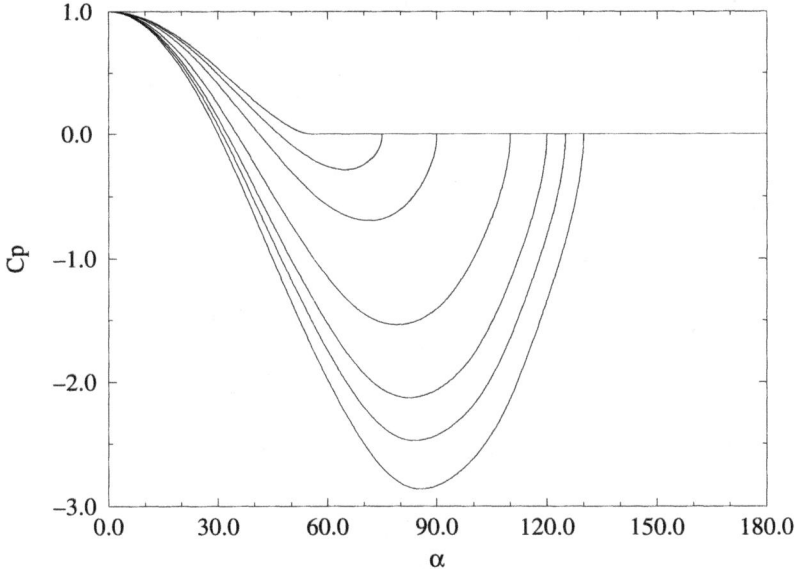

FIG. 6.10. Pressure coefficient $C_p = 1 - q^2$, for Brodetsky's model. $\alpha_s = 55, 75, 90, 110, 120, 125, 130$.

The nature of the singularity which exists at the separation point for $\alpha_s > 55°$ is not apparent from Figure 6.9. Indeed the streamline direction, θ is continuous across the separation point as the flow moves from the body to the free streamline. If we look at the pressure coefficient, C_p, in Figure 6.10, the singularity is immediately evident. In the case $\alpha_s = 55°$ the slope of the pressure coefficient is continuous at α_s but at larger values of α_s, the slope of the pressure coefficient is clearly discontinuous indicating that there will be a singularity in the pressure gradient at the separation point for separation angles larger than the critical value where smooth separation occurs. It is also important to note that in the case of smooth separation, the pressure gradient is everywhere *favourable* and the absence of any adverse pressure gradient has been a cause for reluctance to regard smooth separation as a model for separation of a viscous fluid where separation is usually associated with regions of *adverse* pressure gradient.

The precise form of the singularity will later take on considerable importance so it is necessary to look further into Brodetsky's solution near the separation point.

Let $\sigma = \pi/2 - \epsilon$ and

$$B(\epsilon) = A_1 \sin \epsilon - \frac{1}{3} A_3 \sin 3\epsilon + \frac{1}{5} A_5 \sin 5\epsilon + \cdots . \tag{6.49}$$

This value Brodetsky set to zero and Γ was given by $\Gamma = 1 + B'(0)$. On the body we have already the speed q and direction θ, and the direction is related to the angle from the stagnation point α by $\alpha = \pi/2 - \theta$. The speed is given by

$$q = \sqrt{\frac{1-\sin\epsilon}{1+\sin\epsilon}} e^{-B(\epsilon)}, \qquad (6.50)$$

and the angle from the separation point by

$$\alpha = -A_1 \cos\epsilon + \frac{1}{3} A_3 \cos 3\epsilon - \frac{1}{5} A_5 \cos 5\epsilon + \cdots. \qquad (6.51)$$

Using distance $s = a\alpha$ on the cylinder, the pressure gradient will be

$$\frac{dp}{ds} = -q\frac{dq}{ds} = -\frac{q}{a}\frac{dq}{d\alpha} = -\frac{q}{a}\frac{dq}{d\epsilon}\bigg/\frac{d\alpha}{d\epsilon}. \qquad (6.52)$$

Examining (6.51) for small ϵ, α can be expanded just upstream of the separation point as

$$\alpha \sim \alpha_s - \frac{1}{a}\epsilon^2 + O(\epsilon^3). \qquad (6.53)$$

The expansion of q near $\epsilon = 0$ is more tedious to derive,

$$q \sim 1 - \Gamma\epsilon + \frac{1}{2}\Gamma^2\epsilon^2 + O(\epsilon^3), \qquad (6.54)$$

so that the pressure gradient approaching the separation point is given by

$$\frac{dp}{ds} \sim -\frac{\Gamma}{2\epsilon} = -\frac{\Gamma}{2\sqrt{a}}(\alpha_s - \alpha)^{-1/2}, \qquad (6.55)$$

and is singular as $\alpha \to \alpha_s$ unless Brodetsky's condition (6.43), $\Gamma = 0$, is satisfied. It is important to observe that the generalisation of Brodetsky's solution with $\Gamma \neq 0$ still leaves the flow direction, θ, continuous at the separation point: that is that the flow leaves tangentially; it is the curvature which is discontinuous, not the flow direction as can be seen from the limits of $\sigma \to \pi/2$ in (6.37) and $\rho \to 1$ in (6.35) being the same.

Brodetsky's solution is for an upstream velocity scale of unity about a cylinder of radius a; to normalise the solution for unit upstream velocity about a unit radius cylinder, both z and w have to be divided by a, so the speed q will remain unchanged but the pressure gradient will have to be scaled by a, and we can rewrite the pressure gradient near the separation point for a unit radius cylinder (so α now also provides distance along the cylinder surface) as

$$\boxed{\frac{dp}{ds} \sim \kappa(\alpha_s - \alpha)^{-1/2},} \qquad (6.56)$$

with $\kappa = -\Gamma\sqrt{a}/2$ being a measure of the singularity in the pressure gradient. Calculation of κ using Brodetsky's method is shown in Figure 6.11.

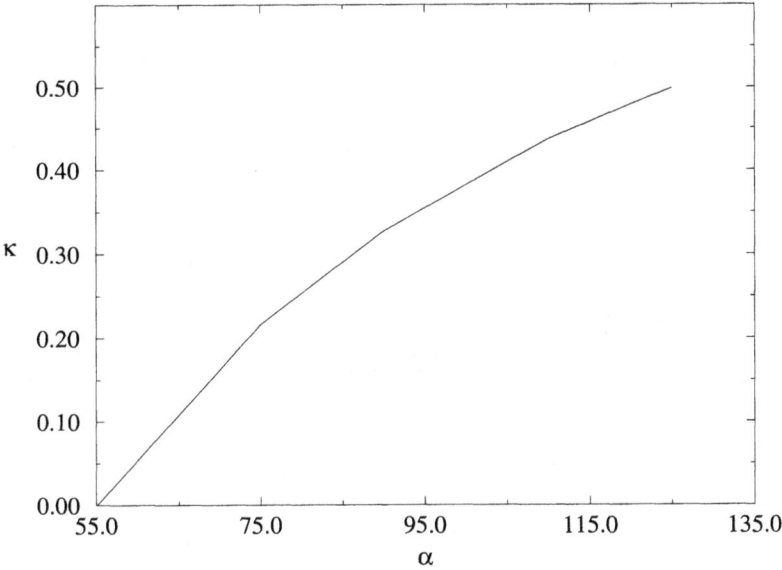

FIG. 6.11. Calculated values from Brodetsky's method of the singularity strength, κ, in the pressure gradient as the separation angle α_s varies from $55°$ to $125°$.

The free streamline shape immediately after separation can be calculated. If in (6.40) we let $\rho = 1 - \varepsilon$ and expand for small ε,

$$\frac{ds}{d\theta} \sim -\frac{2}{\Gamma}\varepsilon. \tag{6.57}$$

Using (6.38) we also have

$$\varepsilon \sim 2\sqrt{\phi - \phi_s}, \tag{6.58}$$

where ϕ_s is the value of ϕ at separation. Since $q = 1$ on the free streamline we can use $\phi - \phi_s$ and distance s *from* the separation point interchangeably and normalising to a unit radius cylinder,

$$\frac{d\theta}{ds} \sim \frac{1}{2}\kappa s^{-1/2} \text{ or } \theta \sim \kappa s^{1/2}. \tag{6.59}$$

The distance of the free streamline from the tangent plane at separation, s_N, relative to distance along the tangent plane, s_T (defined by $s_N = \int \sin\theta ds$ and $s_T = \int \cos\theta ds$) is asymptotically,

$$\boxed{s_N \sim \frac{2}{3}\kappa s_T^{3/2} \quad \text{as} \quad s_T \to 0.} \tag{6.60}$$

The nature of the free streamline far downstream for Brodetsky's solution can be found by examining the limit $t = i\rho \to 0$, so that

$$\Omega = \log \frac{1+i\rho}{1-i\rho} + A(i\rho), \tag{6.61}$$

and thus

$$\theta \sim (2+A_1)\rho - \frac{1}{3}A_3\rho^3 + \cdots, \quad \text{as} \quad \rho \to 0. \tag{6.62}$$

It will also be the case that on the free streamline,

$$\phi \sim \rho^{-2}/4 \quad \text{as} \quad \rho \to 0, \tag{6.63}$$

and since far downstream $\phi \sim x$ and $\theta = dy/dx$, the free streamline must satisfy

$$y \sim (2+A_1)x^{1/2} \quad \text{as} \quad x \to \infty. \tag{6.64}$$

If the flow is about a unit radius cylinder, this must be rescaled to

$$\boxed{y \sim a^{-1/2}(2+A_1)x^{1/2} \quad \text{as} \quad x \to \infty.} \tag{6.65}$$

Of course, this implies that solutions with $A_1 < -2$ are physically unacceptable because the two free streamlines will cross the axis $y = 0$ and hence each other. The data shown in Table 6.1 indicates that the maximum separation angle is just less than $\alpha = 125°$. The result $A_2 \geq -2$ for physically acceptable solutions was first determined by Schmieden (1932).

The drag for Brodetsky's model can be calculated from a slightly modified form of Levi-Civita's formula (6.30),

$$F = -\frac{i}{2}\int_c e^{\Omega(t)} \frac{dw}{dt}(t)dt, \tag{6.66}$$

where c is an anticlockwise contour around the unit circle (the sign change is because the integral over $S'OS$ is already in the anticlockwise direction in the t plane in Brodetsky's model whereas it is in the clockwise direction in Levi-Civita's model). Then

$$\frac{dw}{dt}(t) = -\frac{1}{2}\frac{(t^2+1)(t^2-1)}{t^3}, \tag{6.67}$$

and

$$e^{\Omega(t)} = \frac{1+t}{1-t}[1 + A_1 t + \frac{1}{2}A_1^2 t^2 + O(t^3)] \tag{6.68}$$

so evaluating the residue of the pole at the origin in (6.66), the non-dimensional force on the cylinder is

$$F = \pi(1 + \frac{1}{2}A_1)^2. \tag{6.69}$$

This is completely in agreement with Schmieden's calculation that the streamlines cross when $A_1 = -2$ since the drag goes to zero at that value, as it must to satisfy d'Alemberts prediction of zero drag for a closed wake. It also gives the drag associated with smooth separation since for that case $A_1 = -0.942...$ so that $F = (0.5289..)^2 \pi$.

6.2.7 Woods (1955)

A powerful method for computing general free streamline flows is due to Woods (1954a), Woods (1955) and described comprehensively in Woods (1961). Instead of mapping the flow region to part of the unit disc, Woods proposed mapping the flow region to a strip in such a way that either the real or the imaginary part of Ω would be specified on the boundaries of the strip and the general solution for Ω could then be written down in terms of integrals of the boundary values. The map between the physical plane and the new transformed plane still has to be determined numerically, but Woods found that an iterative process to determine the map seemed to converge very quickly. We shall also look at Woods' method when we deal with channel flows but here the method can be described and applied to symmetric free streamline flow about a cylinder.

Woods defined an auxiliary plane t by

$$w = \phi_s \tanh^2 t, \tag{6.70}$$

where ϕ_s is the potential where the free streamline leaves the cylinder.

If we assume that the speed q is given on the free streamline and the direction θ on the body, then the complex function Ω has its real part, $\Omega_0 = -\log q$ specified on the line $\Im t = \pi/2$, and its imaginary part, θ_0, specified on the line $\Im t = 0$. The solution of this problem can be found by contour integration around the strip $0 \leq \Im t \leq \pi/2$ and Woods (1954a) had shown that solution was

$$\Omega(t) = \frac{1}{\pi} \int_{-\infty}^{\infty} [\theta_0(k)\operatorname{cosech}(k-t) + \Omega_0(k)\operatorname{sech}(k-t)] \, dk. \tag{6.71}$$

It is important to realise that this form of the solution is derived from an integral around a strip which contains the point t, so great care is needed with the limit $t \to \infty$ so that the point t always remains within the contour of integration. As Woods observed, the limits of $t \to \infty$ and the integration in k cannot be interchanged.

Of course, the primary difficulty with using this solution is that θ_0 and Ω_0 are not known as functions of t, rather θ_0 is known in terms of z and Ω_0 either vanishes if $q = 1$ on the free streamline or must be specified some other way.

From a practical viewpoint, having the flow direction on the body at separation specified as $t \to \pm \infty$ presents a potential computational difficulty. Woods dealt with this problem by defining a second auxiliary plane, ζ, with

$$\tanh t = -i \sinh \frac{1}{2}\zeta, \tag{6.72}$$

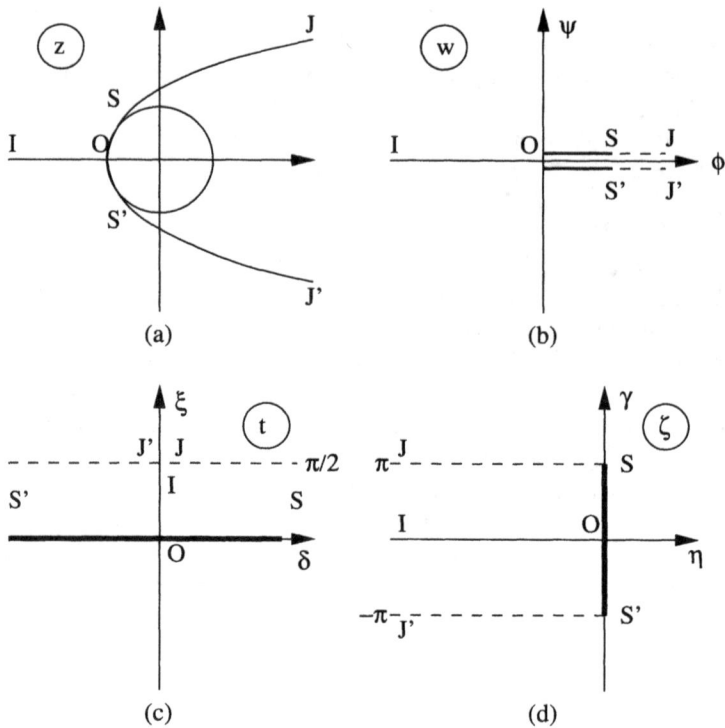

FIG. 6.12. Various planes used in applying Woods' method to a circular cylinder. (a) Coordinate plane z, (b) complex potential w, (c) auxiliary plane t, (d) auxiliary plane ζ.

so that

$$w = -\phi_s \sinh^2 \tfrac{1}{2}\zeta. \tag{6.73}$$

In the $\zeta = \eta + i\gamma$ plane, the body is given by $-\pi \leq \gamma \leq \pi$ with $\eta = 0$ while the free streamlines are given by $\gamma = \pm\pi$, $-\infty < \eta \leq 0$ (see Figure 6.12 (d)) so that to calculate a solution it will eventually be necessary to determine the map from angle from the stagnation point, α to γ so as to obtain $\theta_0(\gamma)$. Using this new variable, the solution is given by

$$\Omega(\zeta) = \frac{1}{2\pi}\int_{-\pi}^{\pi} \frac{\theta_0(\gamma')\cosh\tfrac{1}{2}\zeta}{\sin\tfrac{1}{2}\gamma' + i\sinh\tfrac{1}{2}\zeta}d\gamma' + \frac{1}{\pi}\int_{-\infty}^{0} \frac{\Omega_0(\eta')\cosh\tfrac{1}{2}\zeta\cosh\tfrac{1}{2}\eta'}{\cosh^2\tfrac{1}{2}\eta' + \sinh^2\tfrac{1}{2}\zeta}d\eta' \tag{6.74}$$

We shall compute solutions for two cases; the equivalent of Brodetsky's model where $\Omega_0 = 0$ and a model proposed by Woods to account for the experimentally observed fact that the pressure in a separated region behind a cylinder is less than the pressure far upstream. As we have noted, the direct form of the solution, (6.74) has a difficulty with the limit $\zeta \to \infty$ and this emerges in the computational problem too. To overcome this Woods integrated the first integral in (6.74) by parts, so that for the case where $\Omega_0 = 0$ ($q = 1$ on the free streamline),

$$\Omega(\zeta) = i\theta(-\pi) - \frac{1}{\pi} \int_{-\pi}^{\pi} \log\left(\frac{\sin\frac{1}{4}(\gamma' + i\zeta)}{\cos\frac{1}{4}(\gamma' - i\zeta)}\right) d\theta_0(\gamma'). \tag{6.75}$$

To evaluate this it has to be recalled that θ_0 has a step discontinuity of π at the stagnation point, so the differential $d\theta_0$ will act like π times a delta-function at $\gamma' = 0$ and the differential will otherwise be an even function of γ'. Thus the function $\Omega(\zeta)$ can be written

$$\Omega(\zeta) = -i\theta_s - \log\left(\frac{\sin\frac{1}{4}i\zeta}{\cos\frac{1}{4}i\zeta}\right) - \frac{1}{\pi} \int_0^{\pi} \log\left(\frac{\sin\frac{1}{4}(i\zeta - \gamma')\sin\frac{1}{4}(i\zeta + \gamma')}{\cos\frac{1}{4}(i\zeta - \gamma')\cos\frac{1}{4}(i\zeta + \gamma')}\right) d\theta_0(\gamma'). \tag{6.76}$$

In order to finally solve the problem it is necessary to study (6.76) on the surface of the cylinder, where writing $\zeta = i\gamma$, $0 \le \gamma \le \pi$, and considering only the real part of (6.76),

$$-\log q(\gamma) = -\log\left|\tan\frac{1}{4}\gamma\right| - \frac{1}{\pi}\int_0^{\pi} \log\left|\frac{\sin\frac{1}{4}(\gamma - \gamma')\sin\frac{1}{4}(\gamma + \gamma')}{\cos\frac{1}{4}(\gamma - \gamma')\cos\frac{1}{4}(\gamma + \gamma')}\right| d\theta_0(\gamma'). \tag{6.77}$$

The potential on the surface of the cylinder is given by

$$\phi = \phi_s \sin^2 \frac{1}{2}\gamma, \tag{6.78}$$

the speed by $q = d\phi/ds$ and the radius of the cylinder, a, is constant so that $d\theta_0/ds = -1/a$. Hence as

$$d\theta_0(\gamma') = \frac{d\theta_0}{ds}\frac{ds}{d\phi}\frac{d\phi}{d\gamma}d\gamma', \tag{6.79}$$

then

$$d\theta_0(\gamma') = -\frac{\phi_s}{aq(\gamma')}\sin\frac{1}{2}\gamma'\cos\frac{1}{2}\gamma' d\gamma' = -\frac{\phi_s}{2aq(\gamma')}\sin\gamma' d\gamma'. \tag{6.80}$$

Thus if (6.80) is substituted into (6.77) the result is an integral equation for the surface speed $q(\gamma)$,

$$-\log q(\gamma) = -\log\left|\tan\frac{1}{4}\gamma\right|$$
$$+ \frac{\phi_s}{2a\pi}\int_0^{\pi}\frac{\sin\gamma'}{q(\gamma')}\log\left|\frac{\sin\frac{1}{4}(\gamma - \gamma')\sin\frac{1}{4}(\gamma + \gamma')}{\cos\frac{1}{4}(\gamma - \gamma')\cos\frac{1}{4}(\gamma + \gamma')}\right| d\gamma', \tag{6.81}$$

which once solved enables the whole flow field to be generated by substituting (6.76) into

$$z = \int^{\zeta} e^{\Omega(\zeta')} \frac{dw}{d\zeta'} d\zeta', \qquad (6.82)$$

since the map $w(\zeta)$ is given by (6.73).

Woods (1954a) proposed an iterative method to solve the integral equation (6.77): assume a surface speed distribution and by substitution into the right-hand side of (6.77) a new estimate could be obtained and the iteration continued. The integral is evaluated by quadrature. Suppose the separation angle α_s is given. For some integer N define $\Delta\gamma = \pi/N$, and $\gamma_j = j\Delta\gamma$, $j = 0, \cdots, N$. Let $q_j = q(\gamma_j)$ and q_j^k be the estimate of q_j at the k-th iteration. A set of simple initial values is $q_j^0 = 1$, $\phi_s^0 = \alpha_s$. The integral in (6.77) is evaluated using simple trapezoidal quadrature at the mid-points, $\bar{\gamma}_j = (\gamma_j + \gamma_{j+1})/2$, $j = 0, \cdots, N-1$ with $\bar{q}_j^k = (q_j^k + q_{j+1}^k)/2$ so that if

$$b_{i,j} = \log \left| \frac{\sin\frac{1}{4}(\gamma_i - \bar{\gamma}_j) \sin\frac{1}{4}(\gamma_i + \bar{\gamma}_j)}{\cos\frac{1}{4}(\gamma_i - \bar{\gamma}_j) \cos\frac{1}{4}(\gamma_i + \bar{\gamma}_j)} \right|, \qquad (6.83)$$

then the main iteration is determined by

$$\log q_i^{k+1} = \log|\tan\frac{1}{4}\gamma_i| - \frac{\phi_s^k}{2a\pi} \sum_{j=0}^{j=N-1} b_{i,j} \frac{\sin\bar{\gamma}_j}{\bar{q}_j^k} \Delta\gamma. \qquad (6.84)$$

It is necessary to use the discrete form of the identity

$$s = a\alpha_s = \int_0^{\phi_s} d\phi = \frac{\phi_s}{2a} \int_0^{\pi} \frac{\sin\gamma'}{q(\gamma')} d\gamma', \qquad (6.85)$$

to obtain

$$\phi_s^{k+1} = 2\alpha_s / [\sum_{i=0}^{i=N-1} \frac{\sin\bar{\gamma}_i}{\bar{q}_i^{k+1}} \Delta\gamma]. \qquad (6.86)$$

Once converged values of q_i and ϕ_s are obtained the whole flow field can be determined from (6.76) and (6.82) using numerical integration.

Example computations using $N = 40$ are shown in Figure 6.13 for three separation angles, $\alpha_s = 55°$, $90°$, $120°$. Woods and Brodetsky's methods predict the same speed distribution on the cylinder, an example is shown in Figure 6.16 for $\alpha_s = 90$. The map between the flow direction θ and γ on the body is shown in Figure 6.14 for separation angles between $55°$ and $120°$.

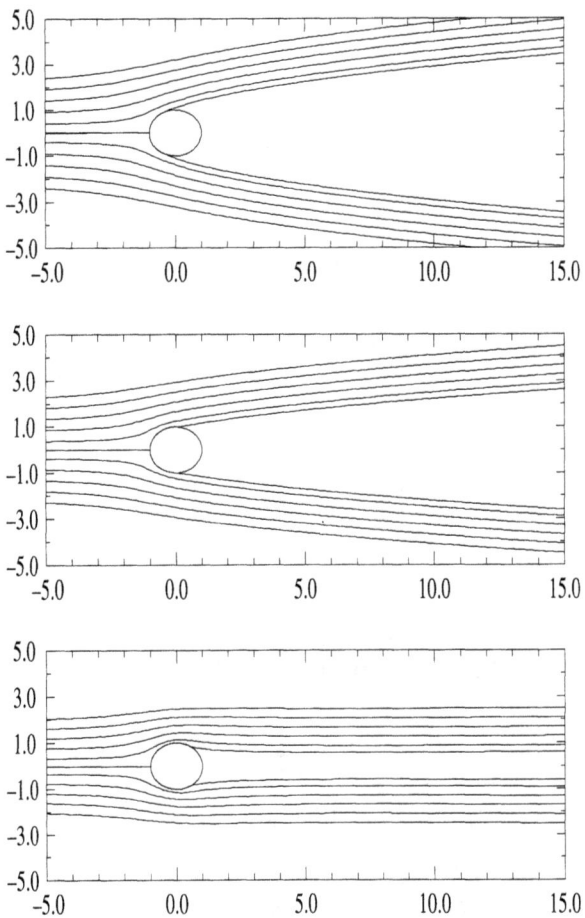

FIG. 6.13. Computed streamlines using Woods' method for free streamline speed $q = 1$, $\alpha_s = 55°$, $90°$, $120°$.

Near the separation point, $\zeta = \pi i$, expressions for the pressure on the body and the angle of the flow can be derived in much the same way as for Brodetsky's solution. On the body, let $\zeta = i(\pi - \epsilon)$, then the potential is given by

$$\phi \sim \phi_s(1 - \frac{1}{4}\epsilon^2), \tag{6.87}$$

and the speed by

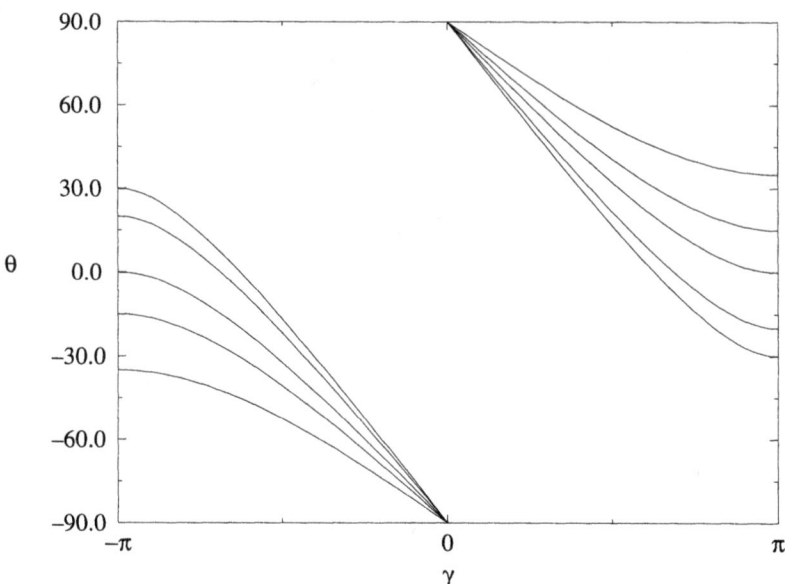

FIG. 6.14. Computed map between flow angle θ and γ on a cylinder for separation angles $\alpha_s = 55°, 75°, 90°, 110°, 120°$.

$$q(\pi - \epsilon) \sim 1 - [\frac{1}{2} - \frac{\phi_s}{\pi a}\int_0^\pi \frac{\sin \frac{1}{2}\gamma'}{q(\gamma')}d\gamma']\epsilon. \qquad (6.88)$$

Using similar algebra as before, on the body

$$\frac{\partial p}{\partial s} \sim \kappa(\alpha_s - \alpha)^{-1/2} \text{ as } \alpha \to \alpha_s, \qquad (6.89)$$

where Woods' model predicts that

$$\boxed{\kappa = \frac{1}{\sqrt{a\phi_s}}[\frac{1}{2} - \frac{\phi_s}{\pi a}\int_0^\pi \frac{\sin \frac{1}{2}\gamma'}{q(\gamma')}d\gamma']} \qquad (6.90)$$

Calculation of Woods' solution shows that this is the same value as computed using Brodetsky's method, see Figure 6.15. The speed on the body, and hence the pressure coefficient, C_p, is also indistinguishable between the two methods, see Figure 6.16.

On the free streamline, far from the body, let $\zeta = \pi i + \eta$, with $-\eta \gg 1$, then the potential behaves like

$$\phi \sim \frac{1}{4}\phi_s e^{-\eta}, \qquad (6.91)$$

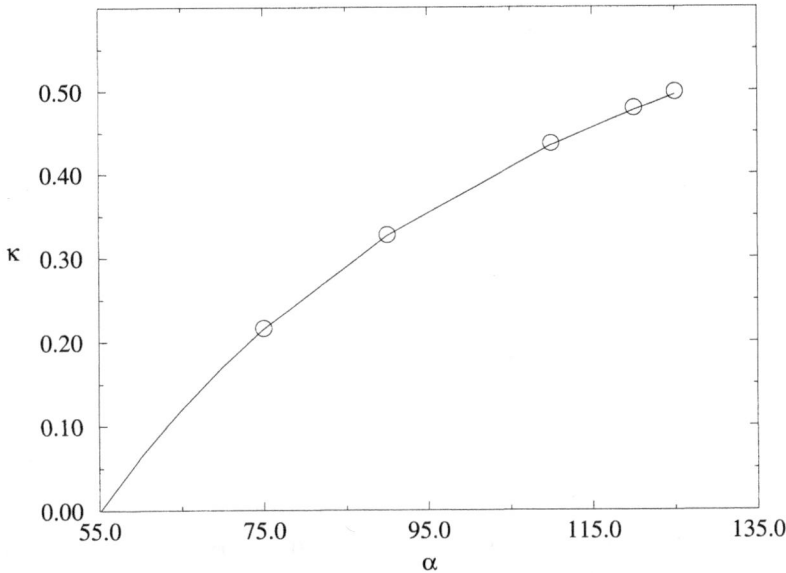

FIG. 6.15. Comparison of Woods' (———) and Brodetsky's (circles) prediction of pressure gradient singularity strength at separation, κ, versus separation angle.

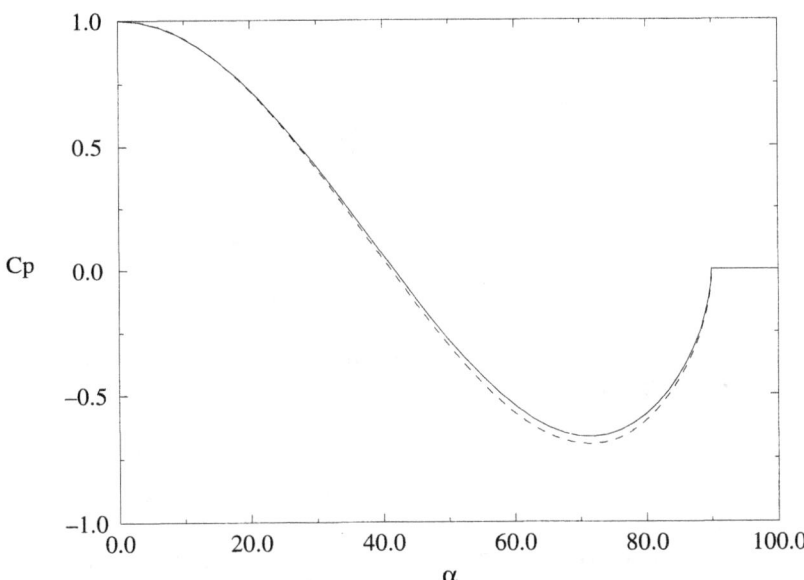

FIG. 6.16. Comparison of Woods' (———) and Brodetsky's (- - - -) solution for pressure coefficient C_p on the surface of a cylinder with $\alpha_s = 90°$.

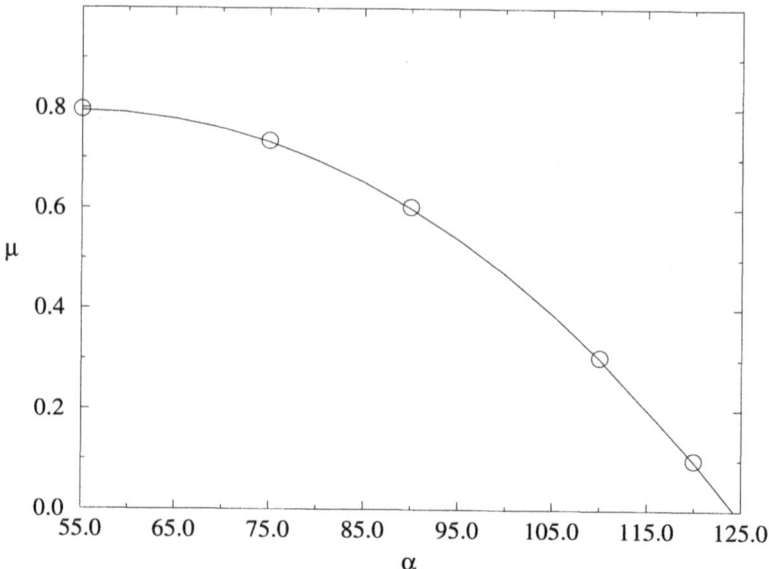

FIG. 6.17. Comparison of Woods' (———) and Brodetsky's (circles) prediction of the coefficient, μ, of free streamline growth versus separation angle.

while the flow direction can be expanded

$$\theta \sim \frac{2}{\pi} \int_0^\pi \theta_0(\gamma') \sin \frac{1}{2}\gamma' \, d\gamma' e^{\frac{1}{2}\eta}, \tag{6.92}$$

so that

$$\theta \sim \frac{1}{\pi} \int_0^\pi \theta_0(\gamma') \sin \frac{1}{2}\gamma' \, d\gamma' \sqrt{\frac{\phi_s}{\phi}}. \tag{6.93}$$

In the far field, $\phi \sim x$, and as $\theta = dy/dx$, this can be integrated to show that the free streamline behaves like

$$y \sim \frac{\sqrt{\phi_s}}{\pi} \int_0^\pi \theta_0(\gamma') \sin \frac{1}{2}\gamma' \, d\gamma' \sqrt{x} = \mu\sqrt{x}, \tag{6.94}$$

where the coefficient of $\sqrt{x}$ is denoted μ.

As expected, calculation of the coefficient in this expression matches the values obtained using Brodetsky's solution, see Figure 6.17. Thus the methods of Brodetsky and of Woods produce exactly the same solution for inviscid flow about a cylinder, provided suitable rescaling is applied to Brodetsky's solution to make the cylinder of unit radius.

Woods' method is however more powerful than this example has so far shown. Both methods can be extended to non-symmetric flow but Woods method can be extended in a way that Brodetsky's method cannot. As we mentioned above, a significant objection to free streamline models for describing finite Reynolds number separation is that the free streamlines diverge, with the width between them growing like $\sqrt{x}$ as $x \to \infty$. When the problem was first posed, we assumed that the flow upstream **and** downstream would be uniform flow. The free streamline solution gives a very different picture: it indicates that we cannot specify uniform flow far downstream (or at least not over the whole downstream region). Woods (1955) proposed that it is possible for the free streamlines not to diverge far downstream if the value of the free streamline *speed* was specified: that is the function Ω_0 not taken to be zero but rather only asymptotically tend to zero, so that the speed asymptotically returns to the upstream value. Woods argued that an advantage in doing this would be that the pressure would also vary along the free streamline and so could accommodate the experimentally observed fact that the pressure coefficient was not zero but negative immediately behind a cylinder with separation. This was an entirely ad-hoc proposal by Woods who noted that a source in a uniform flow has parallel streamlines downstream and the velocity behaves like $q \sim 1 - x^{-1}$, so since (1) $\phi \sim x$ as $x \to \infty$, and (2) $\phi \sim e^{-\eta}$ on $\zeta = \pi i + \eta$ as $\eta \to -\infty$, Woods used

$$\Omega_0(\eta) = -\log q = \frac{-K}{1 + b^2 \sinh^2 \tfrac{1}{2}\eta}, \tag{6.95}$$

on the free streamline.

The two constants K and b cannot be arbitrarily specified, it is necessary that the coefficient of $e^{-\tfrac{1}{2}\eta}$ should vanish in the far downstream expansion of θ so that K and b have to be related by

$$K = \frac{b}{\pi} \int_0^\pi \theta_0(\gamma) \sin \frac{1}{2}\gamma \, d\gamma. \tag{6.96}$$

Using this distribution of speed on the free streamline, the function Ω is given by

$$\Omega(\zeta) = -i\theta_s - \log\left(\frac{\sin \tfrac{1}{4}i\zeta}{\cos \tfrac{1}{4}i\zeta}\right) - \frac{K}{1 + b \cosh \tfrac{1}{2}\zeta}$$
$$- \frac{1}{\pi} \int_0^\pi \log\left(\frac{\sin \tfrac{1}{4}(i\zeta - \gamma') \sin \tfrac{1}{4}(i\zeta + \gamma')}{\cos \tfrac{1}{4}(i\zeta - \gamma') \cos \tfrac{1}{4}(i\zeta + \gamma')}\right) d\theta_0(\gamma'). \tag{6.97}$$

The pressure coefficient at the separation point is

$$C_p = 1 - e^{2K}, \tag{6.98}$$

and Woods found that separation at $\alpha_s = 80°$ with $K = 0.4077$ gave very good agreement with observations for a Reynolds number of $R = 1060$. The flow field corresponding to these conditions is shown in Figure 6.18.

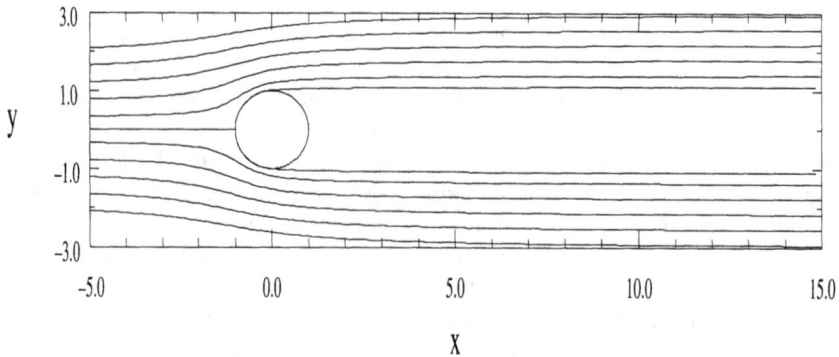

FIG. 6.18. Calculation of streamlines for Woods' extended model with parallel streamlines downstream, $\alpha_s = 80°$, $K = 0.4077$.

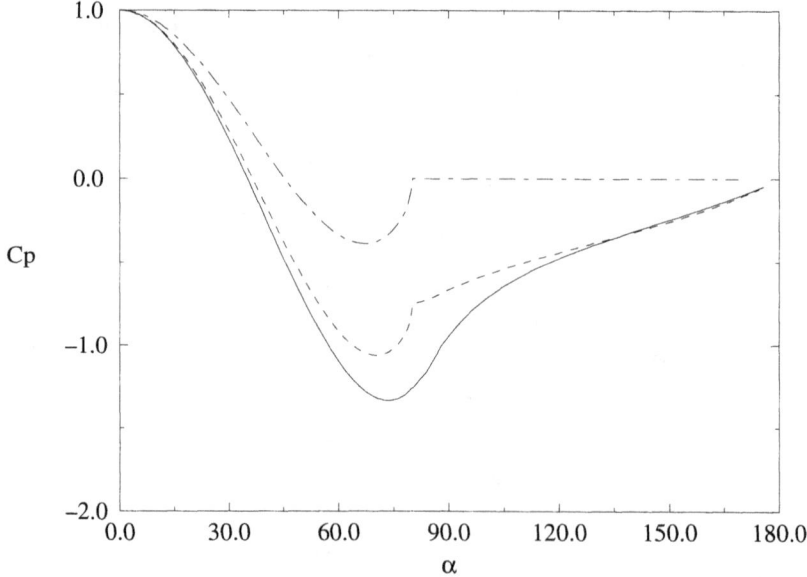

FIG. 6.19. Calculation of pressure coefficient C_p for Woods' extended model for $\alpha_s = 80°$, (- · · · -) $K = 0$, (- - - -) $K = 0.28$, (———) $K = 0.4077$. Note that the angle α is measured from the cylinder centre and $\alpha > 80°$ is on the free streamline and not the cylinder.

The pressure coefficient for Woods' comparison with experimental results is shown in Figure 6.19 and at first sight, this seems to have a remarkable property that the pressure gradient is not singular at the separation point, rather there is a step in the pressure gradient.

Near the separation point on the body, the analysis which led to (6.88) is easily extended to deal with the extra term introduced by allowing the free stream velocity to vary, giving

$$q(\pi - \epsilon) \sim 1 - [\frac{1}{2} - \frac{\phi_s}{\pi a}\int_0^\pi \frac{\sin\frac{1}{2}\gamma}{q(\gamma)}d\gamma - Kb]\epsilon. \qquad (6.99)$$

so that the magnitude of the pressure gradient singularity becomes

$$\boxed{\kappa = \frac{1}{\sqrt{a\phi_s}}[\frac{1}{2} - \frac{\phi_s}{\pi a}\int_0^\pi \frac{\sin\frac{1}{2}\gamma}{q(\gamma)}d\gamma - \frac{K^2\pi}{\int_0^\pi \theta_0(\gamma)\sin\frac{1}{2}\gamma\, d\gamma}].} \qquad (6.100)$$

Thus there will be a value of the pressure behind the cylinder for which the pressure gradient singularity vanishes (only for a further singularity to appear in the second derivative of the pressure coefficient, it should be remembered). It is remarkable that the value of the pressure coefficient used by Woods to match a set of observed results should be the very close to the value which removes the pressure gradient singularity (for $\alpha_s = 80°$, Woods used $K = 0.4077$, $C_p = -1.26$, calculation gives $K = 0.4012$, $C_p = -1.23$). The pressure coefficient calculated by Woods is shown in Figure 6.19 together with the corresponding pressure coefficients for $C_p = 0$ in the wake (i.e. Brodetsky and Woods simpler model) and for an intermediate case, $C_p = -0.75$ at separation. For each of the two latter conditions, the singularity in the pressure gradient is clearly evident. Wu (1962a) later also proposed a model with variable pressure in the wake to consider separation behind an oblique plate.

6.3 Boundary layer with a variable pressure gradient

We now turn to attempts to model flow around a cylinder using boundary layer theory. We shall suppose that a far field velocity U and cylinder radius a have been used to non-dimensionalise the variables. The Navier–Stokes equations in cylindrical coordinates (r, α) (using angular coordinate α from the front stagnation point) with velocities (u_r, u_α) and pressure p, are

$$u_r\frac{\partial u_\alpha}{\partial r} + \frac{u_\alpha}{r}\frac{\partial u_\alpha}{\partial \alpha} + \frac{u_r u_\alpha}{r} = -\frac{1}{r}\frac{\partial p}{\partial \alpha}$$
$$+ R^{-1}[\frac{1}{r}\frac{\partial}{\partial r}(r\frac{\partial u_\alpha}{\partial r}) + \frac{1}{r^2}\frac{\partial^2 u_\alpha}{\partial \alpha^2} + \frac{2}{r^2}\frac{\partial u_r}{\partial \alpha} - \frac{u_\alpha}{r^2}], \qquad (6.101)$$

and

$$u_r\frac{\partial u_r}{\partial r} + \frac{u_\alpha}{r}\frac{\partial u_r}{\partial \alpha} - \frac{u_\alpha^2}{r} = -\frac{\partial p}{\partial r}$$
$$+ R^{-1}[\frac{1}{r}\frac{\partial}{\partial r}(r\frac{\partial u_r}{\partial r}) + \frac{1}{r^2}\frac{\partial^2 u_r}{\partial \alpha^2} - \frac{u_r}{r^2} - \frac{2}{r^2}\frac{\partial u_\alpha}{\partial \alpha}], \qquad (6.102)$$

while continuity is

$$\frac{\partial}{\partial r}(r u_r) + \frac{\partial u_\alpha}{\partial \alpha} = 0. \qquad (6.103)$$

Near the surface of the cylinder, let $r = 1 + \epsilon Y$, where $\epsilon \ll 1$ is to be determined and suppose the velocities can be expanded by

$$u_\alpha = u + u',$$
$$u_r = \epsilon v + v', \qquad (6.104)$$

where $u' = o(u)$ and $v' = o(\epsilon v)$ as $\epsilon \to 0$.

Then the leading order part of the continuity equation is just

$$\frac{\partial u}{\partial \alpha} + \frac{\partial v}{\partial Y} = 0. \qquad (6.105)$$

Leaving aside the scaling for the pressure for just a moment, the tangential momentum equation has leading terms

$$u \frac{\partial u}{\partial \alpha} + v \frac{\partial u}{\partial Y} = -\frac{\partial p}{\partial \alpha} + \frac{1}{\epsilon^2 R} \frac{\partial^2 u}{\partial Y^2} + o(1), \qquad (6.106)$$

while the normal momentum equation has an expansion beginning

$$\frac{\partial p}{\partial Y} = \epsilon u^2 + \cdots. \qquad (6.107)$$

This gives: (a) that we should choose $\epsilon = R^{-1/2}$ and (b) that the pressure can be expanded $p = p_0(\alpha) + o(1)$. Thus the leading order equation for tangential momentum is identical with the boundary layer equation on a flat plate, excepting we can not take the pressure gradient parallel to the wall to be zero. The leading order pressure gradient is determined by the outer potential flow, if the outer flow velocity tangential to the cylinder wall is $u_e(\alpha)$ then $p_0(\alpha) = -u_e^2(\alpha)/2$, The leading order boundary layer equations for flow near the cylinder wall are continuity and

$$u \frac{\partial u}{\partial \alpha} + v \frac{\partial u}{\partial Y} = u_e(\alpha) \frac{du_e}{d\alpha}(\alpha) + \frac{\partial^2 u}{\partial Y^2}. \qquad (6.108)$$

Note that a circular cylinder is a special case, if a body has variable curvature then the transverse pressure gradient may not be negligible at leading order.

This simple set of equations has exercised the finest fluid dynamicists for over half a century and lead to the most famous 'singularity' in fluid dynamics. We shall follow attempts to solve these equations, analytically, approximately and numerically before seeing that in the study of separation, in themselves they lead nowhere useful.

6.3.1 Heimenz (1911)

The earliest attempt to use boundary layer theory to determine the point of separation was by Heimenz (1911). We have already discussed his observations of flow around a cylinder and he used the experimentally observed pressure gradient to obtain an approximate solution of (6.108). Assuming that the observed pressure gradient could be expanded

$$\frac{\partial p}{\partial \alpha} = p_1 \alpha + p_3 \alpha^3 + p_5 \alpha^5 + \cdots, \quad (6.109)$$

(and hence the velocity at the edge of the boundary layer deduced in a form $u_e \sim u_1 \alpha + u_3 \alpha^3 + u_5 \alpha^5 + \cdots$) he expanded the stream-function as

$$\psi \sim \alpha \psi_1(Y) + \alpha^3 \psi_3(Y) + \cdots, \quad (6.110)$$

and substituted these expansions into (6.108) obtaining a set of inhomogeneous differential equations for $\psi_1, \psi_3, \cdots$, and from these, using only three terms in the expansion, the separation point was predicted by

$$\frac{\partial u}{\partial Y}(\alpha, 0) = \alpha \psi_1''(0) + \alpha^3 \psi_3''(0) + \alpha^5 \psi_1''(0) = 0. \quad (6.111)$$

Heimenz found that when he used the measured pressure gradient to calculate solutions for $\phi_1 - \phi_5$ and then substitute the results into (6.111), the observed separation point was predicted accurately, just a little above $\alpha = 81°$.

The result that *given* the observed pressure on the surface, *then* the separation point can be predicted will be repeated; however, it ought to be kept in mind that what we would hope to obtain from a true theory is a prediction of *both* the surface pressure *and* the separation point.

6.3.2 Pohlhausen (1921)

Whilst Karman (1921) had shown the importance of an integral form of the boundary layer equations (see section 1.6 above) and applied an integral form to the boundary layer on disc rotating in its plane, Pohlhausen (1921) applied Karman's ideas to the boundary layer on a cylinder using the same outer velocity as Heimenz had deduced from pressure measurements. In particular, Pohlhausen approximated the velocity across the boundary layer by a quartic distribution which could then be substituted into an integral form of the boundary layer equation to obtain a first order ordinary differential equation for a boundary layer thickness. Pohlhausen's integral formulation predicted essentially the same separation point as had Heimenz.

The integral formulation used by Pohlhausen was developed in three steps, first rewriting the boundary layer equation in terms of a velocity defect, $u_e - u$; then integrating in Y, noting that $u_e - u = 0$ for $Y > \delta$, where δ is the boundary layer thickness; and finally using integration by parts to replace the term involving the v velocity, resulting in

$$2u'_e \int_0^\delta (u_e - u)dY + u_e \frac{d}{d\alpha} \int_0^\delta (u_e - u)dY - \frac{d}{d\alpha} \int_0^\delta (u_e - u)^2 dY = -\frac{\partial u}{\partial Y}(\alpha, 0). \tag{6.112}$$

Polhausen let the velocity across the boundary layer be approximated by

$$u = \sigma Y + bY^2 + cY^3 + dY^4, \tag{6.113}$$

(thus σ is a form of non-dimensional wall shear coefficient) so that the coefficients σ, b, c and d come from the conditions $u = u_e(\alpha)$, $\frac{\partial u}{\partial Y} = 0$ and $\frac{\partial^2 u}{\partial Y^2} = 0$ at $Y = \delta$ and $\frac{\partial^2 u}{\partial Y^2} = -u_e u'_e(\alpha)$ at $Y = 0$. Pohlhausen used an auxiliary variable, $\lambda = u'_e \delta^2$ and coefficients

$$\sigma = \frac{(12+\lambda)u_e}{6\delta}, \quad b = -\frac{1}{2}u_e u'_e, \quad c = \frac{(\lambda-4)u_e}{2\delta^3}, \quad d = \frac{(6-\lambda)u_e}{6\delta^4}. \tag{6.114}$$

This enables the integrals which occur in (6.112) to be evaluated as functions of u_e and δ, resulting in

$$\frac{d}{d\alpha}\delta^2 = \frac{-4 + \frac{232}{315}u'_e\delta^2 - \frac{1}{3780}[79(u'_e)^2 + 8u_e u''_e]\delta^4 - \frac{1}{2268}u'_e[(u'_e)^2 + u_e u'''_e]\delta^6}{-\frac{37}{315}u_e + \frac{1}{315}u_e u'_e \delta^2 + \frac{5}{9072}u_e(u'_e)^2 \delta^4}. \tag{6.115}$$

This first order differential equation needs to be integrated from $\alpha = 0$, but there is a potential difficulty in that the denominator vanishes with u_e at $\alpha = 0$, so Pohlhausen required that the boundary layer thickness have a non-zero value at $\alpha = 0$ so as to provide a zero in the numerator too. It is also clear that separation will occur when $\lambda = u'_e \delta^2 = -12$ since the coefficient σ will change sign at that point.

In order to calculate the solution for a cylinder, Pohlhausen used Heimenz's velocity distribution

$$u_e(\alpha) = 1.8348\alpha - 0.2742\alpha^3 - 0.0478\alpha^5, \tag{6.116}$$

which requires $\delta(0) = 1.9605$ and gives $\frac{d\delta^2}{d\alpha}(0) = 0$. Pohlhausen used a graphical method to solve (6.115); using a Runge–Kutta method the same results are obtained, the boundary layer thickness δ grows steadily until some form of 'singularity' occurs a little above $\alpha = 80°$, see Figure 6.20.

If the coefficient σ is plotted against α, Figure 6.21, then the solution predicts separation just above $\alpha = 81°$, the same value as had been obtained by Heimenz.

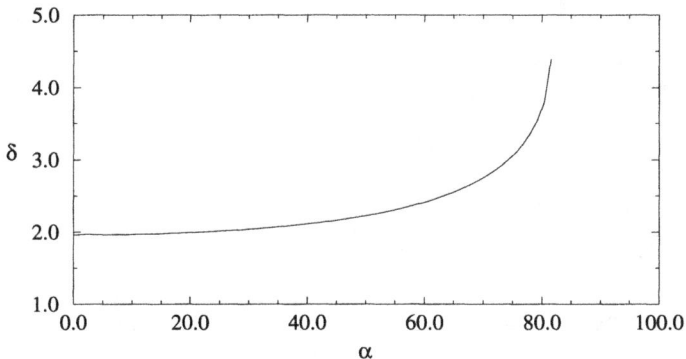

FIG. 6.20. Non-dimensional boundary layer thickness, δ, for flow around a cylinder, calculated using Pohlhausen's method for Heimenz's surface velocity distribution.

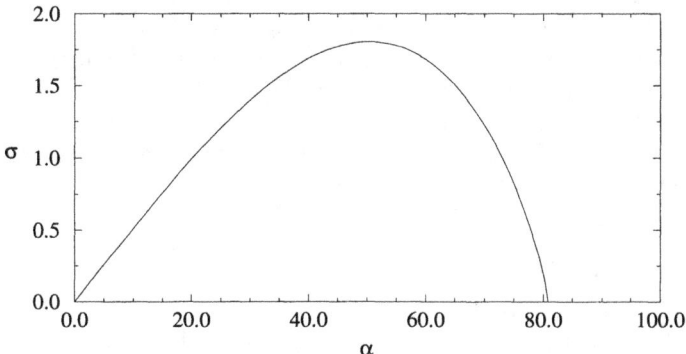

FIG. 6.21. Non-dimensional boundary layer shear, σ, for flow around a cylinder, calculated using Pohlhausen's method for Heimenz's surface velocity distribution. Separation should occur when $\sigma = 0$.

One example which Pohlhausen did not examine is the case of a potential outer flow. If the tangential velocity on the cylinder is taken from the potential solution

$$\psi = (r - \frac{1}{r}) \sin \alpha, \qquad (6.117)$$

then the surface speed is $q = 2\sin\alpha$. Computation of Pohlhausen's method for

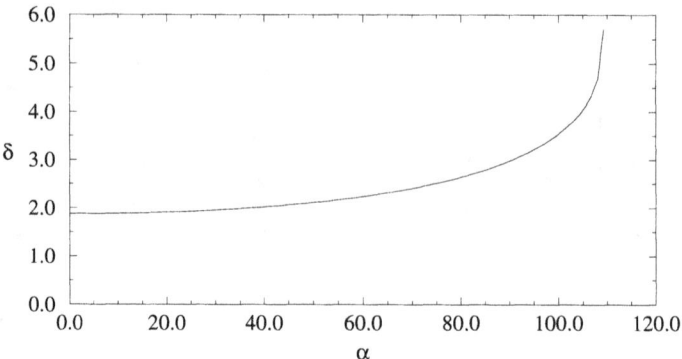

FIG. 6.22. Non-dimensional boundary layer thickness, δ, for flow around a cylinder, calculated using Pohlhausen's method for a potential surface velocity distribution.

this case, using $\delta(0) = 1.8778$ predicts separation just above $\alpha = 106°$, a value much in excess of the observed separation angle. The boundary layer thickness and non-dimensional wall shear for this case are shown in Figures 6.22 and 6.23. Once again the calculation of the boundary layer thickness cannot be continued much beyond the predicted separation point. We shall examine the results of methods which combine free streamline flow with boundary layer calculations in the next section but it seems that combining an 'unseparated' potential outer flow with a boundary layer solution will not work.

6.3.3 Bairstow (1925), Maccoll (1930)

The boundary layer formulation for a cylinder which lead to (6.108) is a special case of the formulation for a curved body given by Bairstow (1925). While he did not give any solution of the boundary layer equations for a curved body and seemed unaware of the work of Heimenz he did indicate that efforts were being made to solve the boundary layer equation using measured pressure values (although he did not give any idea of what method was being attempted). This work was at a time when developments in boundary layer theory in Germany were less well known in Britain than later, after Maccoll (1930) had returned to Britain from working in Germany and given a comprehensive account of progress there.

6.3.4 Goldstein (1930)

The idea used by Heimenz, that given a Taylor expansion for the pressure gradient and assuming a polynomial expansion for the velocity u, then the subsequent flow development could be determined was very attractive. Indeed it held out the promise of a step by step method for calculating solutions to the boundary

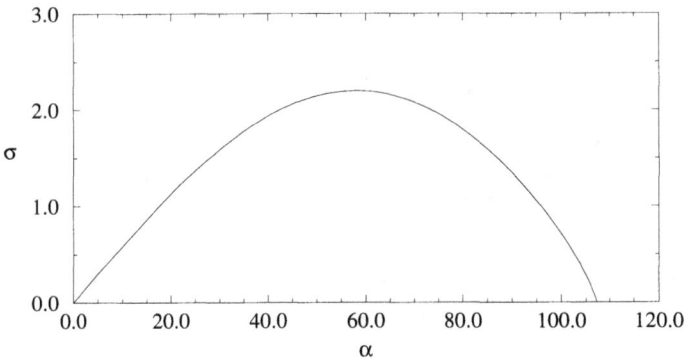

FIG. 6.23. Non-dimensional boundary layer shear, σ, for flow around a cylinder, calculated using Pohlhausen's method for a potential surface velocity distribution. Separation should occur when $\sigma = 0$.

layer equations: given u and p_α at one α point, the expansions could predict u at a later α point and the whole process repeated to give the solution at further α points. This idea and others (see for instance, the discussion about the wake of a flat plate in Chapter 2) were considered by Goldstein (1930) in a seminal study on boundary layer theory. In the parts of his paper relevant to this section, Goldstein showed was that while it might seem plausible that given a pressure gradient, the boundary layer development could be taken from an initial Taylor series expansion of u, the reality was that the coefficients of the pressure gradient expansion and the Taylor series for the velocity u should to be related in a particular way.

Thus if an initial velocity profile were given by

$$u(0, Y) = a_1 Y + a_2 Y^2 + a_3 Y^3 + \cdots, \tag{6.118}$$

with stream-function

$$\psi(0, Y) = \frac{1}{2} a_1 Y^2 + \frac{1}{3} a_2 Y^3 + \cdots, \tag{6.119}$$

(with $a_1 \neq 0$) and the pressure gradient by

$$\frac{\partial p}{\partial \alpha} = p_0 + p_1 \alpha + p_2 \alpha^2 + \cdots, \tag{6.120}$$

then the coefficients $\{a_m\}$ and $\{p_n\}$ cannot be independently specified. Goldstein presented the conditions between the coefficients as ones '*for the absence of singularities in the solution*' without derivation. I find it easiest to see them as consistency conditions near the point where the velocity profile is specified. If the u velocity is written

$$u(\alpha, Y) \sim A_1(\alpha)Y + A_2(\alpha)Y^2 + A_3(\alpha)Y^3 + \cdots, \qquad (6.121)$$

then the v velocity is

$$v \sim -\frac{A_1'(\alpha)Y^2}{2} - \frac{A_2'(\alpha)Y^3}{3} - \frac{A_3'(\alpha)Y^4}{4} - \cdots, \qquad (6.122)$$

then when these are substituted into the boundary layer equation, and powers of Y equated to zero, we obtain successively,

$$2A_2(\alpha) - p_0 - p_1\alpha - p_2\alpha^2 - \cdots = 0, \qquad (6.123)$$

$$A_3(\alpha) = 0, \qquad (6.124)$$

$$12A_4 + \frac{A_1 A_1'}{2} = 0, \qquad (6.125)$$

$$20A_5 - \frac{2}{3}A_1 A_2' = 0, \qquad (6.126)$$

$$30A_6 - \frac{2}{3}A_2 A_2' = 0, \qquad (6.127)$$

and so on. If we now consider $\alpha = 0$, and using $A_2'(0) = p_1/2$ from (6.123), then consistency conditions from (6.123), (6.124), (6.126), (6.127) are respectively

$$2a_2 - p_0 = 0, \ a_3 = 0, \ 5!a_5 - 2a_1 p_1 = 0, \text{ and } 6!a_6 + 2p_0 p_1 = 0. \qquad (6.128)$$

The condition (6.125) gives $A_1'(0) = 4!a_4/a_1$, and a_1, a_4 can be specified arbitrarily. Considering the coefficient of higher powers of Y leads to further relations between the coefficients $\{a_m\}$ and $\{p_n\}$. Goldstein thought that this process cumbersome and likely to result in numerical errors with increasingly large factorials occurring in the numerators of expressions for coefficients as the process continued.

Goldstein did, however, continue his study by showing that starting from (6.121) and defining $\eta = Y/\alpha^{1/3}$, the stream-function would have to be given by an expansion

$$\psi = \alpha^{2/3}[f_0(\eta) + \alpha^{1/3} f_1(\eta) + \alpha^{2/3} f_2(\eta) + \cdots]. \qquad (6.129)$$

Of course, Goldstein had to determine the functions $f_1, f_2, \cdots$ analytically, a process which became more tedious the further it was carried and even Goldstein thought that determining f_5 to be '*out of the question*'. In finding the first few functions $\{f_r\}$ Goldstein obtained their behaviour for large η, for example,

$$f_0 \sim \frac{1}{2}a_1\eta^2, \quad f_1 \sim \frac{1}{3}a_2\eta^3 + \frac{3^{1/3}\Gamma(\frac{1}{3})}{2\Gamma(\frac{2}{3})}\frac{p_0 + 2a_2}{a_1}\eta - \frac{p_0 + 2a_2}{a_1}, \qquad (6.130)$$

The genius of Goldstein is illustrated by his realisation that (6.129) is entirely unsatisfactory when η is large. Remember that the expansion of the velocity (and hence the stream-function) as $\alpha \to 0$ is one of integral powers of Y yet the expansion (6.129) shrinks to a point as $\alpha \to 0$. Goldstein saw that a separate expansion would be necessary for $Y > 0$, $\alpha \to 0$ and two decades before matched asymptotic analysis was developed, he realised that this new expansion would have to correspond to the asymptotic expansion of (6.129) with its terms rearranged so as to be an asymptotic sequence as $Y \to 0$.

If the expansion of (6.129) for large η is written down using (6.130), the terms rearranged and η replaced in terms of α and Y, then

$$\psi \sim \frac{1}{2}a_1Y^2 + \frac{1}{3}a_2Y^3 + \cdots + \alpha^{2/3}[\frac{3^{1/3}\Gamma(\frac{1}{3})}{2\Gamma(\frac{2}{3})}\frac{p_0 + 2a_2}{a_1}Y + \cdots] + \cdots, \qquad (6.131)$$

and the leading order term (in the α expansion) gives precisely the initial velocity profile, (6.119). Thus Goldstein took the expansion for this region to be

$$\psi \sim \psi_0(Y) + \alpha^{2/3}\psi_1(Y) + \alpha\psi_2(Y) + \alpha\log\alpha\,\psi_4(Y) + \cdots, \qquad (6.132)$$

(the logarithms arise in the large η expansion of f_2) and it is this expansion which can be matched to the outer velocity as well as the initial velocity profile.

In addition to the case $a_1 \neq 0$, Goldstein considered the case where the wall shear is zero at the starting point, $a_1 = 0$. In that case the most general subsequent development was quite different, the similarity variable which had to be used was $\eta = Y\alpha^{-1/4}$ and the stream-function had to be expanded

$$\psi = \alpha^{3/4}[f_0(\eta) + \alpha^{1/4}f_1(\eta) + \alpha^{1/2}f_2(\eta) + \cdots], \qquad (6.133)$$

with a correspondingly different expansion to match the sublayer expansion to the outer velocity but we shall defer considering this particular case until we come to Goldstein's later proposal of a singularity in the boundary layer solution at separation.

6.3.5 Falkner & Skan (1931)

Falkner and Skan (1931) realised that if the pressure gradient varied because the tangential outer velocity was changing, then the special case

$$u_e(\alpha) = \alpha^m, \qquad (6.134)$$

would lead to a similarity solution of the boundary layer equation.

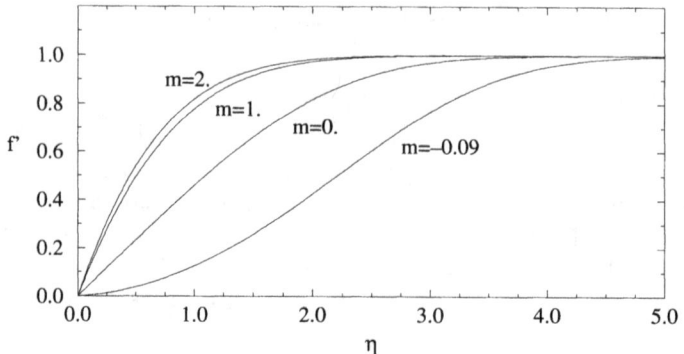

FIG. 6.24. Solution of Falkner–Skan equation.

If

$$\eta = \sqrt{\frac{m+1}{2}} \frac{Y}{\alpha^{(1-m)/2}}, \tag{6.135}$$

and

$$\psi = \frac{2}{\sqrt{m+1}} \alpha^{(m+1)/2} f(\eta), \tag{6.136}$$

then with $\beta = 2m/(m+1)$, the boundary layer equation becomes

$$f''' + ff'' + \beta(1 - f'^2) = 0. \tag{6.137}$$

This equation is now called the Falkner–Skan equation and the flat plate Blasius solution is recovered when $\beta = 0$. The boundary conditions are $f(0) = f'(0) = 0$ and $f'(\eta) \sim \sqrt{m+1}$ as $\eta \to \infty$. A similar equation occurs in compressible flow but as one of a pair of simultaneous equations; see for example Rogers (1969).

Falkner & Skan solved this equation using a combination of a power series and a graphical method; it is nowadays straight forward to solve the equation numerically, and computed profiles for f' corresponding to those calculated by Falkner & Skan are shown in Figure 6.24. If the index m is positive then the outer flow is being accelerated, if the index is negative then the outer flow is being decelerated. In that case Falkner & Skan thought it not possible to find solutions for $m < -0.09$, a value for which $f''(0) = 0$ and beyond which the flow would stop moving forward everywhere, having to have a layer of backward moving fluid *everywhere* along the wall.

The Falkner–Skan equation was, for a while, thought to be one of the keys in the study of separation so we shall look deeper into this solutions of this equation

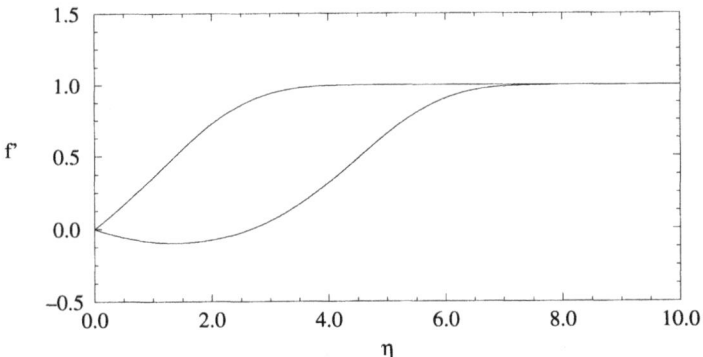

FIG. 6.25. An example of the multiple solutions of the Falkner–Skan equation (for $m = -0.05$) discovered by Stewartson.

out of historical sequence before returning to how attempts were made to predict separation from boundary layer theory.

Hartree (1937) looked to solve the equation using a 'differential analyser', an analog computer. He realised that the formulation of the problem, a differential equation with two boundary conditions at $\eta = 0$ and the condition $f' \to 1$ as $\eta \to \infty$ was not sufficient to determine a unique solution. The example problem he gave is still relevant today. If a second order linear differential equation is required to vanish at infinity and has one condition imposed at the origin, then the solution is only unique if one of the two solutions to the homogeneous equation does **not** vanish at infinity. If both fundamental solutions vanish at infinity then the solitary condition at the origin is not sufficient to determine a unique solution. Hartree's view was that an additional condition which would give a unique solution to the Falkner–Skan equation would be $f' \to 1^-$, and as fast as possible. The solutions which satisfied this condition were only obtained for $\beta > \beta_0 = -0.1988$ ($m = -0.0904$).

The question of lack of uniqueness was then partially resolved by Stewartson (1954). First, he described how Hartree's concern about the far boundary condition not providing uniqueness could be partially resolved by first applying the boundary condition at a fixed point and then allowing that point where the boundary condition was applied to become large. He was able to show analytically that if $\beta < \beta_0$ then it must be the case that $f' > 1$ somewhere and he regarded such solutions, with a velocity in excess of the outer velocity as unphysical. If however, $\beta_0 < \beta < 0$, then there must be two solutions, both satisfying $f' < 1$ but one having $f''(0) < 0$, or reversed flow everywhere along the wall. Additional material on the Falkner–Skan equation can also be found in Coppel (1960)

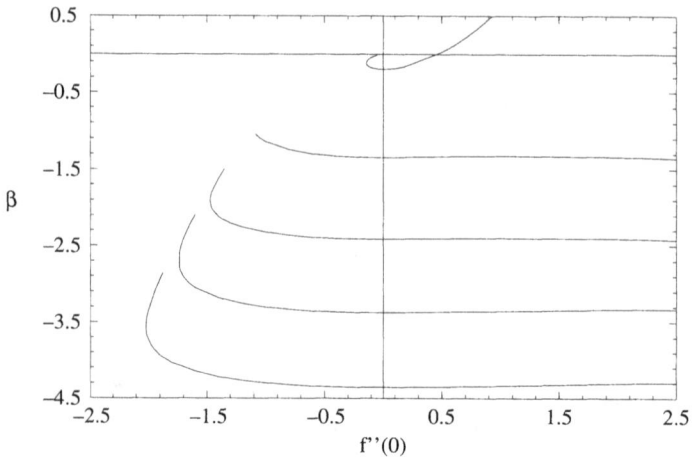

FIG. 6.26. Map between $f''(0)$ and β for the Falkner–Skan equation for which solutions exist, first determined by Libby & Liu, showing multiple solutions for either fixed $f''(0)$ or fixed β.

An example calculation of the two solutions is shown in Figure 6.25 for $m = -0.05$, $\beta = 0.105$ and it is clear that Hartree's condition requiring the most rapid convergence to $f' = 1$ would eliminate the solution with $f''(0) < 0$.

The full complexity of solutions to the Falkner–Skan equation emerged in the work of Libby and Liu (1967) who deduced that there would be an infinite number of solutions to the equation for $\beta < 0$ which converge exponentially to the $f' = 1$ at infinity (solutions which might have an algebraic decay to $f' = 1$ are discarded as unphysical, see the discussion in chapter 2 about this point). The map they determined showed that in addition to the solutions Stewartson had found for $\beta_0 < \beta < 0$, there were other solutions for $\beta < \beta_0$ all of which had, as predicted by Stewartson, $f' > 1$. The map is shown in Figure 6.26; the uppermost curve for $f''(0) > 0$ was first calculated by Falkner & Skan and is the solution Hartree thought physically relevant, the rest of the upper curve was first found by Stewartson and the general structure of the diagram was fist given by Libby & Liu although they did not calculate the left-hand turning points on the lower three curves. Curiously the second curve from the top does not seem to have any form of turning point, ending as shown in the diagram. The easiest examples to calculate are for $f''(0) = 0$ and β negative and the graphs of f' shown in Figure 6.27 were first calculated by Libby & Liu. The belief that the Falkner–Skan equation would give solutions which might be part of a description of separation was probably always flawed. Indeed Stewartson (1954) recognised

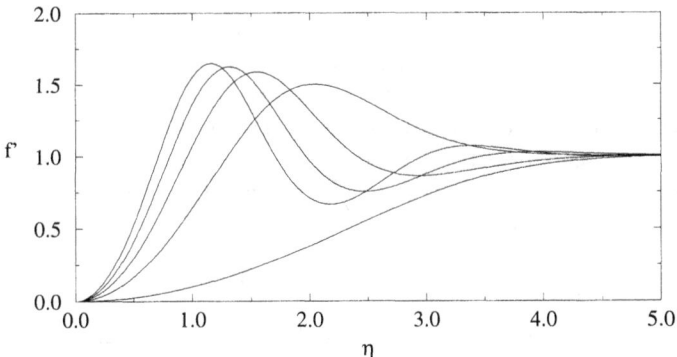

FIG. 6.27. Multiple solutions of the Falkner–Skan equation discovered by Libby & Liu, calculated for $f''(0) = 0$ and representing values $\beta = -0.1988, -1.347, -2.416, -3.375, -4.357$ as the velocity changes from monotonic to oscillatory.

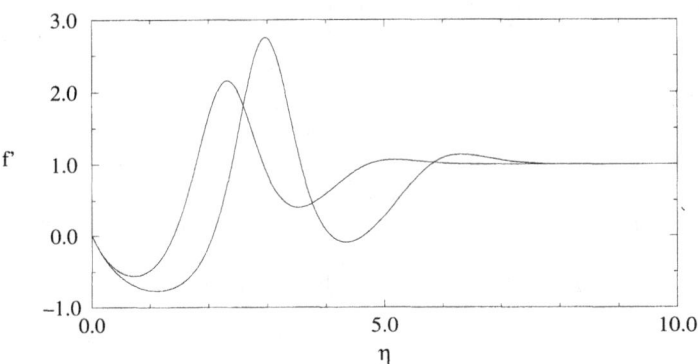

FIG. 6.28. Multiple solutions of the Falkner–Skan equation for $f''(0) = -1.7$.

this when he wrote of the solutions, 'as $x \to \infty$ some of them may be asymptotic boundary-layer profiles if the main stream velocity is asymptotically proportional to x^m, **provided that in the meantime separation has not occurred**'. On the other hand, the utility of the Falkner–Skan equation in a favourable pressure gradient is widely accepted. Goldstein (1965) was also clear about the applicability (only to converging passages) and proposed a more rigorous statement of the outer condition, that $Y^N(f'-1) \to 0$ for any positive integer N, as $Y \to \infty$.

6.3.6 Howarth (1938)

An important contribution to boundary layer theory came from Howarth (1938) who introduced a form of problem for separation which was to play a central role in thinking for thirty years. As we have seen, separation was observed to occur a short distance after the pressure gradient became adverse. Howarth formalised this by considering the solution of the boundary layer equations for a simple adverse pressure gradient: that where the external velocity was a linearly decreasing function of distance,

$$u_e(\alpha) = b_0(1 - c\alpha), \tag{6.138}$$

where b_0 and c are constants. Using this external velocity, the pressure gradient was

$$\frac{\partial p}{\partial \alpha} = c b_0^2 (1 - c\alpha), \tag{6.139}$$

so that the pressure gradient is positive (there is an implicit assumption that $\alpha < 1/c$) and as main flow is being decelerated, separation might be expected to occur. Howarth saw this problem as part of a wider approach to solving the boundary layer equations where the outer velocity would be replaced by a sequence of piecewise functions (in the α direction), each of which is solved using the previous one as an initial condition (hence his reason for not putting $b_0 = 1$ since the outer velocity need not have that value at the point where the linearly decreasing velocity begins). Nevertheless, the problem he specified, that of the response of a boundary layer to a linearly varying pressure gradient, came to be viewed as one of the key problems for separation of a laminar boundary layer.

Howarth first looked to solve for the boundary layer by using a series expansion with similarity variable $\eta = \sqrt{b_0} Y / 2\sqrt{\alpha}$ and stream-function expansion

$$\psi \sim \sqrt{b_0 \alpha} [f_0(\eta) - 8c\alpha f_1(\eta) + (8c\alpha)^2 f_2(\eta) - \cdots], \tag{6.140}$$

Substituting (6.140) into the boundary layer equation gave a series of differential equations to the functions $\{f_r\}$ and Howarth found the solutions to be all positive and slowly decreasing with increasing r, so the series expansion being the sum of an alternating sequence should have a radius of convergence near $c\alpha \approx 1/8$. Unfortunately he also found that this expansion predicted that the wall shear would vanish near this value and that convergence of the series was very slow near separation. He combined his series expansion, approximations for f_r when r was large and Pohlhausen's integral method to calculate that separation should occur when $c\alpha \approx 0.120$. Direct application of Pohlhausen's method gave a much larger value, $c\alpha \approx 0.156$. Of course, the expansion (6.140) which has singularities at its origin was not altogether suitable for the step by step approach which he was proposing, so he turned to the integral form of the equations and considered the problem as one where the boundary layer and momentum thickness would be specified at discrete α points and the integral momentum equation solved for a piecewise specified pressure gradient, including the case where the outer velocity distribution might be given on a segment by a quadratic. The idea of solving for the boundary layer in this way was shortly overtaken by other ideas, particularly numerical methods and Thwaites (1949) approximate method but the view remained that a linearly decreasing external velocity was an important aspect in boundary layer separation.

6.3.7 Thwaites (1949)

The application of Karman's integral formulation to separation reached something of an apogee with the work of Thwaites (1949) who proposed a very general formula for predicting separation from an amalgam of previous results. If the Karman integral, (1.46) is written in the form

$$\frac{d\theta}{d\alpha} + [H+2]\frac{\theta}{u_e}\frac{du_e}{d\alpha} = \frac{1}{R}\frac{A}{\theta u_e^2}\frac{\partial u}{\partial Y}(0), \tag{6.141}$$

then the Reynolds number can be removed from the equation by letting

$$\theta = R^{-1/2}\Theta. \tag{6.142}$$

Thwaites defined new variables by

$$m = -\Theta^2 u_e', \text{ and } l = \frac{A}{u_e}\frac{\partial u}{\partial Y}(0). \tag{6.143}$$

Rewriting (6.141), assuming that $l = l(m)$, $H = H(m)$,

$$u_e \frac{d}{d\alpha}\frac{m}{u_e'} = 2(m[H+2] + l(m)), \tag{6.144}$$

a differential equation for $m(\alpha)$. Of course, the functions $H(m)$ and $l(m)$ still have to be determined. Thwaites let the right-hand side of this equation be a function $L(m) = 2(m[H+2] + l(m))$ and by examining a number of examples, proposed that for practical purposes, there was a universal $L(m)$,

$$L(m) = 0.45 + 6m, \tag{6.145}$$

and that separation should occur when $m = 0.082$ Using (6.145) then (6.144) becomes

$$u_e(\alpha)\frac{d}{d\alpha}\Theta^2 = 0.45 - 6u_e'(\alpha)\Theta^2, \tag{6.146}$$

with solution

$$\Theta^2 = \frac{0.45}{u_e^6}\int_0^\alpha u_e(\alpha')d\alpha'. \tag{6.147}$$

If this is combined with the condition $m = -0.082$ then separation is predicted when

$$\boxed{-\frac{0.45 u_e'}{u_e^6}\int_0^\alpha u_e^5(\alpha')d\alpha' = 0.082.} \tag{6.148}$$

In the case of potential flow around a cylinder, we have already seen that $u_e = 2\sin\alpha$ so that (6.148) gives

$$-0.45\cos\alpha_s\left(\frac{8}{15} - \cos\alpha_s + \frac{2}{3}\cos^3\alpha_s - \frac{1}{5}\cos^5\alpha_s\right) = 0.082\sin^6\alpha_s, \qquad (6.149)$$

and the predicted separation angle is given by $\alpha_s \approx 102°$; as with all the predictions we have examined which are based around potential outer flow about a cylinder, the separation angle is too large by a considerable amount.

6.4 Combined boundary layer – free streamline models

We have seen that if we use free streamline theory then the separation point can be specified somewhere within the range between that for smooth separation where streamlines cease to cut into the body and that point further round the body where a free streamline would cross the free streamline from the other side of the body. Within this range all solutions are equally plausible, excepting that other than for the point of smooth separation, a singularity will be evident in the pressure gradient at the point of separation. We have also seen that attempting to use boundary layer theory leads to problems near the point of separation. What is also true is that neither theory predicts the separation point accurately, excepting when the outer pressure gradient (or equivalently, velocity) is specified from experiments. A natural question is then: if free streamline theory is used to specify an outer flow, can it be combined with a boundary layer approximation in such a way that the two theories consistently specify the same separation point?

6.4.1 Burgers (1922), Squire (1934)

An approximate solution for viscous flow about a body had been developed by Burgers (1922, 1930) which fell into neither boundary layer theory nor free streamline theory but rather was an attempt to apply Oseen's approximation to the vorticity equation. The essence of the method was to envisage the flow as the result of a series of point forces, to write the vorticity as a convolution integral of the fundamental solution with the point force strengths and then to use the known velocity on the body surface to determine the point force strengths (see also the discussion in Chapter 2 on drag of a plate).

Starting from the equation for the vorticity, ω, in coordinates (x, y)

$$u\frac{\partial\omega}{\partial x} + v\frac{\partial\omega}{\partial y} = \frac{1}{R}\nabla^2\omega, \qquad (6.150)$$

rewritten in terms of the potential ϕ and stream-function ψ,

$$\frac{\partial\omega}{\partial\phi} = \frac{1}{R}\left(\frac{\partial^2\omega}{\partial\phi^2} + \frac{\partial^2\omega}{\partial\psi^2}\right), \qquad (6.151)$$

becomes with a boundary layer type approximation, $\psi \sim R^{-1/2}\Psi$, $\omega = R^{1/2}\omega_0 + o(R^{1/2})$,

Combined boundary layer – free streamline models 165

$$\frac{\partial \omega_0}{\partial \phi} = \frac{\partial^2 \omega_0}{\partial \Psi^2}, \qquad (6.152)$$

that is, a diffusion equation.

Burgers had given a solution for this form of approximate equation. Assuming that the vorticity vanishes upstream of the front stagnation point, taken at say $\phi = \phi_0$, then the vorticity could be written as

$$\omega_0(\phi, \Psi) = \int_{\phi_0}^{\phi} \frac{f(\phi')}{\sqrt{\pi(\phi - \phi')}} \exp \frac{-\Psi^2}{4\pi(\phi - \phi')} d\phi', \qquad (6.153)$$

where f has to be determined. If the velocity in the boundary layer parallel to the surface is denoted q and Y is the normal boundary layer coordinate, then since $q \sim \Psi_Y$, $\omega_0 \sim -q_Y$

$$q \sim -\int_0^\Psi \frac{\omega}{q} d\Psi, \qquad (6.154)$$

Burgers then made an approximation, that in the integral of (6.154) the velocity q might be replaced by the outer velocity u_e. In that case, substituting (6.153) and changing the order of integration,

$$q \approx -\frac{1}{u_e} \int_{\phi_1}^{\phi} f(\phi') \mathrm{erf}[\frac{1}{2} \frac{\Psi}{\sqrt{\phi - \phi'}}] d\phi'. \qquad (6.155)$$

Then letting $\Psi \to \infty$ with $q \to u_e$,

$$u_e^2 \approx \int_{\phi_1}^{\phi} f(\phi') d\phi', \text{ or } f(\phi) \approx \frac{du_e^2}{d\phi} = 2 \frac{du_e}{d\alpha}, \qquad (6.156)$$

where we have returned to using α to denote distance on the body surface and have used $d\phi = u_e d\alpha$. With the distribution f known, the vorticity can be calculated from (6.153) and separation from where $\omega_0 = 0$.

Burgers applied this method to an outer velocity given by potential flow around a cylinder, where if the cylinder has unit radius, the potential is given by

$$\phi = -(r + \frac{1}{r})\cos\alpha, \qquad (6.157)$$

so that $\phi_0 = -2$, and the surface velocity is given by $u_e = 2\sin\alpha = 2\sqrt{1 - \phi^2/4}$. In this case the force distribution f is

$$f(\phi) = -2\phi, \qquad (6.158)$$

and thus the surface vorticity is

$$\omega_0 = \int_{-2}^{\phi} \frac{2\phi'}{\sqrt{\pi(\phi - \phi')}} d\phi' = \frac{8}{3\sqrt{\pi}}(1-\phi)\sqrt{\phi+2}. \qquad (6.159)$$

This distribution of surface vorticity vanishes at $\phi = 1$ or $\alpha = 120°$, see Figure 6.29. Hence when combined with a potential outer flow, Burgers' method, in

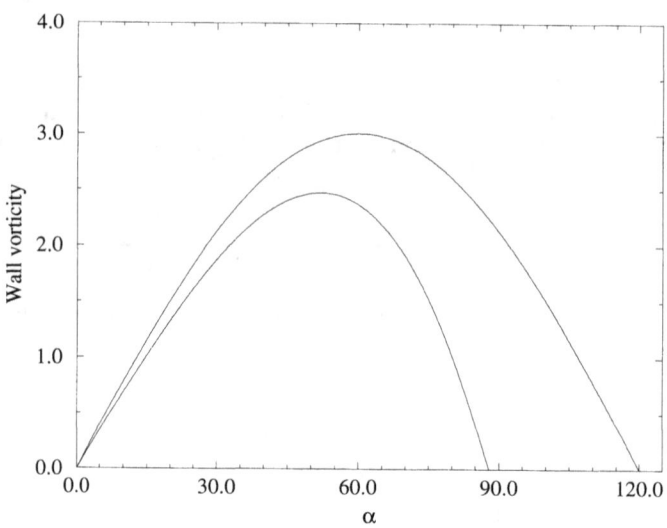

FIG. 6.29. Wall vorticity versus angle on the surface of a cylinder calculated by Burgers' approximate method using (a) potential outer flow, separation at $\alpha = 120°$ and (b) Heimenz's experimental outer velocity, separation at $\alpha \approx 88°$.

common with other attempts we have described, gave a separation angle for a cylinder which was far too great. Burgers recognised this and attributed it to 'the difference is caused by the calculation of convection, which here keeps a finite velocity up to the surface so that' the separation point is swept 'along too far by the flow'. Burgers either did not know of Heimenz's measurements of pressure on the surface of a cylinder or did not try to use them; had he used the outer velocity deduced from Heimenz's measurements he would have predicted a separation angle only a little above the observed value. The result of substituting Heimenz's velocity profile as the outer velocity in (6.153) and letting $\Psi = 0$ to obtain the wall vorticity is shown in Figure 6.29 and compared with the wall vorticity predicted for potential outer flow (6.159). The predicted separation point is still too great, probably for the same reason Burgers thought the prediction using a potential outer flow was in error.

Squire (1934) realised that Burgers' method could be extended to include other outer velocity distributions, in particular the outer flow given by the free streamline solution determined by Brodetsky and extended by Schmieden. We have already seen that Brodetsky's method leads to a surface velocity given by (6.36) and on the surface of the body where $t = e^{i\sigma}$ the surface potential is related to σ by $\phi = \sin^2 \sigma$ with $\phi_0 = 0$ at the forward stagnation point so we have all the information necessary to apply Burgers' method. There is, however, a problem near the separation point. In the expansion for the surface velocity near the stagnation point, (6.54), we saw that with $\sigma = \pi/2 - \epsilon$,

$$u_e \sim 1 - \Gamma\epsilon + \frac{1}{2}\Gamma^2\epsilon^2, \qquad (6.160)$$

and it is easy to show that $\epsilon \sim \sqrt{1-\phi}$ near the separation point so that the point force strength needed in Burgers' method will have

$$f(\phi) = \frac{\mathrm{d}u_e^2}{\mathrm{d}\phi} = \frac{\Gamma}{\sqrt{1-\phi}} + g(\phi), \qquad (6.161)$$

where $g(\phi)$ has no singularities for $0 \leq \phi \leq 1$. Thus the leading order vorticity was given by

$$\omega_0 = \int_0^\phi \frac{\Gamma}{\sqrt{\pi(1-\phi')(\phi-\phi')}}\mathrm{d}\phi' + \int_0^\phi \frac{g(\phi')}{\sqrt{\pi(\phi-\phi')}}\mathrm{d}\phi'. \qquad (6.162)$$

The second term was regular on $0 \leq \phi \leq 1$ but the first term had a logarithmic singularity near $\phi = 1$,

$$\omega_0 \sim \frac{\Gamma}{\sqrt{\pi}} \log(1-\phi) + O(1), \quad \text{as } \phi \to 1. \qquad (6.163)$$

Now of course we have seen that the value of Γ varies from zero at angles greater than those for smooth separation and is positive. Squire observed that if the separation angle were to be greater than that for smooth separation ($\alpha_s \approx 55°$) then (6.163) indicated that the vorticity must tend to $-\infty$ at the assumed separation angle and so the vorticity would have passed through zero at a smaller angle. Hence separation would actually have occurred at that smaller angle. If the free streamline flow were then calculated for that smaller angle, separation would still have to have occurred at yet a smaller angle until eventually Γ went to zero. The only conclusion Squire could come to was that in the limit of vanishing viscosity, separation must occur where free streamline theory indicated smooth separation, $\Gamma = 0$. This is a very powerful conclusion and has played an important part in how the inviscid flow associated with $R \to \infty$ has been viewed. Squire was also very blunt about the use of measured pressure distributions in boundary layer calculations, '*any deductions of the position of break-away are meaningless, since that is fixed when the pressure distribution is known*'.

6.4.2 Imai (1953), Kawaguti (1953)

Whereas Squire had combined Burgers' approximate model with free streamline theory, Imai (1953) and Kawaguti (1953a) combined a boundary layer integral model with a free streamline model. Imai argued in much the same way as Squire, if the separation angle was taken to be greater than that for smooth separation then the singularity in the pressure gradient which is associated with the free streamline outer flow would always force the separation point further upstream until it reached the point of smooth separation. These ideas were quantified by Kawaguti for a circular cylinder who used the boundary layer integral result for the separation point (6.148) proposed by Thwaites,

$$\frac{u'_e(\alpha)}{u_e^6(\alpha)} \int_0^\alpha u_e^5 d\alpha \approx -0.018. \tag{6.164}$$

After setting an initial separation angle, he calculated the surface speed, $u_e(\alpha)$, from Brodetsky's method and then solved (6.164) to obtain a next iterate for the separation angle and repeated the process. He found that this procedure lead to a converged separation angle of $58°$, just above the value for smooth separation and possibly due to only having solutions for the free streamline case at a small number of angles. Nevertheless, this too seemed strong evidence that the only consistent model of the flow for $R \to \infty$ would be the free streamline, smooth separation case.

6.4.3 Vortex closure – Prandtl–Batchelor flows

In a brief but seminal paper, Batchelor (1956b) questioned whether the inviscid solution which came from irrotational free streamline theory could be the correct large Reynolds number limit of a viscous separated flow. This followed an earlier work, Batchelor (1956a) dealing with some aspects of the theory of closed streamline flow. Batchelor appreciated clearly the role that an inviscid solution might play, *'it may well be that a knowledge of the steady flow in the limit of infinite Reynolds number would allow the determination by some kind of asymptotic expansion, of the flow at the upper end of the range of Reynolds numbers at which the flow is stable, more readily than by an expansion valid in the neighbourhood of zero Reynolds number'*. None of the free streamline models which had been examined in detail provided a closed recirculation region, and that was a substantial problem since all observations of viscous separation had a closed recirculation region. There had been free streamline flows with a finite region of separation, for instance Riabouchinsky (1920) had proposed putting a second plate downstream of a plate so that the free streamline went from the edge of one plate to the edge of the downstream plate creating a 'cavity' between the plates but such models were highly contrived in the context of separation.

In addition to his introduction of a boundary layer approximation (that is neglect of longitudinal second derivatives in the Navier–Stokes equations), Prandtl (1904) had also observed that *also ergibt sich hier das einfache Resultat, daš innerhalbeines Gebietes von geschlossenen Stromlinien der Wirbel einen konstanten Wirt annimmt*, roughly translated 'gives the simple result that the vorticity inside a closed streamline takes a constant value'. In his paper Prandtl acknowledged that the essential mathematics behind this, that the vorticity might be treated as a function of the stream-function, $\omega = \omega(\psi)$, came from Helmholtz. Prandtl's 1904 paper is known mostly for the introduction of the boundary layer into fluid mechanics, it was a remarkable paper in that it also contained this important result for inviscid rotational flow. Consequently flows of the type envisaged by Batchelor as the more likely high Reynolds number limit than just irrotational flow are called Prandtl–Batchelor flows. Such flows contain finite regions of constant vorticity embedded in an overall potential flow. There is a growing literature concerning these flows and we will return to them at the end of the next chapter.

Southwell and Vaisey (1946) and Lighthill (1949) had obtained closed wakes from free streamline theory which had a cusped shape by varying the pressure within the separation region but these seemed less plausible as models for viscous separation. Batchelor approached the question very much from a physical plausibility view point, looking at the inviscid solution as the double limit, first $t \to \infty$ in the unsteady Navier–Stokes equations and then as $R \to \infty$. A cornerstone of his thinking was that because the steady solution was an infinite time solution, it would be very difficult to neglect the action of small viscous stresses over such a long time on the fluid within the separation region. Hence the velocity field within the separation region should play a part in the global description and was unlikely to be just zero. A second part of his thinking was more philosophical. Is the problem correctly specified by demanding uniform flow **upstream** and **downstream** or only **upstream**? Batchelor inclined to the former specification of the problem and so thought that the infinitely wide wake predicted by free streamline theory, while undoubtedly a possible limit for the Navier–Stokes equations, was nevertheless not the correct limit for viscous separated flow. Batchelor proposed that a physically plausible model would have two features, first that the wake would be closed and second that the action of viscosity over an infinite time would have to leave some vorticity in the flow and mathematically the only inviscid flow with vorticity and closed streamlines would have uniform **vorticity** within the separation region with potential flow outside separated regions.

The question of what is the correct inviscid model to match with large Reynolds number laminar flow remains partially unresolved although recent computations by Fornberg (1985) and McLachlan (1990, 1991b) give preliminary computational evidence that regions of constant vorticity do begin to emerge in solutions for separated flow at larger Reynolds numbers. We shall return to this question in discussion in Chapter 7.

6.5 Goldstein's hypothesis of a boundary layer singularity

As we have seen, attempts to calculate a boundary layer flow during a specified deceleration of the outer flow using Karman's integral method cannot be carried on too far in the streamwise direction. Numerical experiments by Howarth (1938) (repeated by Hartree), for a linearly decreasing outer velocity had also shown difficulty near the point where the wall shear vanished and in particular the numerical experiments indicated that approaching the separation point, the wall shear behaved like distance from the separation point to a power somewhere between 1/4 and 1. The significance of this was that if the shear had a power dependence with power index less than 1, then that would provide a singularity

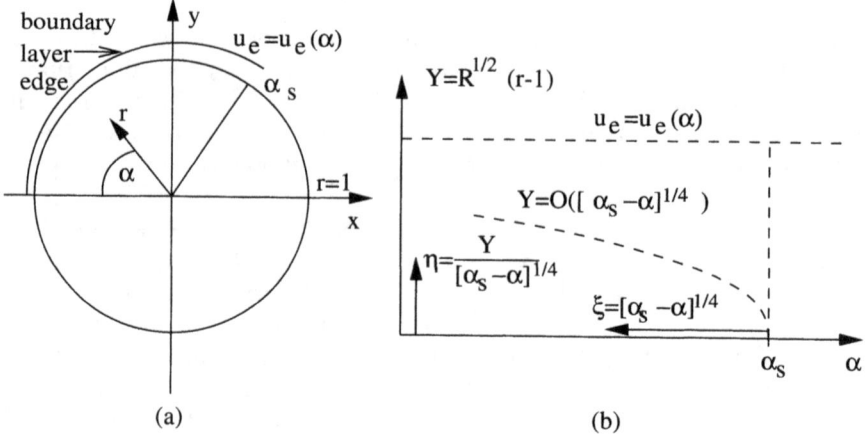

FIG. 6.30. Coordinate definitions for Goldstein's singularity at separation. (a) Rectangular coordinates for cylinder. (b) Boundary layer and sublayer coordinates near separation.

in the transverse velocity at the separation point. Goldstein (1948) used his previous ideas from 1930 to analyse how a boundary layer would develop in the approach to a point of zero shear and in particular the nature of the singularity in the boundary layer solution.

We shall consider Goldstein's singularity at separation specifically in the context of a cylinder, consequently some definitions are a little different from those in Goldstein (1948) but not in any essential way. If separation occurs at angle α_s then Goldstein used a local length scale $u_e(\alpha_s)/u'_e(\alpha_s)$ and local velocity scale $u_e(\alpha_s)$ and a Reynolds number based on these. We shall retain the definition of the Reynolds number based on cylinder radius and upstream velocity although this leads to slightly more intricate expressions for the functions which determine the boundary layer solution than those derived by Goldstein. The coordinate systems are shown in Figure 6.30, both in the physical plane of the cylinder, where the tangential velocity at the edge of the boundary layer, u_e is assumed specified and in the (α, Y) plane. Define new local coordinates near the separation point by

$$\xi = (\alpha_s - \alpha)^{1/4} \quad \text{and} \quad \eta = \frac{Y}{(\alpha_s - \alpha)^{1/4}} = \frac{Y}{\xi}. \tag{6.165}$$

The essence of Goldstein's work is that the expansion of the stream-function (or velocities) has to be developed in the main part of the oncoming boundary layer, (ξ, Y) variables, and in a sublayer, (ξ, η) variables and that the matching of these two regions leads to a singularity across the whole boundary layer at $\xi = 0$, or $\alpha = \alpha_s$. This singularity is called Goldstein's singularity at separation. There is no singularity at separation of a real fluid, the singularity arises only in the boundary layer formulation.

As with Goldstein's earlier work, suppose that near separation the outer velocity is given by
$$u_e(\alpha) = b_0(1 - c\alpha), \tag{6.166}$$
so that the outer velocity at separation is $U_1 = b_0(1 - \alpha_s)$, (assumed to have $U_1 > 0$) and the pressure gradient near separation is
$$\frac{\partial p}{\partial \alpha} = b_0^2 c(1 - c\alpha) = b_0 c U_1 + b_0^2 c^2 \xi^4. \tag{6.167}$$
Additionally, an expansion for the streamwise velocity at separation is assumed to be of the form
$$u(\alpha_s, Y) \sim b_0 c U_1 [a_2 Y^2 + a_3 Y^3 + a_4 Y^4 + \cdots], \tag{6.168}$$
where the coefficients a_r (with $a_2 \neq 0$) have to satisfy Goldstein's compatibility conditions, (6.128), so for instance $a_2 = 1/2$ and $a_3 = 0$.

6.5.1 *Sublayer expansion*

In the sublayer, let the stream-function be expanded,
$$\psi \sim b_0 c U_1 \xi^3 \{f_0(\eta) + \xi f_1(\eta) + \xi^2 f_2(\eta) + \cdots\}, \tag{6.169}$$
with velocities
$$u \sim b_0 c U_1 \xi^2 [f_0'(\eta) + \xi f_1'(\eta) + \xi^2 f_2'(\eta) + \cdots\}, \tag{6.170}$$
and
$$v \sim \frac{1}{4} b_0 c U_1 \xi^{-1} [3f_0 + \eta f_0' + \xi(4f_1 + \eta f_1') + \xi^2(5f_2 + \eta f_2') + \cdots]. \tag{6.171}$$
If these are substituted into the boundary layer equation, the functions f_r have to satisfy
$$f_0''' - \frac{3}{4} f_0 f_0'' + \frac{1}{2} f_0'^2 = 1, \tag{6.172}$$
and for $r > 0$
$$f_r''' - \frac{3}{4} f_0 f_r'' + \frac{1}{4}(r+4) f_0' f_r' - \frac{1}{4}(r+3) f_0'' f_r = G_r, \tag{6.173}$$
where Goldstein defined functions $G_r(\eta)$ which here are: $G_1 = 0$ and
$$G_r(\eta) = \frac{1}{4} \sum_{s=1}^{r-1} (r-s+3) f_s'' f_{r-s} - (r-s+2) f_s' f_{r-s}'] + p_r, \ r \geq 2. \tag{6.174}$$
The pressure coefficients p_r all vanish when the pressure gradient is given by (6.167) except $p_4 = b_0 c / U_1$.

The boundary conditions are no slip at the wall, $f_r = f'_r = 0$ at $\eta = 0$ and to match with the form of velocity, (6.168), $f'_r \sim a_{r+2}\eta^{r+2}$ as $\eta \to \infty$.

The sublayer solution allows the wall shear to be calculated, if we denote $f''_r(0) = \beta_r$,

$$\left.\frac{\partial u}{\partial Y}\right|_{Y=0} \sim b_0 c U_1 [\beta_0 \xi + \beta_1 \xi^2 + \beta_2 \xi^3 + \cdots]. \tag{6.175}$$

Goldstein was able to determine the first five functions, although the complexity of the solutions increases rapidly after the first three functions. The first function is

$$f_0(\eta) = \frac{1}{6}\eta^3, \tag{6.176}$$

giving $\beta_0 = 0$, while the second is

$$f_1(\eta) = \frac{1}{2}\beta_1 \eta^2, \tag{6.177}$$

satisfying the compatibility condition $a_3 = 0$. The third function is

$$f_2(\eta) = \frac{1}{2}\beta_2 \eta^2 - \frac{1}{240}\beta_1^2 \eta^5, \tag{6.178}$$

so that $a_4 = -\beta_1^2/48$. In solving for the fourth function, f_3 the next constant β_3 is introduced and β_2 related to β_1, a pattern which Goldstein believed could be continued to all further functions. Since Goldstein was trying to demonstrate the existence of a singularity by constructing an appropriate asymptotic expansion, everything hinged around success in showing the existence of the unknown constants β_r with functions f_r which had a physically plausible solution. Goldstein conjectured that the expansion would, in the end, have a single unknown constant β_1 and that this constant would be determined by the condition on u as $Y \to \infty$ and $\xi \to 0$.

In order to understand the difficulty Goldstein faced, it is necessary to know more about general solutions of (6.173). There are three independent solutions to the homogeneous equation: η^2 and two can be denoted $j_r(\eta)$ and $k_r(\eta)$ where the later two functions satisfy the boundary conditions $j_r(0) = 0$, $j'_r(0) = 1$, $k_r(0) = 1$ and $k'_r(0) = 0$. By themselves neither j_r or k_r satisfies the condition of a double zero at $\eta = 0$. Goldstein was able to determine both j_r and k_r explicitly as a power series expansion and rewrite this as an integral of a hypergeometric function. Both j_r and k_r could have an exponentially large behaviour for $\eta \to \infty$, **except** for certain values of r. If m were an integer and $r = 4m + 1$ then k_r was exactly a finite power series, if $r = 4m + 2$ then j_r was a finite power series. For $r = 4m$ or $r = 4m + 3$ the expansions containing exponentially large terms were the same excepting a multiplicative factor and it was possible to choose a suitable linear combination of the two functions so that the exponentially large parts cancelled and the solution could then satisfy the condition that $f'_r \sim \eta^{r+2}$

as $\eta \to \infty$. If this had not been possible then an exponentially growing solution would not be physically plausible and the expansion (and hence the existence of a singularity at separation of the form Goldstein was proposing) would have failed. Thus everything hinged around the cases $r = 4m + 1$ and $r = 4m + 2$ which first occurred when $r = 5$ and $r = 6$.

The equation for f_3 was

$$f_3''' - \frac{1}{8}\eta^3 f_3'' + \frac{7}{8}\eta^2 f_3' - \frac{3}{2}\eta f_3 = -\frac{5}{8}\beta_1\beta_2\eta^2 - \frac{1}{3}\beta^1\eta^5, \qquad (6.179)$$

and the solution satisfying the boundary conditions at $\eta = 0$ is

$$f_3 = \frac{1}{2}\beta_3\eta^2 + \beta_1\beta_2(\eta - j_3(\eta)) - \frac{1}{3}\beta_1^3(1 + \frac{1}{16}\eta^4 - k_3(\eta)). \qquad (6.180)$$

Goldstein showed that the exponentially large part of the solution would vanish identically if

$$\beta_2 = \frac{\pi^{3/2}}{2^{5/4}5(\frac{1}{4}!)^3}\beta_1^2. \qquad (6.181)$$

The equation for f_4 was similar and its solution, with j_4 and k_4 would have the required algebraic behaviour for large η provided

$$\beta_3 = \frac{\pi^3}{400(\frac{1}{4}!)^6}(\frac{35}{8} - \sqrt{2})\beta_1^3. \qquad (6.182)$$

Thus by explicitly constructing the solutions Goldstein had shown that his proposal of a singularity was plausible for the first five terms (and for further terms where $r = 4m$ or $r = 4m + 3$).

However, when he turned to the case $r = 4m + 1$ (starting with $r = 5$) then the function k_5 did not have exponential growth so the trick of balancing the two functions to remove the unwanted behaviour could not be used. However it is always the case that the coefficient β_{r-1} only multiplied a factor of the form $\eta - j_r(\eta)$ since that is the only term which gives the correct η^2 behaviour in G_r. hence the solution would always be a combination of $\beta_1\beta_{r-1}$ times j_r and other multiples of j_r which depend only on known constants, β_s, $s < r - 1$ and so β_{r-1} can be chosen to eliminate the exponential growth in the solution as $\eta \to \infty$. Thus, in principle, the cases $r = 4m + 1$ had a plausible solution.

The cases $r = 4m + 2$ were more problematical. This first occurred in the case $r = 6$ when the function G_6 could be written

$$G_6 = -\beta_1\beta_5\eta^2 + p_1 l(\eta) + \bar{G}_6(\eta), \qquad (6.183)$$

where $l(\eta)$ was a known polynomial and $\bar{G}_6$ was potentially going to cause a problem. Goldstein was able to show that any exponentially large terms in the $\eta \to \infty$ expansion of f_6 would vanish identically with the integral

$$\int_0^\infty (\eta^2 - \frac{1}{20}\eta^6 + \frac{1}{2880}\eta^{10})\bar{G}_6(\eta)\exp(-\frac{\eta^4}{32})\mathrm{d}\eta. \tag{6.184}$$

Goldstein could not prove that this integral vanished, and although he did have some numerical evidence (see discussion of Jones (1948) below) that it might be small, his conjecture of a singularity at separation in the boundary layer formulation rested uneasily on the requirement that the integral (6.184) vanished.

6.5.2 Outer part of the boundary layer

Using the large η expansions of the functions f_r as a guide, Goldstein proposed that in the boundary layer outside of the sublayer, the stream-function should have an expansion

$$\psi \sim b_0 c U_1 [g_0(Y) + \xi^2 g_2(Y) + \xi^3 g_3(Y) + \xi^4 (g_4(Y) + h_4(Y)\log \xi) + \cdots]. \tag{6.185}$$

As separation is approached, $\xi \to 0$ and the first term in the expansion must satisfy (6.168),

$$g_0'(Y) \sim \frac{1}{2}Y^2 - \frac{1}{48}\beta_1^2 Y^4 + \cdots, \tag{6.186}$$

and substituting the expansion into the boundary layer equation so as to determine the subsequent functions, most importantly that the next term did not vanish,

$$g_2(Y) = \frac{1}{2}\beta_1 g_0'(Y). \tag{6.187}$$

This brings matters to the core consequence of Goldstein's singularity. If $\beta_1 \neq 0$, then the transverse velocity across the boundary layer (in the case of a cylinder, the radial velocity u_r) is given by

$$u_r = -R^{-1/2}\frac{\partial \psi}{\partial \alpha} \sim \frac{\beta_1}{4}R^{-1/2}b_0 c U_1 g_0'(Y)(\alpha_s - \alpha)^{-1/2}, \text{ as } \alpha \to \alpha_s^-, \tag{6.188}$$

and this is singular across the whole boundary layer at separation. It is also characteristic of Goldstein's solution that the wall shear, given by (6.175) will have slope with respect to angle α (or distance in a more general circumstances) which becomes infinite at separation, the shear behaving like $(\alpha_s - \alpha)^{1/2}$. We have only looked at a linearly decreasing outer velocity, if the pressure gradient had further terms of order ξ^{4m} where m is some integer, the start of the expansion and the final conclusion of a singularity would be unaltered. Goldstein seemed to see the outer part of the boundary layer as only an opportunity to verify the algebraic detail of the sublayer functions and he seemed not to have considered the perturbation to the outer potential flow which a singularity in the transverse velocity would cause.

This seemed an impenetrable singularity; Goldstein showed that there was no real solution downstream of the singularity (effectively, the equivalent constant to β_1 would be imaginary) and although the existence of a singularity was only postulated by Goldstein, there was some numerical evidence to support Goldstein's conclusion.

Jones (1948) attempted to integrate Goldstein's equations for f_r, (6.173), directly using a Taylor series expansion. At each η value the differential equation, and higher derivatives of the equation were used to estimate the coefficients in a Taylor series expansion about η so as to determine the functions and their derivatives at a further η point. Jones felt that this was better suited to the laborious tabular methods then in use rather than finite differences. Jones integrated both from $\eta = 0$ and using asymptotic expansions, from large η back towards smaller values. He successfully calculated the functions $f_0 - f_5$ but found considerable difficulty in evaluating the integral (6.184). In the end he found the integral to be small and non-zero but with a possible error in his calculations of the same size as the estimated value, so he cautiously took the integral to be zero and incorrectly confirmed Goldstein's hypothesis.

Our plan is to look first at how Stewartson put Goldstein's conjecture on a secure theoretical footing and some discussion of that work, before coming back to see how the study of this singularity was gradually taken up by direct numerical integration of the boundary layer equations as computing power increased and became widely available.

6.5.3 Stewartson (1958)

Goldstein's formulation of the flow near separation relied on developing a power series expansion in ξ and on being able to relate the unknown constants β_r to each other. It was not clear that this was possible and as we have seen, everything revolved around the integral (6.184) vanishing. Stewartson (1958) showed that the later problem was related to the assumption of a power series expansion in ξ and that if the expansion involved not only powers of ξ but also logarithms of ξ then there would be no difficulty in choosing constants so that a slightly modified version of Goldstein's integral condition (6.184), vanished and hence that the expansion which gave a singularity would be properly established.

Stewartson relied on his earlier work, Stewartson (1957), to propose that the form of the expansion of the stream-function should be

$$\psi \sim b_0 c U_1 \xi^3 \{ f_0(\eta) + \xi f_1(\eta) + \xi^2 f_2(\eta) + \xi^3 f_3(\eta) + \xi^4 f_4(\eta) + \\ \xi^5 [f_5(\eta) + h_5(\eta) \log \xi] + \xi^6 [f_6(\eta) + h_6(\eta) \log \xi] + O(\xi^7 \log \xi) \}. \quad (6.189)$$

The functions $f_0 - f_5$ were as determined by Goldstein. Stewartson showed that

$$h_5(\eta) = \gamma_5 \eta^2, \quad (6.190)$$

and

$$h_6(\eta) = \gamma_6 \eta^2 - \frac{1}{15\sqrt{2}} \beta_1 \gamma_5 (\eta^2 - \frac{1}{336} \eta^9), \qquad (6.191)$$

where γ_5 and γ_6 were new constants. The integral condition for the absence of exponentially large terms in f_6 as $\eta \to \infty$ was then

$$\int_0^\infty \left(\bar{G}_6 - \frac{\beta_1 \gamma_5}{2\sqrt{2}} [\eta^2 - \frac{\eta^6}{20} + \frac{\eta^{10}}{1440}] \right) \left(\eta^2 - \frac{\eta^6}{20} + \frac{\eta^{10}}{2880} \right) \exp(-\frac{\eta^4}{32}) d\eta = 0. \qquad (6.192)$$

It was then straightforward to choose γ_5 so the integral vanished. However, it did mean that β_5 cold not be determined. The constant γ_6 would then be determined along with f_7. The continuation of this process resulted in $\beta_1, \beta_5, \beta_9, \cdots$ remaining undetermined from consideration of the sublayer alone, so although the expansion (6.189) put the singularity on a proper footing it also brought an infinite number of unknowns to be determined by matching, presumably with the upstream oncoming boundary layer. Additionally, Stewartson was concerned that the expansion (6.189) might imply a potentially unphysical logarithmic singularity in u as $\alpha \to \alpha_s$ for $Y > 0$.

This last point was clarified in Brown and Stewartson (1969) where it was noted that cancellation between logarithmic terms in different stages of the expansion would mean that there would not be a logarithmic singularity in u as $\xi \to 0$. They also considered the general question of the occurence of a singularity, using the observation in Curle (1962) that the wall shear satisfied

$$\frac{\partial}{\partial \alpha} \left(\frac{1}{2} \left(\frac{\partial u}{\partial Y} \Big|_{Y=0} \right)^2 \right) = \frac{\partial^4 u}{\partial Y^4} \Big|_{Y=0}, \qquad (6.193)$$

and thus if u_{YYYY} were non-zero at separation, then the wall shear would have to behave like $(\alpha_s - \alpha)^{1/2}$ whereas were $u_{YYYY} = 0$ at separation then there need be no singularity. There was a sense in which this was just a restatement of Goldstein's singularity, since in Goldstein's expansion $u_{YYYY} = -\beta_1^2/2$ at separation and $u_Y = \beta_1(\alpha_s - \alpha)^{1/2}$ which satisfies (6.193).

The correctness of Stewartson's expansion was confirmed by Terrill (1960) who used a Runge–Kutta integration to determine the functions f_r and in doing so, demonstrated that Jones' hope that the errors in his integration would cancel out the value of the integral (6.184) was false, rather the errors added and the correct value was close to twice Jones estimate and non-zero. Terrill was able to calculate a value for γ_5 and thus verified the practicality of computing all the functions in Stewartson's modified form of Goldstein's expansion.

Further reading about the expansion for Goldstein's singularity and the nature of the solution near separation can be found in Part 2, Chapter 3 of Kaplun (1967).

6.6 Direct numerical solution of boundary layer equations

While Goldstein and Stewartson had approached the question of flow near the separation point by looking at the form of the *upstream* expansion it should be clear that they left the practical question of how to compute the flow approaching separation on one side. The boundary layer equations, having parabolic character, ought to be able to be computed by direct numerical integration using finite differences. Of course, there are some technical difficulties in carrying this out. Explicit schemes which march along the surface relating values at $\alpha + \Delta\alpha$ to those at α may be unstable, implicit schemes which try to calculate all the values at $\alpha + \Delta\alpha$ simultaneously result in a set of coupled non-linear equations. Alternately the equations can be discretised in α and the resulting non-linear ordinary differential equations in Y integrated using Runge–Kutta and indeed existing attempts to compute boundary layer flows outside of Karman's integral formulation relies on a power series expansion in the streamwise coordinate to provide ordinary differential equations across the boundary layer using either Y coordinate or a similarity coordinate. What should be clear is that each of these methods require considerable computing power and direct finite difference integration the most onerous requirement.

6.6.1 Thom (1928)

Thom (1928a) discretised the equations on a rectangular grid and proposed what nowadays is called a cell vertex method. The velocities were kept at the vertices of a cell of size $\Delta\alpha, \Delta Y$ and the non-conservative form of the boundary layer discretised using a cell velocity which was the average of the values at the cell vertices. In this way a given velocity at each of the four cell vertices could be used to give a better estimate of the velocity at the $(\alpha + \Delta\alpha, Y + \Delta Y)$ vertex, a form of Picard iteration. One of the technical difficulties in solving the boundary layer equations is dealing with the pressure, see Chapter 4, Thom unwittingly avoided this by using a prescribed pressure distribution, $c_p = 1 - 3.5\sin^2\alpha$, Thom then iterated through the mesh a number of times to obtain a solution. He used step sizes of $\Delta\alpha = 2°$ and $\Delta Y = 0.24$ (with ten cells across the boundary layer) and used the calculation for angles up to $\alpha = 20°$. At larger angles he used as an approximate form of solution the external speed times a function of Y and did not continue his finite difference calculation to separation but this was nevertheless a remarkable attempt to compute boundary layer solutions directly.

6.6.2 Leigh (1955)

While Thom had lacked computing power to tackle direct numerical solution properly, Leigh (1955) was able to use the Cambridge Mathematical Laboratory EDSAC machine to calculate one of the first direct finite difference solutions. He chose a linearly decreasing outer velocity profile as introduced by Howarth,

$$u_e(\alpha) = 1 - \frac{1}{8}\alpha, \qquad (6.194)$$

and adopted a finite difference method proposed by Hartree and Womersley (1937). The method first discretised the equations in the α direction to obtain a set of ordinary differential equations and then discretised those in the transverse Y direction using central differences.

Letting $U_i(Y)$ be the approximation for $u_i(Y) = u(i\Delta\alpha, Y)$ and P'_i the pressure gradient at $\alpha_i = i\Delta\alpha$, then the Hartree–Womersley discretisation of the boundary layer equations was a central difference scheme about $(i + \frac{1}{2})\Delta\alpha$,

$$(U_i + U_{i+1})\frac{U_{i+1} - U_i}{\Delta\alpha} - (U'_{i+1} + U'_i)\int_0^Y \frac{U_{i+1} - U_i}{\Delta\alpha}dY$$
$$= -P'_i - P'_{i+1} + U''_i + U''_{i+1}. \qquad (6.195)$$

Rather than try to determine $U_{i+1}(Y)$ directly from this equation, Leigh supposed $W(Y)$ were an approximation for $2U_{i+\frac{1}{2}} = U_i + U_{i+1}$ for which the discrete boundary layer equation, (6.195) could be rewritten,

$$W'' + \frac{W'}{\Delta\alpha}\int_0^Y (W(s) - 2U_i(s))ds - \frac{W - 2U_i}{\Delta\alpha}W = P'_i + P'_{i+1}. \qquad (6.196)$$

Leigh adopted a semi-implicit scheme for the kth iterate $W^{(k)}$,

$$\Delta\alpha\frac{d^2W^{(k)}}{dY^2} - W^{(k-1)}W^{(k)} + \frac{dW^{(k-1)}}{dY}\int_0^Y W^{(k)}dY = P'_i + P'_{i+1}$$
$$- 2U_iW^{(k-1)} - \frac{dW^{(k-1)}}{dY}\int_0^Y 2U_i dY, \qquad (6.197)$$

and then discretised this in Y using central differences and trapezoidal quadrature. Together with the boundary conditions, $u = 0$ at $Y = 0$ and $u \to u_e$ as $Y \to \infty$ (this condition applied at the outer edge of the computation domain), the resulting matrix equation for the discrete values $W_j^{(k)} = W^{(k)}(j\Delta Y)$ was solved using a Choleski factorisation and Leigh found that the solution converged in relatively few k-iterations. Hartree had previously tried to integrate (6.197) from $Y = 0$ from a given starting shear $W'(0)$ but had found great difficulty in isolating the correct initial conditions which correspond to the velocity at the boundary layer edge. Leigh's method had no such difficulty since the outer boundary condition was included in his matrix formulation. Leigh computed solutions for two step sizes, ΔY and $2\Delta Y$, and used Richardson extrapolation to obtain a final solution.

As an example of Leigh's method, consider the solution for the outer velocity, (6.194), when the initial velocity profile is

$$u(0, Y) = \begin{cases} \sin Y, & Y < \pi/2, \\ 1, & Y \geq \pi/2. \end{cases} \qquad (6.198)$$

In that case the solution comes to a singularity near $\alpha = 0.873$ and Leigh's method has been applied with 400 steps in the α direction and 80 steps of size $\Delta Y = 0.1$ across the boundary layer. The results are shown in Figure 6.31. If the u velocity is plotted there is no evidence of singular behaviour. The wall shear also

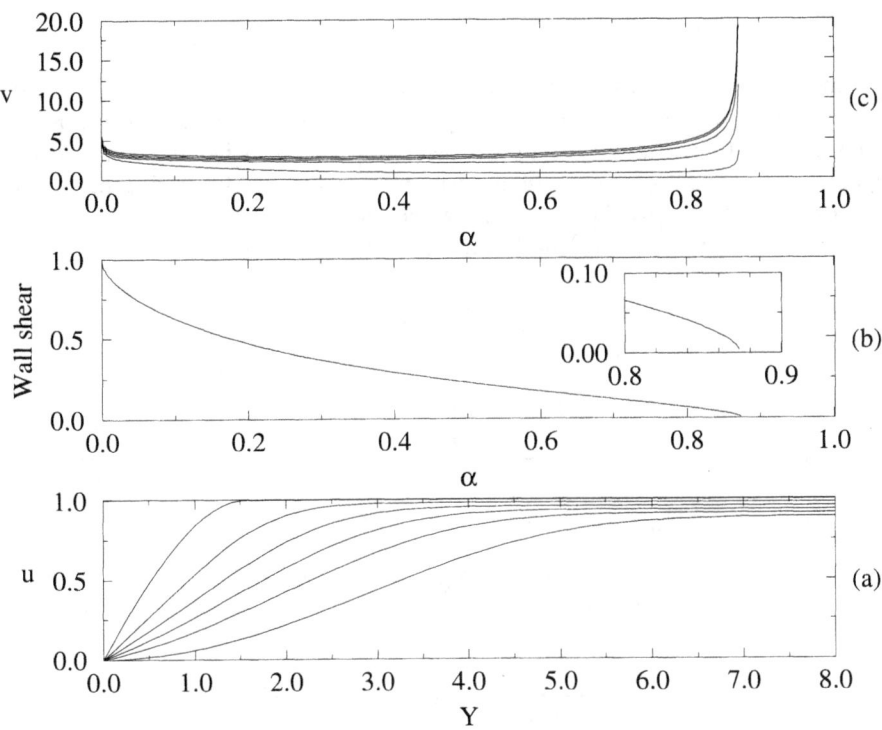

FIG. 6.31. Computed solution for Leigh's problem with initial velocity profile (6.198). (a) u velocity profiles at equal α steps from $\alpha = 0$ to separation at $\alpha = 0.873$, (b) Wall shear showing detail of shear near separation point, (c) v velocity versus α at equal Y steps between $Y = 0$ and $Y = 8$, showing catastrophic nature of singularity at separation

does not look particularly singular at separation although close inspection near that point (see inset diagram) shows the rapid curving downwards indicating a fractional power dependence on distance from the separation point. If the v velocity is examined, it is immediately apparent that disaster strikes at the separation point (as was predicted by Goldstein, see (6.188)).

Leigh's main purpose in carrying out the calculations was to check that a singularity of the form $(\alpha_s - \alpha)^q$ existed for the wall shear by estimating both α_s and the power index q. As might be expected from the wall shear curve in Figure 6.31, inset to (b), this was a difficult problem. Leigh started from a solution Hartree had calculated up to $\alpha = 0.8$ and found separation at $\alpha = 0.9585$ (this is different to the case calculated for Figure 6.31 because of different starting profiles) but the best he was able to say about q was that it was close to Goldstein's predicted value of $1/2$.

6.6.3 Terrill (1960)

In addition to calculation of the functions in Goldstein's expansion, Terrill (1960) developed a direct numerical solution using a formulation in terms of the potential and a similarity variable. Terrill interest was in flows with suction and he developed a numerical solution to deal with this feature; in discussing his work we shall restrict matters to the case with no suction.

The potential at the edge of the boundary layer was defined by

$$\phi = \int_0^\alpha u_e(\alpha) d\alpha, \qquad (6.199)$$

and a similarity variable by

$$\eta = \frac{u_e(\alpha) Y}{\sqrt{2\phi}}. \qquad (6.200)$$

Then the stream-function

$$\psi = \sqrt{2\phi} F(\phi, \eta), \qquad (6.201)$$

gave

$$u = u_e F_\eta, \quad v = -\frac{u_e}{\sqrt{2\phi}}[F + 2\phi F_\phi + (b(\phi) - 1)\eta F_\eta], \qquad (6.202)$$

where the function $b(\phi) = 2\phi u_e'/u_e^2$. When these were substituted into the boundary layer equation, Terrill obtained,

$$F_{\eta\eta\eta} + F F_{\eta\eta} + b(\phi)[1 - F_\eta^2] = 2\phi[F_\eta F_{\eta\phi} - F_\phi F_{\eta\eta}]. \qquad (6.203)$$

This equation has boundary conditions $F_\eta(\phi, 0) = F(\phi, 0) + 2\phi F_\phi(\phi, 0) = 0$ and $F_\eta \to 1$ as $\eta \to \infty$. There is also a sensible initial condition; assuming $b(0) = 0$, then at $\phi = 0$, $F_{\eta\eta\eta} + F F_{\eta\eta} = 0$, that is F has a Blasius profile at $\phi = 0$. This improved upon Leigh's formulation which could not deal with the boundary layer starting from a leading edge or from a stagnation point unless there was some other method to provide a velocity profile to start the calculation. The computational problem formulated by Terrill is illustrated in Figure 6.32, for the simplified case where there is no suction and $b(0) = 0$ where the outer boundary condition is applied at some suitably large value of η.

Terrill solved the non-linear equation (6.203) using the same method as Leigh. First writing $v = F_\eta$, then (6.203) can be rewritten

$$v_{\eta\eta} + v_\eta \int_0^\eta [v + 2\phi v_\phi] d\eta - 2\phi v v_\phi = -b(\phi)(1 - v^2). \qquad (6.204)$$

The potential was discretised by steps $\Delta\phi$ with $\phi_i = i\Delta\phi$, and Terrill applied Hartree & Womersley's discretisation in ϕ using W as for Leigh, $W = 2v_{i+\frac{1}{2}} = v_i + v_{i+1}$ and k to represent iteration of a semi-explicit method, to obtain

Direct numerical solution of boundary layer equations

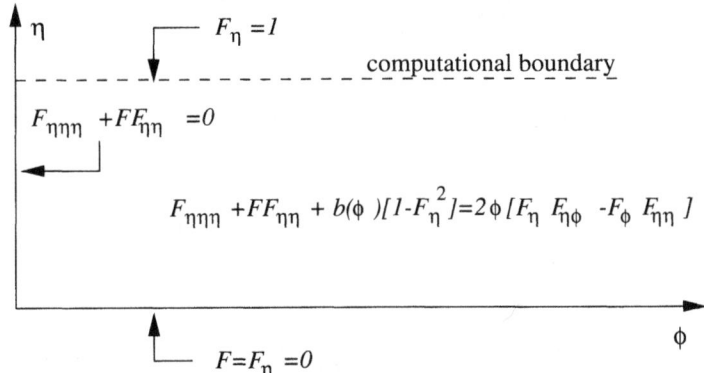

FIG. 6.32. Simplified version of Terrill's boundary layer problem.

$$\Delta\phi W_{\eta\eta}^{(k)} + W_\eta^{(k-1)} \int_0^\eta (\frac{1}{2} + 2\phi_{i+\frac{1}{2}})W^{(k)})\mathrm{d}\eta - 2\phi_{i+\frac{1}{2}}W^{(k-1)}W^{(k)} =$$
$$4\phi_{i+\frac{1}{2}}W_\eta^{(k-1)} \int_0^\eta v_i \mathrm{d}\eta - 4\phi_{i+\frac{1}{2}}W^{(k-1)}v_i$$
$$-\Delta\phi[b_i(1-v_i^2) + b_{i+1}(1-(W^{(k-1)}-v_i)^2)]$$
(6.205)

This equation was discretised in η using central differences and trapezoidal quadrature to obtain a matrix equation for $W^{(k)}$ as with Leigh's method and that matrix equation solved using Choleski factorisation. If the coefficient matrix is denoted A then Terrill found that the order of factorisation $A = LU$ resulted in a stable scheme whereas the order $A = UL$ (U and L being upper and lower triangular matrices respectively) was much less stable. The factorisation $A = LU$ is particularly simple as the upper triangular matrix U is banded to the diagonal and off diagonal elements. However, even using Terrill's factorisation, the iteration scheme is prone to instability unless the transverse step, $\Delta\eta$ is small. This seemed to be related to the maximum value of η rather than just the step size. There was also an α direction instability which is evident for small α in some of the results we show. In order to be able to compute to reasonable outer values of η we have modified Terrill's procedure to use a fairly severe relaxation (relaxation parameter 0.05–0.1) and this seemed to overcome problems with the location of the outer boundary. Terrill also started his solution from small positive values of α when the outer flow had a stagnation point at the origin $\alpha = 0$. This seems unnecessary if the value $b(0)$ is well defined since the problem on the initial axis $\alpha = 0$ for $v_0(\eta) = v(0,\eta) = F_\eta(0,\eta)$ is

$$v_{0\eta\eta} + v_{0\eta} \int_0^\eta v_0(s)\mathrm{d}s + b(0)[1-v_0^2] = 0, \qquad (6.206)$$

with boundary conditions $v_0(0) = 0$, $v_0(\eta) \to 1$ as $\eta \to \infty$. This is of course the Falkner–Skan equation and we have already discussed its solution by Runge–

182   *Separated Flow about a Cylinder*

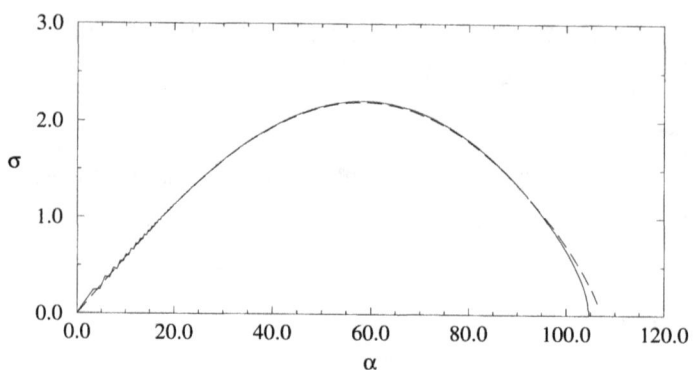

FIG. 6.33. Comparison of wall shear for potential outer flow calculated by (a) (——) Terrill's method, (b) (- - -) Pohlhausen's method.

Kutta integration. This non-linear equation can also be solved using a semi-implicit method with minor modification to the routine which implements Terrill's procedure. Even though the physical velocity $u_e(0)v_0$ is zero the values of v_0 can be used to start Terrill's marching procedure in the α direction.

Terrill described the result of applying his numerical technique to potential flow around a cylinder, both with and without suction at the wall, although we have only considered the situation with no suction here. We shall apply Terrill's formulation to two cases, potential flow around a cylinder and flow around a cylinder using Heimenz's observed pressure distribution to determine the outer velocity. Both the cases below were calculated with $\Delta \eta = 0.07$, 80 steps across the boundary layer and 400 steps between $\alpha = 0$ and $\alpha = \alpha_s$.

6.6.3.1   Terrill's solution for potential outer flow   If the outer flow is potential flow around a cylinder then

$$u_e(\alpha) = 2\sin\alpha, \qquad (6.207)$$

and so

$$\phi = 2(1 - \cos\alpha), \quad \alpha = \cos^{-1}(1 - \phi/2), \qquad (6.208)$$

and

$$b(\alpha) = \frac{\cos\alpha}{\cos^2\tfrac{1}{2}\alpha}. \qquad (6.209)$$

Terrill computed solutions for this case with separation being near $\alpha_s = 104°$. This wall shear for this solution is shown in Figure 6.33 and compared with the result of applying Pohlhausen's integral method. The two results are remarkably close, with the integral method predicting separation at a slightly higher angle.

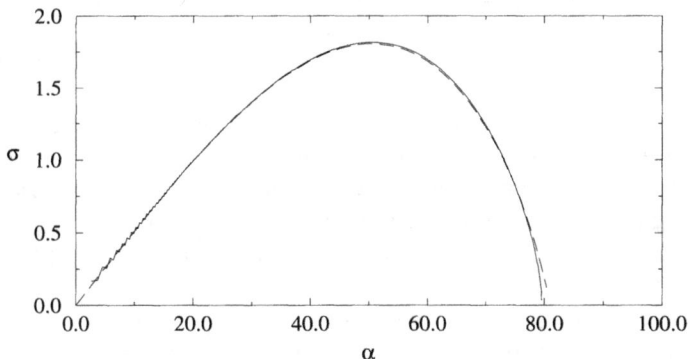

FIG. 6.34. Comparison of wall shear for Heimenz's outer flow calculated by (a) (——) Terrill's method, (b) (- - -) Pohlhausen's method.

6.6.3.2 *Terrill's solution for Heimenz's observed outer flow* If we use the outer velocity Heimenz deduced from his observed pressure, (6.116), then it is straightforward to integrate the polynomial distribution to determine $\phi(\alpha)$; the inversion to obtain $\alpha(\phi)$ can easily be carried out numerically. Since for small α, u_e is linear in α and ϕ is quadratic, the function $b(\alpha)$ is regular with value $b(0) = 1$ and the integration can be started from $\alpha = 0$. If this is done the wall shear is as shown in Figure 6.34 and separation is predicted at an angle just below $\alpha_s = 80°$. The results of using Pohlhausen's method are also shown, again there is remarkable agreement between the approximate integral method and direct numerical solution.

6.7 Reprise

There has been a great deal of technical detail in the last few sections; it is useful here to summarise important points and issues which come from the those sections before moving on to see how some of them have been resolved.

In describing separation there are two fundamental problems. One is a local problem: at which point on a surface will separation occur? In steady two-dimensional flow that is the same as where does the wall shear or vorticity vanish. The second problem is much more difficult: what is the global structure of a steady separated flow? In this case not only is the point of separation of interest but also what happens to fluid beyond separation: how is the wake continued and what is the flow field far behind the body?

If inviscid free streamline theory is used then a number of difficulties arise. When the pressure on the free streamline is the same as the far upstream pressure then the separation region must grow without bound, the separation region is not closed and the wake cannot be specified across its entirety as having the same velocity as the oncoming flow. Further, the separation point can be specified arbitrarily between two limits, one where smooth separation occurs and that where separation lines would cross each other. Within that range, for all points beyond smooth separation, there will be a singularity in the pressure gradient at separation. If the separation is smooth, then there is no singularity but the pressure gradient is everywhere favourable, hence no apparent reason for the flow to separate. For flows at finite Reynolds number the pressure within the separated region is always observed to be lower than the upstream pressure. If a decreased base pressure is applied in an ad-hoc manner the spreading of the streamlines in the wake can be countered, as even could the singularity in the pressure gradient at separation but there appears no rational way of doing this.

If boundary layer theory is applied together with a specified outer flow then the local separation point can be computed correctly if the outer flow comes from observation but not, apparently otherwise. In all cases where boundary layer theory is used, as separation is approached the boundary layer not only thickens but thickens catastrophically, resulting in a singularity at separation right across the boundary layer and boundary layer calculations cannot be continued through the separation point.

Note that although both free streamline theory and boundary layer theory lead to singularities at separation, the type of singularity is very different in each case and the distinction should be kept clear between either a boundary layer singularity (Goldstein's) or a free streamline singularity at separation.

If boundary layer theory is combined with free streamline outer flow, any assumed separation point beyond that for smooth separation leads to a prediction of a separation point closer to that for smooth separation. This gives a conclusion that the only rational infinite Reynolds number solution must be that of smooth separation with its associated wake difficulties and lack of any adverse pressure gradient.

The calculations shown reinforce the assertion earlier that given the outer flow from observation, then numerous methods verify the location of the separation point but if not given the correct outer flow, none of the methods we have examined are close to predicting the separation point.

6.8 Numerical solution of Navier–Stokes equations

Since the Navier–Stokes equations are non-linear, numerical solutions for the full equations fall into a small set of techniques. The first choice is whether to solve in primitive variables (velocity, pressure) or if the calculation is in two-dimensions, stream-function – vorticity. Attempts to extend the idea of stream-function-vorticity to three dimensions have mostly not been widespread. The velocity can be represented as the sum of the curl operator acting on a vector of 'stream-functions' and the gradient of a potential but the overall complexity of the resulting problem makes the primitive variable formulation equally attractive. Primitive variable methods solve directly for required unknowns but in the process a difficult pressure equation with mostly derivative boundary conditions has to be solved. Additionally satisfying continuity can be a problem. However generalisation of primitive variable implementations from two to three dimensions is clear and that may be a great advantage. The stream-function – vorticity formulation in two-dimensions automatically satisfies continuity and is straight forward to program although the pressure has to be calculated from a separate integration.

The next choice is usually whether to solve the steady equations or to solve an unsteady formulation and look for a time evolved steady solution. If considering the steady equations then an iterative method has to be used and these may be explicit Picard iteration or some form of semi-implicit iteration, for instance reducing the non-linear terms to a product of a known iteration and an unknown iteration and solving a matrix equation for the unknown iteration. Alternately a Newton iteration can be applied from a linearisation about a known iteration. Some methods combine a few Picard type iterations with subsequent Newton type iterations hoping to use the rapid convergence of Newton iterations without having a problem of being too far from the solution for Newton iteration to converge quickly. If the unsteady equations are used then reliability of time convergence is a serious problem as is stability of the iteration. It will also be the case that if the flow has a time periodic solution and an unstable steady solution then time marching schemes will calculate the former and not the latter.

Finally one has to decide on a detailed numerical algorithm: early solutions were all based on some form of finite difference formulation, nowadays finite element formulations are highly successful and through the development of *a posteriori* error analysis, hold out the prospect of automated satisfaction of error criteria and automated mesh adaptation and local refinement.

Our plan in this section is to consider the history of numerical solutions to the full Navier–Stokes equations for flow around a cylinder up to the early 1970s and to try to see what guidance was available about what limit the solutions for laminar viscous flow might tend to as the Reynolds number becomes infinite. We are particularly interested in results which might resolve either the counter intuitive aspects of free streamline theory or the failure of the boundary later solutions near separation. In the next chapter we will look at more recent computations of the full Navier–Stokes equations.

6.8.1 Thom (1928)

The earliest numerical solution of the Navier–Stokes equations for flow around a cylinder is due to Thom (1928b, 1933) who implemented an iteration method for the steady stream-function vorticity form of the equations. The hand calculation was at first applied on a rectangular grid but because of difficulty in using the method with curved boundaries, then applied to transformed equations where the transformation gave a rectangular computational area. The iteration process was in two steps, one which used cell vertex values to calculate a cell centred update, and then a second step which used cell centred values to carry out a cell vertex update. The calculation was carried out for a Reynolds number of $R_d = 10$

and showed a recirculation region behind the cylinder. Having found the stream-function and vorticity, the pressure was obtained by a further integration.

Letting $z = x + iy$ and $t = \xi + i\eta$, Thom used two transformations, $t \sim \log z$ and $t \sim z + z^{-1}$. We shall consider his method only for transformations of the latter type, which with $h(\xi,\eta) = |dt/dz|$ transforms the Navier–Stokes equations for the stream-function ψ and vorticity ζ to

$$\frac{\partial \psi}{\partial \xi}\frac{\partial \zeta}{\partial \eta} - \frac{\partial \psi}{\partial \eta}\frac{\partial \zeta}{\partial \xi} = \frac{1}{R}\left(\frac{\partial^2 \zeta}{\partial \xi^2} + \frac{\partial^2 \zeta}{\partial \eta^2}\right), \tag{6.210}$$

and

$$\zeta = -h^2\left(\frac{\partial^2 \psi}{\partial \xi^2} + \frac{\partial^2 \psi}{\partial \eta^2}\right). \tag{6.211}$$

In order to transform a unit cylinder to the t-plane where the surface of the upper half of the cylinder is located on $-1 \le \xi \le 1$, $\eta = 0$, let

$$t = \frac{1}{2}\left(z + \frac{1}{z}\right), \tag{6.212}$$

and discretise the domain in steps $\Delta\xi, \Delta\eta$ with $\psi_{i,j} \approx \psi(i\Delta\xi, j\Delta\eta)$ and $\zeta_{i,j} \approx \zeta(i\Delta\xi, j\Delta\eta)$. Then Thom's scheme set $\Delta\xi = \Delta\eta$ and used a Taylor expansion about $(i+\frac{1}{2})\Delta\xi, (j+\frac{1}{2})\Delta\eta$ to write

$$\zeta_{i+\frac{1}{2},j+\frac{1}{2}} = \frac{1}{4}(\zeta_{i,j} + \zeta_{i+1,j} + \zeta_{i,j+1} + \zeta_{i+1,j+1}) - \frac{1}{8}\Delta\eta^2 \nabla^2 \zeta\big|_{i+\frac{1}{2},j+\frac{1}{2}}, \tag{6.213}$$

and

$$\psi_{i+\frac{1}{2},j+\frac{1}{2}} = \frac{1}{4}(\psi_{i,j} + \psi_{i+1,j} + \psi_{i,j+1} + \psi_{i+1,j+1}) - \frac{1}{8}\Delta\eta^2 \nabla^2 \psi\big|_{i+\frac{1}{2},j+\frac{1}{2}}. \tag{6.214}$$

The Navier–Stokes equation (6.210) and stream-function – vorticity equation, (6.211) was then be used to replace the Laplace terms. Writing ζ_M and ψ_M for the average of the corner values in (6.213) and (6.214) and using central differences about the centre point to evaluate the inertial terms, this gave the compact formulae,

$$\zeta_{i+\frac{1}{2},j+\frac{1}{2}} = \zeta_M - \frac{R}{16}((\psi_{i+1,j+1} - \psi_{i,j})(\zeta_{i+1,j} - \zeta_{i,j+1}) +$$
$$(\psi_{i+1,j} - \psi_{i,j+1})(\zeta_{i,j} - \zeta_{i+1,j+1})), \tag{6.215}$$

and

$$\psi_{i+\frac{1}{2},j+\frac{1}{2}} = \psi_M + \frac{\Delta\eta^2}{8h^2_{i+\frac{1}{2},j+\frac{1}{2}}}\zeta_{i+\frac{1}{2},j+\frac{1}{2}}. \tag{6.216}$$

Numerical solution of Navier–Stokes equations

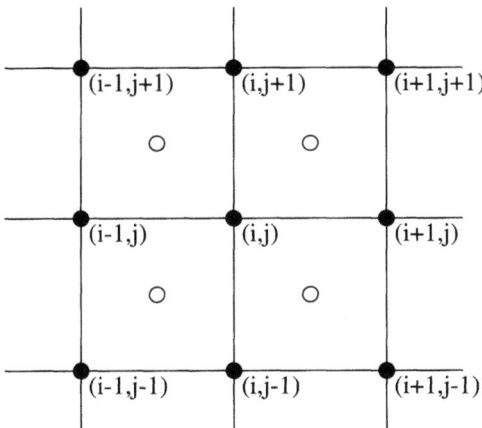

FIG. 6.35. Thom's scheme. Vertex or corner values (closed circles) are used to update cell centres (open circles) which in turn are used to update corner values.

Having found updated values for the cell centres at $(i+\frac{1}{2}, j+\frac{1}{2})$, those in turn were used in an identical manner to calculate updated values for the cell vertices or corners (i,j) from $(i+\frac{1}{2}, j+\frac{1}{2})$, $(i+\frac{1}{2}, j-\frac{1}{2})$, $(i-\frac{1}{2}, j-\frac{1}{2})$ and $(i-\frac{1}{2}, j+\frac{1}{2})$, see Figure 6.35. In this way all the interior points are all dealt with. On the boundaries, the stream-function and vorticity were given, $\psi = 2\eta$ and $\zeta = 0$, except on the cylinder surface where the stream-function vanished and the vorticity was determined at each iteration from a Taylor expansion about $\eta = 0$,

$$\zeta_{i,0} = -\frac{2h_{i,0}^2}{\Delta\eta^2}\psi_{i,1}, \quad i \text{ on cylinder.} \qquad (6.217)$$

The procedure was then iterated by hand calculation. In addition to calculating two-dimensional flow, Thom and Orr (1931) applied this scheme to problems in elasticity. As the solution to his scheme evolved, Thom used a certain amount of judgement about whether particular vertex values had converged and so did not need further iteration in order to reduce the computational effort.

The result of a calculation using Thom's scheme for $R_d = 20$ to calculate the stream-function is shown in Figure 6.36. There is a vortex pair behind the cylinder with separation at an angle $\alpha = 134°$. The vortex length (measured from the centre of the cylinder) is $L_s = 2.85$, a little higher than Thom calculated, but the scheme has had 50000 iterations, far greater than that used by Thom. The pressure coefficient, c_p is also shown in Figure 6.37; there has been a small problem calculating the coefficient at the leading edge and the pressure minimum is much lower than that given by Thom

In order to calculate the pressure around the cylinder, Thom used the following form for the Navier–Stokes equations,

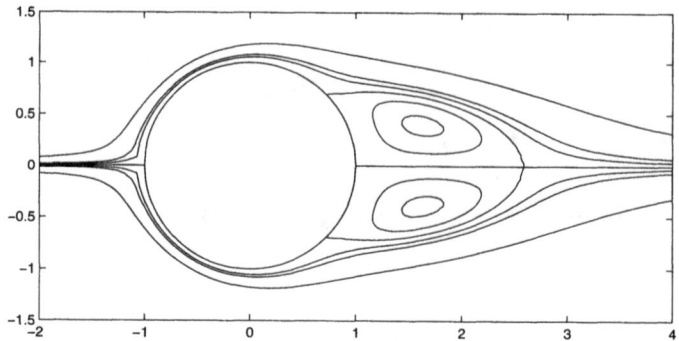

FIG. 6.36. Calculation of streamlines at $R_d = 20$ using Thom's scheme.

$$\nabla(p + \frac{1}{2}q^2) = \frac{1}{R}(-\frac{\partial \zeta}{\partial y}, \frac{\partial \zeta}{\partial x}) + \zeta(v, -u), \tag{6.218}$$

which in transformed coordinates became

$$\frac{\partial}{\partial \xi}(p + \frac{1}{2}q^2) = -\frac{1}{R}\frac{\partial \zeta}{\partial \eta} - \zeta\frac{\partial \psi}{\partial \xi},$$
$$\frac{\partial}{\partial \eta}(p + \frac{1}{2}q^2) = \frac{1}{R}\frac{\partial \zeta}{\partial \xi} - \zeta\frac{\partial \psi}{\partial \eta}. \tag{6.219}$$

On the line $\eta = 0$ where $\psi_\xi = 0$, this gave the pressure coefficient

$$c_p(\xi) = 1 - q^2 - \frac{2}{R}\int_{-\infty}^{\xi} \zeta_\eta d\xi. \tag{6.220}$$

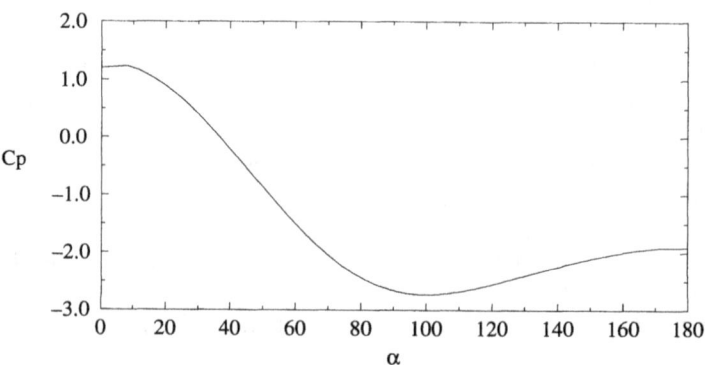

FIG. 6.37. Calculation of pressure coefficient c_p at $R_d = 20$ using Thom's scheme.

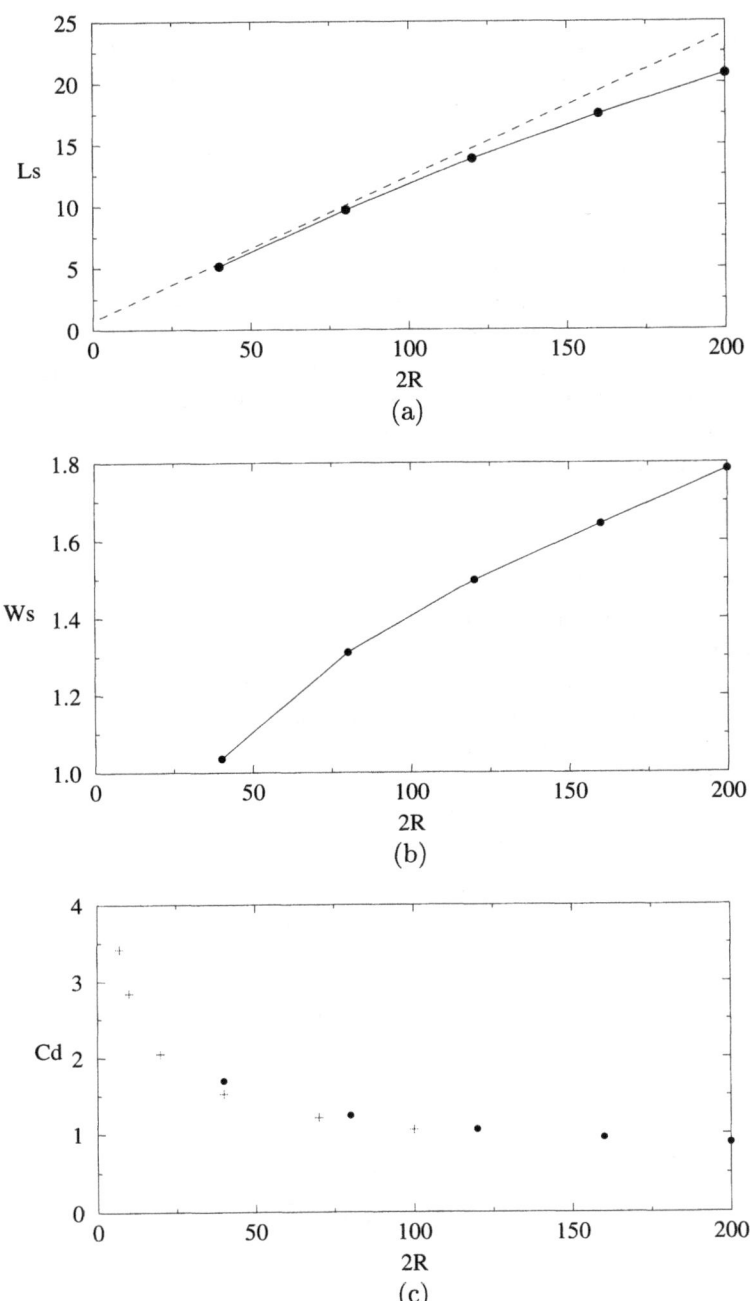

FIG. 6.38. Calculated values of (a) separation length L_s, (b) separation width W_s (c) drag coefficient C_D, using Thom's method shown by filled circles. Results from Dennis and Chang (1970) shown by +, separation length from Coutanceau and Bouard (1977), (6.6) by (- - -).

It was straightforward to convert this back to physical coordinates, x, y, or against angle, α, on the cylinder surface. The overall pressure and viscous drag coefficients, C_p, C_v, were also calculated from

$$C_p = 2\int_0^\pi c_p(\alpha)\cos\alpha\, d\alpha, \quad C_v = \frac{4}{R}\int_0^\pi \zeta\sin\alpha\, d\alpha, \tag{6.221}$$

where the dimensional drag per unit width of cylinder is $\frac{1}{2}\rho U^2 a(C_p + C_v)$.

In modern terms, Thom's calculation techniques may appear rather simplistic but the effectiveness of his method in calculating flow about a cylinder can be seen from Figure 6.38 where the predicted separation length, width and drag are shown together with a comparison with the separation length from Coutanceau and Bouard (1977) as given in (6.6) and with a comparison with the drag coefficient from Dennis and Chang (1970). It is only at high Reynolds numbers that the calculated separation length deviates substantially from the prediction from observation. Recall also that the experimental prediction is based on observations only for $2R \leq 40$.

6.8.2 Kawaguti (1953)

The second calculation of the full Navier–Stokes equations for flow about a cylinder was by Kawaguti (1953b) who also used hand calculation of an iterative finite difference scheme. Kawaguti transformed the equations into inverse cylindrical coordinates and then used a cell-centred difference formulation for the transformed equations to calculate the flow.

Kawaguti defined transformed coordinates

$$\xi = \frac{1}{r} \text{ and } \eta = \frac{2\alpha}{\pi}, \tag{6.222}$$

so the computational domain was $0 \leq \xi \leq 1$ and $0 \leq \eta \leq 2$. The surface of the cylinder was transformed to part of the line $\xi = 1$ while the far field transformed to part of the line $\xi = 0$.

If

$$\psi = y + \Psi, \tag{6.223}$$

then the Navier–Stokes equations transformed to

$$\frac{2\xi^2}{\pi}[\Psi_\eta\zeta_\xi - \Psi_\xi\zeta_\eta] + \xi\zeta_\xi\cos\frac{\pi\eta}{2} + \frac{2}{\pi}\zeta_\eta\sin\frac{\pi\eta}{2} =$$
$$\frac{1}{R}(\xi^3\nabla^2\zeta - (\xi^3 - \frac{4\xi}{\pi^2})\zeta_{\eta\eta} + \xi^2\zeta_\xi), \tag{6.224}$$

and

$$\zeta = -\xi^4\nabla^2\Psi + (\xi^4 - \frac{4\xi^2}{\pi^2})\Psi_{\eta\eta} - \xi^3\Psi_\xi. \tag{6.225}$$

The boundary conditions were $\Psi = \zeta = 0$ on $\eta = 0, 2$, $\Psi = 0$ on $\xi = 1$, $\zeta = 0$ on $\xi = 0$. On the cylinder, $\xi = 1$ the vorticity has to be found from (6.225) and on

$\xi = 0$, the far field, Kawaguti used a result from Imai (1951) to set $\Psi = -\frac{1}{4}C_d\eta$ where C_d was the overall drag of the cylinder.

Kawaguti used a hand calculation to compute the flow at $R_d = 40$ using a 11×21 mesh. His solution for streamlines and pressure coefficient compared well with experimental results which were available and with Thom's calculation at a lower Reynolds number, $R_d = 20$.

In view of his own work on combining free streamline theory with boundary layer theory, Kawaguti (1953a), also introduced a new and revealing plot, that of the contour where the fluid had unit speed and the streamline from smooth separation (on which the speed would also have unit value). This was a remarkable insight supporting the theory that the proper limit for steady separated flow was indeed that given by smooth separation.

Kawaguti should have a special place as one of the pioneers of computational fluid dynamics. Anyone tired of waiting for their computer to complete a calculation might think on his simple statement *'the numerical integration took about one year and a half with twenty working hours every week, with a considerable amount of labour and endurance'*.

Following Kawaguti's calculation a number of other computations were published, particularly as digital computers became available.

6.8.3 Allen & Southwell (1955)

Allen and Southwell (1955) introduced exponential functions locally to try to model solutions of a convection–diffusion equation. Their calculations were at Reynolds numbers up to 1000 but predicted separated regions which were too small and decreased in size as the Reynolds number increased.

Allen & Southwell used a modified form of Thom's mesh, with the normal variable η scaled by $R^{1/2}$. Considering a one-dimensional convection diffusion vorticity equation written in the form

$$\frac{\partial^2 \zeta}{\partial \eta^2} - v\frac{\partial^2 \zeta}{\partial \eta^2} = F, \quad (6.226)$$

then assuming constant v and F a local solution is

$$\zeta = -\frac{F}{v}\eta + b_1 + b_2 e^{v\eta}, \quad (6.227)$$

where b_1 and b_2 are constants. If this is applied to a finite difference formulation on a uniform mesh of size $\Delta\eta$ with vorticity at mesh points ζ_{j-1}, ζ_j and ζ_{j+1}, the constants b_1 and b_2 can be eliminated, resulting in

$$\lambda \coth \lambda \frac{\zeta_{j+1} - 2\zeta_j + \zeta_{j-1}}{\Delta\eta^2} - v\frac{\zeta_{j+1} - \zeta_{j-1}}{2\Delta\eta} = F, \quad (6.228)$$

where $\lambda = v\Delta\eta/2$ is half the mesh Peclet number. The difference to a normal scheme for this equation is the factor $\lambda \coth \lambda$ which is approximately one if λ is small. Their method involved splitting the vorticity equation into two convection diffusion equations, first in one direction and then the other direction to obtain a suitable formula on the two-dimensional mesh neglecting local variation in the velocities and the 'source' terms in each equation. Kawaguti (1959) thought that the failure of Allen & Southwell's method to predict the correct length for the separation region was due the either inaccurate accounting for the far field or because of the coarseness of their mesh. Neither of these seems sufficient to explain such a fundamental structural problem with the solution. Certainly the length of the separation region can be restricted by too close a downstream boundary condition requiring parallel flow but even that does not explain why the length of the separation region should decrease with increasing Reynolds number. Surprisingly, the real success of their idea has been in finite element calculations where their local exponential variation is the basis for the Hemker test functions in Petrov–Galerkin approximation; see for instance Morton (1996).

6.8.4 Payne (1958), Apelt (1958)

Payne (1958) calculated solutions at Reynolds numbers $R_d = 40$ and 100 using a time marching scheme. He used an exponential stretching of a radial mesh,

$$\xi = \log r, \tag{6.229}$$

and polar angle to transform from rectangular coordinates and was the first to solve the conservative form of the unsteady vorticity equation (using τ for time),

$$\frac{\partial \zeta}{\partial \tau} + \frac{\partial (u\zeta)}{\partial x} + \frac{\partial (v\zeta)}{\partial y} = \frac{1}{R}\left(\frac{\partial^2 \zeta}{\partial x^2} + \frac{\partial^2 \zeta}{\partial y^2}\right), \tag{6.230}$$

a form which is much to be preferred for calculation over the non-conservative form. Instead of using a stream-function he calculated the velocity field directly as an integral of the vorticity field. His solution for $2R = 40$ appeared to be converging to Kuwaguti's solution.

At the same time as Payne, Apelt (1958) obtained solutions of the steady equations for $2R = 40$ and $2R = 44$ also using an exponentially stretched mesh. Apelt was very careful to take into account a number of potential improvements over earlier schemes, using Woods (1954b) method to calculate vorticity at a surface, a fourth order difference scheme and relaxation in vorticity updates. The pressure on the surface was calculated using the equivalent of (6.219) for his transformed coordinate system. These solutions agreed quantitatively with the earlier solutions of Thom and Kawaguti.

6.8.5 Other computations to 1970

In a later work using a computer, Kawaguti and Jain (1966), solutions were found using time evolution of the unsteady Navier–Stokes equations. Continuing with stream-function – vorticity, the equations were transformed using an exponential stretching in the radial direction and then discretised using an explicit central difference scheme. The procedure for time marching was to calculate the vorticity at a new time step at internal points from the explicit finite difference form of the vorticity equation, use over-relaxation to calculate the stream-function from a Poisson equation and then to update the vorticity on the cylinder surface. Time evolved solutions were calculated on a 46 × 31 mesh for Reynolds numbers between 10 and 50 and showed substantially the same features as had been calculated earlier by Thom and by Kawaguti. The greatest uncertainty about the quantitative results was whether the solution was properly time converged.

Dennis and Chang (1969, 1970) computed separated flows using an exponential stretching in the radial direction combined with a Fourier series expansion in the angle. In the case of a circular cylinder, they defined a coordinate $\xi = \log r$ so that the Navier–Stokes equations for the stream-function and vorticity became

$$\frac{\partial \psi}{\partial \theta}\frac{\partial \omega}{\partial \xi} - \frac{\partial \psi}{\partial \xi}\frac{\partial \omega}{\partial \theta} = \frac{1}{R}\left[\frac{\partial^2 \omega}{\partial \xi^2} + \frac{\partial^2 \omega}{\partial \theta^2}\right], \qquad (6.231)$$

and

$$\frac{\partial^2 \psi}{\partial \xi^2} + \frac{\partial^2 \psi}{\partial \theta^2} = -e^{2\xi}\omega. \qquad (6.232)$$

These equations had then to satisfy the appropriate boundary conditions, $\psi = \psi_\xi = 0$ on $\xi = 0$, $\psi = \omega = 0$ on $\theta = 0, \pi$ and ψ, ω tending to uniform flow as $\xi \to \infty$. As with polar coordinates and assuming symmetry about the centreline, the computational region transformed to a strip, $0 \leq \theta \leq \pi$, $\xi \geq 0$. The interesting point of their solution was that while the vorticity transport equation, (6.231), was solved using a finite difference method, the stream-function equation, (6.232), was solved by using a Fourier series representation

$$\psi \sim \sum_{j=1}^{\infty} f_j(\xi)\sin j\theta. \qquad (6.233)$$

so that the functions $f_j(\xi)$ were to be determined by solving an ordinary differential equation in ξ,

$$f_j'' - j^2 f_j = -\frac{2e^{2\xi}}{\pi}\int_0^\pi \omega(\xi,\theta)\sin j\theta d\theta, \qquad (6.234)$$

with initial conditions $f_j(0) = f_j'(0) = 0$. While considerable care had to be taken with some of the details of implementing the method so as to correctly deal with the conditions at $\xi = 0$ for the vorticity and to match the far field, this method gave results which were in very good agreement with experimental observations.

A similar expansion for the stream-function to (6.233), excepting in polar coordinates, had been used by Underwood (1969) except he had adopted a series truncation method. The Fourier expansion for the stream-function and the vorticity was substituted into the vorticity transport equation and terms involving $\sin j\theta$ were collected to obtain non-linear ordinary differential equations for the coefficient functions, the set of which was truncated after a given number of terms. Underwood considered up to a fifth order truncation and he too obtained a linear variation in the separation length as the Reynolds number varied up to $2R = 40$.

Some other numerical solutions for steady and unsteady flow around a circular cylinder can be found in Cheng (1969), Takami and Keller (1969), Jain and Sankara Rao (1969) and Thoman and Szewczyk (1969).

6.8.6 *Summary*

The picture which emerged from these early calculations was mostly in agreement with experiments. The leading edge pressure coefficient decreased towards one with increasing Reynolds number in accordance with low Reynolds number (Oseen approximation) predictions, the length of the separation region increased approximately linearly with Reynolds number, the pressure coefficient over the body agreed with observations and the calculated drag coefficient was reasonably close to observed values. More importantly, such calculations showed no particular difficulty near separation, the regions in which there were vortices were calculated accurately and without any inkling that there should be some form of singularity at separation.

However, there had been contrary calculations such as Allen & Southwell's which predicted a decrease in separation length as the Reynolds number increased. These are now thought to have been inaccurate calculations but they engendered an atmosphere of uncertainty about the correct calculation of separated flow around a cylinder which seems to have persisted to the present even though computational resources are nowadays magnitudes greater than was the case in the past.

6.9 Attempts to resolve Goldstein's singularity

The publication of Catherall and Mangler (1966) began the resolution of Goldstein's boundary layer singularity. Not that the singularity could be eliminated, but rather that it came about through asking the wrong question. Whereas Goldstein had considered the response of a boundary layer to an imposed pressure gradient and so obtained a singularity at separation, he had also indicated that he believed that in reality, the pressure gradient adjusted so that a boundary layer singularity would not arise, *Another possibility is that ... experimentally, whatever we may do, the pressure variations near separation will always be such that no singularity will occur* (Goldstein (1948), page 50) but this had not been followed up until Catherall & Mangler showed that if instead of specifying the external speed (or equivalently, the pressure gradient), the displacement thickness was specified, then the boundary layer equations could be integrated past a point of separation.

Catherall and Mangler approached a boundary layer problem in an ingenious way. Normally, the body is a fixed element in a problem formulation. Instead of this they took the displacement surface as a fixed rather than the body surface. Of course, in the limit of infinite Reynolds number the two are the same, but at finite Reynolds number the body surface was then defined implicitly and located at an unknown position. Part of the algorithm had to provide a solution for the body position in a transformed plane.

Physical coordinates $z = x + iy$ were mapped to transformed coordinates $\tilde{z} = \tilde{x} + i\tilde{y}$ where the displacement surface was given by $\tilde{x} = 0$ and the body position by $\tilde{y} = \tilde{y}_B(\tilde{x}, R)$. Corresponding to this transformation there was also a mapping

$$\tilde{z} = \tilde{z}(w, R), \tag{6.235}$$

where $w = \phi + i\psi$ was the complex potential. The flow speed q was determined as usual by

$$q = \left|\frac{dw}{dz}\right|.$$

The Navier–Stokes equations for the stream-function ψ and vorticity ω transformed to

$$\frac{\partial \psi}{\partial \tilde{y}}\frac{\partial \omega}{\partial \tilde{x}} - \frac{\partial \psi}{\partial \tilde{x}}\frac{\partial \omega}{\partial \tilde{y}} = \frac{1}{R}\left(\frac{\partial^2 \omega}{\partial \tilde{y}^2} + \frac{\partial^2 \omega}{\partial \tilde{x}^2}\right), \tag{6.236}$$

and

$$\frac{\partial^2 \psi}{\partial \tilde{y}^2} + \frac{\partial^2 \psi}{\partial \tilde{x}^2} = -J\omega, \tag{6.237}$$

where

$$J = \left|\frac{dz}{d\tilde{z}}\right|^2 \tag{6.238}$$

was the Jacobian of the transformation. Catherall & Mangler noted that their formulation would have advantages in developing high order solutions and although they presented only the leading order solution the following discussion tries to draw out the general formulation they were proposing.

We considered a high Reynolds number expansion for this problem where in the bulk of the flow region

$$\tilde{z} = \sqrt{2w} + O(R^{-1/2}), \tag{6.239}$$

giving

$$\phi \sim \frac{1}{2}(\tilde{x}^2 - \tilde{y}^2) + O(R^{-1/2}),$$

$$\psi \sim \tilde{x}\tilde{y} + O(R^{-1/2}),$$
$$\omega = O(R^{-1/2}),$$
$$J \sim \frac{\tilde{x}^2 + \tilde{y}^2}{q^2} + O(R^{-1/2}). \qquad (6.240)$$

It should be clear that the transformation (6.239) needs an order $R^{-1/2}$ correction since for finite Reynolds number, the stream-function does not vanish on $\tilde{y} = 0$, but on the body and it is only asymptotically for large Reynolds number when the body and the displacement surface are the same that the stream-function will vanish there. Note also that this transformation has the effect of making the flow region $\tilde{y} \geq \tilde{y}_B(\tilde{x})$ with $\tilde{y}_B(\tilde{x}) < 0$.

While the expansion (6.239) provided the leading order outer flow, the expansion for the flow near the body had to be sought in terms of a stretched boundary layer coordinate transverse to the surface.

Define new coordinates by

$$\tilde{x} = X \quad \text{and} \quad \tilde{y} = R^{-1/2}Y, \qquad (6.241)$$

and expand the stream-function, vorticity and Jacobian according to

$$\psi \sim R^{-1/2}\psi_0 + o(R^{-1/2}),$$
$$\omega \sim R^{1/2}\omega_0 + o(R^{1/2}),$$
$$J \sim J_0 + O(R^{-1/2}), \qquad (6.242)$$

then the leading order terms of the Navier–Stokes equations become

$$\frac{\partial \psi_0}{\partial Y}\frac{\partial \omega_0}{\partial X} - \frac{\partial \psi_0}{\partial X}\frac{\partial \omega_0}{\partial Y} = \frac{\partial^2 \omega_0}{\partial Y^2} \qquad (6.243)$$

and

$$\frac{\partial^2 \psi_0}{\partial Y^2} = -J_0\omega_0. \qquad (6.244)$$

In order to match with the potential region when $Y \to \infty$ and $\tilde{y} \to 0$ the Jacobian can be expanded

$$J \sim \frac{\tilde{x}^2 + \tilde{y}^2}{q^2} + O(R^{-1/2}) \sim \frac{X^2}{q(X,0)} + O(R^{-1/2}), \qquad (6.245)$$

so that

$$J_0 = J_0(X) = \frac{X^2}{q(X,0)}; \qquad (6.246)$$

that is, the Jacobian is approximately constant across the boundary layer and only reflects longitudinal stretching as the boundary layer develops. Strictly speaking we should also have written $q = q(X, Y; R)$ and formally expanded

q for high Reynolds number before taking $Y = 0$. The boundary conditions were that

$$\psi_0(X, Y_B) = \frac{\partial \psi_0}{\partial Y}(X, Y_B) = 0, \qquad (6.247)$$

$$\psi_0(X, Y) \sim XY + \text{exponentially small terms as } Y \to \infty. \qquad (6.248)$$

The implicit definition of the surface position comes from noting that the leading order external velocity across the surface is just $u_e \sim X$ so a displacement thickness can be represented by the identity

$$\int_{Y_B}^{\infty} (X - \frac{\partial \psi_0}{\partial Y}) \mathrm{d}Y = -XY_B. \qquad (6.249)$$

To compute solutions for these equations, Catherall & Mangler supposed that the computational domain for Y would extend to $Y = Y_E$ and defined a new transverse coordinate

$$\eta = \frac{Y - Y_B(X)}{Y_E - Y_B(X)}, \qquad (6.250)$$

so that the computational domain was $0 \leq \eta \leq 1$, $X \geq 0$. In addition to the two boundary conditions (6.247) which move to $\eta = 0$, the far field boundary conditions were changed to

$$\psi_0 = XY_E, \quad \omega_0 = \frac{\partial \psi_0}{\partial \eta} = 0, \text{ at } \eta = 1. \qquad (6.251)$$

The convection–diffusion equation for the vorticity, (6.243), was solved by marching in the X direction using a Crank–Nicholson difference scheme. This gave two second order ordinary differential equations (one non-linear) for the streamfunction and vorticity at the next X station so an iterative scheme similar to that used by Leigh was applied. There were, however, two extra unknowns, the speed $q(X, 0)$ and the scaled displacement thickness, $Y_B(X)$. As Catherall & Mangler had proposed five boundary conditions, one of these last two unknowns could be calculated if the other was given.

Catherall & Mangler used two cases to examine the fundamental behaviour of their system. In one case they specified the speed $q(X, 0)$ as a quadratic in X which increased, reached a maximum and began to decrease to simulate an adverse pressure gradient. The solution could not be continued to separation and generally behaved as might have been expected in the approach to Goldstein's singularity. In a second case, they specified the displacement thickness $Y_B(X)$ by a steadily increasing quadratic in X. In that case they were able to integrate to separation and beyond. In a third calculation with a cubic form for the displacement thickness which increased and then decreased, they were able to calculate a separation bubble where the flow separated smoothly and then reattached further downstream.

6.10 Summary

It was the case that studies would shortly appear which confirmed the result of Catherall & Mangler that if the outer pressure was not specified then Goldstein's singularity would not appear in a boundary layer formulation. However, we have reached the point where the triple deck appeared and so we shall finish this chapter at that time. The growing ease with which numerical solutions of the full Navier–Stokes equations could be obtained must have lead many to question whether there was much to be gained by continuing to study boundary layer theory.

Of course, what has happened has been that the great number of computed solutions of the full equations have not lead to an equivalent increase in our understanding of the structure of solutions. Rather computational limits such as stability, accuracy and spatial resolution have for long hindered attempts to apply computational fluid mechanics to probe the **structure** of solutions. In fact the greater developments in our understanding have continued to come from theoretical work, albeit work which can only proceed hand in hand with computation of solutions and asymptotics which have become progressively more complex.

Thus we can move on to consider the consequences of the development by Stewartson and Williams (1969) and Neiland (1969) of a new model for separation in a compressible boundary layer using a triple deck structure and to see how its application to incompressible flow by Sychev (1972), Messiter and Enlow (1973) and Smith (1977b) can give much understanding of separation from a cylinder.

7

PREDICTION OF SEPARATION FROM A CYLINDER

7.1 Introduction

The length scales needed to resolve the singularity at the trailing edge of plate were simultaneously applied to separation in a compressible flow by Stewartson and Williams (1969) and Neiland (1969). In compressible flow separation can occur ahead of the place where an oblique shock interacts with a wall and the spontaneous nature of this upstream separation was something of a puzzle since the mechanism by which a disturbance could be propagated upstream of the wall-shock interaction in a supersonic flow was unclear, see, for example, Reyhner and Flügge-Lotz (1968). However, building on the work of Lighthill (1953a,b) and Gadd (1957) it was possible to construct an analogous structure to the trailing edge triple deck using a longitudinal scale of order $R^{-3/8}$ together with three decks: a deck outside the boundary layer with inviscid flow satisfying the Prandtl–Glauert equation, a middle deck within the boundary layer where streamlines were just displaced outwards and an inner deck of thickness $R^{-5/8}$ where the flow equations were identical to an *incompressible* boundary layer. Connecting the three decks was a simple relation between the pressure and the displacement of the streamlines in the middle deck. It was the solution of this system which showed that a free interaction between the pressure and the displacement could occur upstream leading to separation and that under those circumstances, the boundary layer equation was not singular at the separation point.

While compressible flow was more complicated to set out than incompressible flow since density and temperature variation had to be included, the flow in the outer inviscid deck could be rather simpler than for incompressible flow. Using the same notation as for the triple deck of (X, W) for the outer deck coordinates and P_2 for pressure perturbation, then P_2 satisfied the Prandtl–Glauert equation

$$(M_\infty^2 - 1)P_{2XX} - P_{2WW} = 0, \tag{7.1}$$

where $M_\infty > 1$ was the upstream Mach number. Requiring waves to be only propagated downstream gave

$$P_2(X, W) = p_2(X - \sqrt{M_\infty^2 - 1}W), \tag{7.2}$$

where the function p_2 still had to be found. However, as $W \to 0$,

$$P_2(X, W) \sim p_2(X) - \sqrt{M_\infty^2 - 1}W p_2'(X), \tag{7.3}$$

and matching this to the middle deck gave p_2 as the pressure in the inner 'incompressible' boundary layer. The displacement function $-A_1(X)$ for the middle deck (suitably non-dimensionalised) was related to the boundary layer pressure $p_2(X)$ by

$$p_2(X) = -A_1'(X). \tag{7.4}$$

Putting these together, the problem for the inner boundary layer with suitable velocities u, v and inner coordinates (X, Z) was

$$uu_X + vu_Z = A_1''(X) + u_{ZZ}, \quad u_X + v_Z = 0, \tag{7.5}$$

together with boundary conditions

$$u = v = 0 \text{ on } Z = 0, \quad u \to Z \text{ as } X \to -\infty, \quad u \to Z + A_1(X) \text{ as } Z \to \infty. \tag{7.6}$$

These equations obviously have the trivial solution $u = v = A_1 = 0$. The earlier work of Lighthill had suggested that a disturbance upstream in the flow might be grow exponentially according to

$$A_1(X) \sim -a_1 e^{cX}, \quad u \sim Z - a_1 e^{cX} f'(Z), \quad v \sim ca_1 e^{cX} f(Z), \tag{7.7}$$

where $f(0) = f'(0) = 0$ and $f \to 1$ as $Z \to \infty$. When these were substituted into the boundary layer equation with $e^{cX} \ll 1$, then

$$f(Z) = -\frac{c^{5/3}}{\text{Ai}'(0)} \int_0^Z \int_0^s \text{Ai}(c^{1/3} t) \, dt \, ds, \tag{7.8}$$

and so

$$c = \left[\frac{-\text{Ai}'(0)}{\int_0^\infty \text{Ai}(t) dt}\right]^{3/4} = 0.8272.... \tag{7.9}$$

Stewartson & Williams then found that the velocity gradient at the wall and the pressure would be given by

$$u_Z(X, 0) \sim 1 - 1.910 a_1 e^{cX}, \quad p_2(X) \sim ca_1 e^{cX}, \tag{7.10}$$

so the pressure should rise resulting in an adverse pressure gradient and the velocity gradient at the wall should decrease towards separation.

In order to conclusively demonstrate that self induced separation was possible without the appearance of a singularity in the boundary layer equation, Stewartson & Williams computed a solution by representing the stream-function and its derivatives as a finite series of Chebychev polynomials in the Z direction and applied a form of Box method to discretise the equation in the X direction. Their results can also be obtained directly from the numerical methods already

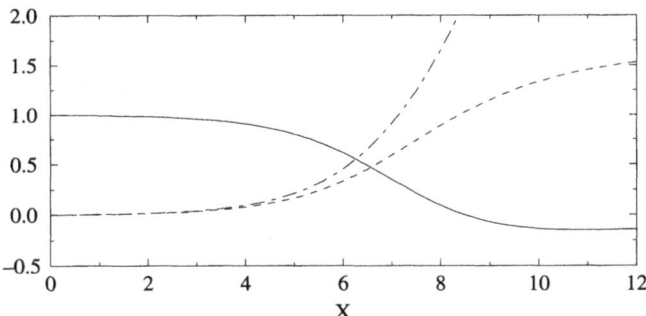

FIG. 7.1. Self-induced separation computed according to Stewartson & Williams' model: (—) velocity gradient at the wall, $u_z(X,0)$, (- - -) pressure $p_2(X)$, (– · –) displacement function $-A_1(X)$.

described for a triple deck problem, replacing the Hilbert integral connecting the pressure and displacement function with the much simpler compressible flow relation (7.4). A solution can be obtained by perturbing the pressure to have a small value at the start of the calculation. The value of X at the origin of the calculation is immaterial, Stewartson & Williams shifted the values of X relative to the point of separation, in the results shown here the calculation was started at $X = 0$ with $p_2(0) = 0.004$. In that case the pressure and the displacement rise and the wall velocity gradient falls until separation occurs without any singularity. The leading order term in (7.10) would give separation at $X = 5.6$ after this starting pressure perturbation, the full boundary layer equation gives a somewhat larger value, separation occurring at $X \approx 8.5$ after the calculation started, see Figure 7.1

To further demonstrate the smooth integration past the separation point, streamlines are shown in Figure 7.2. Note that the flow in the separated region is very slow and the difference between values of the stream-function shown there are one tenth the difference outside of the separation region. The extent to which the calculation can be continued past separation appears to depend only on the lateral extent of the computation region, but of course the flow is moving very rapidly away from the wall once separation occurs so that the computational region must increase correspondingly rapidly beyond separation. See Stewartson and Williams (1973) for further discussion relating to problems calculating a solution beyond separation.

The successful prediction of separation via a free interaction using a triple deck analysis did not of course fully resolve all questions about separation upstream of shock impingement since there was no linking of the solution downstream of separation with re-attachment and hence no prediction of how far upstream of the shock that separation would occur. However, these are questions for compressible flow theory, here our interest is in the consequences for incompressible flow but before going on to discuss this it is important to bring

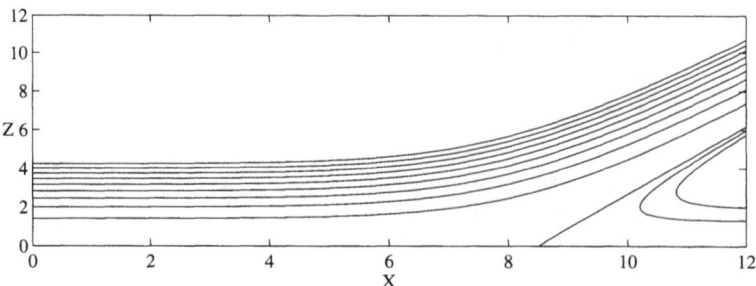

FIG. 7.2. Streamlines for self-induced separation computed according to Stewartson & Williams' model: streamlines for positive ψ at intervals $\Delta\psi = 1$, streamlines for negative ψ at intervals $\Delta\psi = 0.1$.

out one feature of the compressible solution for this particular problem. In this compressible flow, the limiting solution as $R \to \infty$ is straightforward: the hyperbolic Euler equations apply, there is no separation, there is no upstream influence and the pressure computed for the free interaction is regular. This should help to understand why incompressible separation can be so much more difficult: then the limit as $R \to \infty$ is the solution of an elliptic equation and there is a choice between attached potential flow and some form of separated flow such as described by free streamline theory with the problem that brings of the pressure being an irregular function (that is, not differentiable at separation).

Following the application of triple deck theory to separation in a compressible flow it was natural that incompressible separation should be examined in the same style to see if Goldstein's singularity might be resolved by an appropriate scaling of a region near separation and with the pressure remaining regular at separation.

Stewartson (1970a) tried and failed to incorporate a triple deck structure around a regular adverse pressure gradient. He considered a flat plate with nondimensional coordinates (x, y) and boundary layer coordinate $y = R^{-1/2} Y$ for which the velocity profile $U_0(Y)$ at $x = 0$ was about to separate, having $U_0(0) = U_0'(0) = 0$, $U(Y) \to 1$ as $Y \to \infty$, and a linear pressure distribution about $x = 0$. This was of course a similar formulation to the one already seen by Goldstein in section 6.5. Instead of taking Goldstein's formulation right up to $x = 0^-$, Stewartson supposed that new expansions would be needed on a length scale δ about $x = 0$. As with the triple deck theory for the free interaction, this length scale would be long compared with the boundary layer thickness, $R^{-1/2}$, but still a scale which would go to zero as $R \to \infty$. To match the oncoming double layer structure of Goldstein's solution Stewartson needed another length scale, $\mu R^{-1/2}$ to define the sublayer or inner deck where $\mu \to 0$ as $R \to \infty$.

Thus Stewartson considered the three lateral scalings:

$$\textbf{Outer deck}: \quad x = \delta X, \quad y = \delta W, \qquad (7.11)$$

Introduction

Main deck: $x = \delta X, \quad y = R^{-1/2}Y,$ (7.12)

Inner deck: $x = \delta X, \quad y = \mu R^{-1/2}Z.$ (7.13)

Since Goldstein's expansion for his outer layer (6.185), began

$$R^{1/2}\psi \sim g_0(Y) + (-x)^{1/2}g_2(Y) + O((-x)^{3/4}), \qquad (7.14)$$

Stewartson proposed that the expansion in the main deck should begin

$$R^{1/2}\psi \sim g_0(Y) + \delta^{1/2}\psi_1(X,Y) + \delta^{3/4}\psi_2(X,Y)$$
$$+ \delta \log \delta \psi_3(X,Y) + \delta \psi_4(X,Y) + \cdots, \qquad (7.15)$$

to match with Goldstein's expansion as $X \to -\infty$. Since $g_0(Y)$ is the stream-function at the instant of separation, $g_0 \sim Y^3$ as $Y \to 0$, and to match (7.15) the inner expansion would have to start

$$R^{1/2}\psi \sim \mu^3 \Psi_0(X,Z) + \cdots. \qquad (7.16)$$

In the inner region the leading order equation would be just a non-linear boundary layer equation and the inertial terms in the inner region were of order $\mu^4 \delta^{-1}$ whereas the pressure gradient and viscous terms was $O(1)$ and hence the scaling for μ had to be

$$\mu = \delta^{1/4}. \qquad (7.17)$$

If the main deck was examined for $Y \to \infty$ then ψ_1, ψ_2 and ψ_3 were all independent of Y (all being of the form $-A_i(X)g_0'(Y)$ with $g_0'(Y) \to 1$) whereas ψ_4 was proportional to Y This meant that in the outer region, the second term in the expansion there had to match with the $O(\delta^2)$ term in the main deck. Essentially the outer flow perturbation had to match with a slip velocity at the boundary layer edge. This lead Stewartson to expand the stream-function in the outer region as

$$\psi \sim \delta W + \delta^2 \psi_1^o(X,W) + \delta^{5/2}\psi_2^o(X,W) + \cdots \qquad (7.18)$$

and since $\psi_1^o \sim W$ as $W \to 0$, the transverse velocity at the boundary layer edge had to come from the next term giving a velocity

$$v \sim \delta^{3/2}\psi_{2X}^o. \qquad (7.19)$$

However, the transverse velocity at the boundary layer edge also had to be of order $R^{-1/2}\delta^{-1/2}$ so the scaling parameter δ would be given by

$$R^{-1/2}\delta^{-1/2} = \delta^{3/2}, \quad \text{or} \quad \delta = R^{-1/4}, \qquad (7.20)$$

and $\mu = R^{-1/16}$. This gave the lateral extend of the three layers as $R^{-9/16}$ (inner deck), $R^{-1/2}$ (main deck) and $R^{-1/4}$ (outer deck). While these scales were unusual and indeed unexpected they were the only ones which provided a rational asymptotic structure near the separation point.

Once the asymptotic framework was established, Stewartson was able to consider the interaction between the pressure and the displacement function. The pressure in the main deck was expanded

$$p \sim R^{-1/2} p_2(X) + R^{-3/4} p_3(X) + \cdots, \qquad (7.21)$$

and the first term was just the imposed adverse pressure, $p_2(X) = X$ whereas the second could be matched with both the outer deck and the inner deck.

Matching with the outer deck gave the same equation as was found in the triple deck,

$$p_3(X) = \frac{1}{\pi} \int_{-\infty}^{\infty} \frac{A_1'(\zeta)}{X - \zeta} d\zeta. \qquad (7.22)$$

Matching with the inner deck gave a second equation relating p_3 and A_1,

$$A_1^2(X) + c_1 X = -c_2 \int_{-\infty}^{X} \frac{p_3'(\zeta)}{(X - \zeta)^{1/2}} d\zeta, \qquad (7.23)$$

where c_1, c_2 were known positive constants. Eliminating the pressure perturbation, p_3, gave Stewartson

$$A_1^2(X) + c_1 X = c_2 \int_{X}^{\infty} \frac{A_1''(\zeta)}{(\zeta - X)^{1/2}} d\zeta. \qquad (7.24)$$

If the flow downstream of separation were to be able to be matched to some form of boundary layer then (7.21) implies that the worst behaviour for p_3 as $X \to \infty$ would be $p_3 \sim X^{3/2}$ (otherwise the downstream expansion would have to involve terms behaving like a positive power of the Reynolds number R as $R \to \infty$) and the corresponding behaviour of A_1 for $X \to \infty$ would have to be between a constant and $X^{5/2}$. All such cases lead to a contradiction in (7.24). Thus Stewartson concluded that it was not possible to avoid Goldstein's singularity if the external pressure gradient had a specified adverse value near separation.

The work in Stewartson (1970a) appeared to close a chapter in boundary layer research with a very pessimistic note but shortly Sychev's hypothesis would provide the key to finally unlock an asymptotic structure for separation in the context of a boundary layer. Some years later Smith and Daniels (1981) would also show that there were circumstances in which Goldstein's singularity may be removed for a special case where the pressure and displacement were suitably related.

7.2 Sychev's hypothesis for separation

In one of the significant contributions to boundary layer theory, Sychev (1972) proposed as a hypothesis that separation might be described not by trying to remove any singularity in the asymptotic structure at separation but by accepting that the leading terms in the outer expansion would have to match the free streamline singularity at finite Reynolds number. However, he proposed that the singularity strength should vanish as $R \to \infty$. A very similar argument was put forward by Messiter and Enlow (1973) in another important contribution where they argued from a very different viewpoint, that of allowable descriptions of the outer inviscid flow field. Both papers reached the same scaling relating the Reynolds number to the strength of a singularity in the pressure gradient. The development of the theory is set out very clearly in Sychev et al. (1998) and the description here has been suitably abbreviated.

Recall from our discussion of free streamline theory that smooth separation would occur without any singularity in the flow at separation but since there was no adverse pressure gradient then there was no reason for a fluid to have separated in the first place. If separation occurred beyond the point of smooth separation then the pressure $p(x)$ on the streamline at the wall and along the free streamline after separation at $x = 0$ with $p(0) = 0$ behaved like [see for instance (6.56) or Figure 6.10]

$$p \sim \begin{cases} -\kappa(-x)^{1/2} & : \quad x < 0, \\ 0 & : \quad x \geq 0. \end{cases} \tag{7.25}$$

Smooth separation occurs when $\kappa = 0$, negative values of κ would cause the streamline to intersect the body surface. Sychev's hypothesis was disarmingly simple: the flow outside the boundary layer at separation should be governed by (7.25) but with

$$\kappa = \kappa(R) \to 0 \text{ as } R \to \infty. \tag{7.26}$$

Thus the limiting inviscid flow structure was indeed smooth separation to overcome the argument that inviscid separation beyond smooth separation as a high Reynolds number limit would always be contradictory because then separation should have occurred closer to the point of smooth separation. At finite Reynolds number the flow could support separation beyond the point of smooth separation which would agree more closely with observations.

Sychev argued that if the singularity structure occurred over a longitudinal scale δ then perturbations from a pressure change of order $\kappa \delta^{1/2}$ would show up in a sublayer of thickness $\mu R^{-1/2}$ in the boundary layer with velocity scale u_s and balances

$$\text{Inertia}: u_s^2 \delta^{-1}, \text{ Pressure}: \kappa \delta^{-1/2}, \text{ Viscous}: u_s \mu^{-2}. \tag{7.27}$$

In the viscous layer the velocity should be approximately linear in distance from the wall so that $u_s \sim \mu$ whence

$$\delta \sim \kappa^6 \text{ and } \mu \sim \kappa^2. \tag{7.28}$$

In order to complete determination of the scales, Sychev argued that the sublayer would induce a displacement in the main part of the boundary layer of order

$\mu R^{-1/2}$ so that the curvature, of order the displacement divided by δ^2, would match the free streamline curvature (see for instance (6.59)) provided

$$\frac{\mu R^{-1/2}}{\delta^2} \sim \kappa \delta^{-1/2} \qquad (7.29)$$

Substituting (7.28) gave the scale for κ:

$$\boxed{\kappa = \kappa_0 R^{-1/16}} \qquad (7.30)$$

where κ_0 was an order one constant which still needed to be evaluated. This also gave that the streamwise length scale for the structure about separation was $O(R^{-3/8})$ and the sublayer thickness $O(R^{-5/8})$, results which while identical to the triple deck scalings, Sychev considered this similarity entirely coincidental.

An alert reader should see immediately that if Sychev's hypothesis is correct, then only given the value κ_0, it is possible to predict a separation angle α_s for the cylinder problem because we already have the variation $\kappa(\alpha_s)$ with angle beyond smooth separation from Brodetsky's and Woods' work, see Figures 6.11 & 6.15. In principle we have two relations, $\kappa = \kappa(\alpha_s)$ from free streamline theory and $\kappa = \kappa(R)$ from triple deck theory. By eliminating κ we have a relation $\alpha_s = \alpha_s(R)$ between separation angle and Reynolds number. We shall come to this shortly when we discuss Smith (1979b) but before that the constant κ_0 has to be determined along with a evidence that Sychev's hypothesis has a solution at separation. Even a simple correspondence of κ will still leave a question about the pressure at separation: in deriving κ from Brodetsky's model the pressure is taken to be the same as the upstream pressure rather than the observed lower pressure.

7.3 Smith's solution near separation

The non-linear boundary layer equation which comes from Sychev's hypothesis has to be solved numerically and that was first done by Smith (1977b). In order to formulate the non-linear boundary layer equation the flow upstream of separation had to be re-examined to provide the asymptotic conditions which defined the flow about the separation point. Suppose the boundary layer has velocity gradient at the wall upstream of separation $u_y = R^{1/2}\lambda$ and that separation occurs at $x = 0$.

The flow upstream of separation will start developing under the action of an adverse pressure gradient of order $(-x)^{-1/2}$ so we might anticipate a wall layer which is a perturbation of an oncoming uniform velocity gradient. Further, we can try to determine a similarity form for the perturbation. Once again using $y = R^{-1/2}Y$ to determine distance on the oncoming boundary layer scale

$$R^{1/2}\psi \sim \frac{1}{2}\lambda Y^2 + (-x)^a F_1(\eta) + \cdots, \quad \eta = \frac{Y}{(-x)^b}, \qquad (7.31)$$

then balances in the boundary layer equation are:

$$uu_x \sim Y(-x)^{a-b-1} = \eta(-x)^{a-1}, \ p_x \sim (-x)^{-1/2}, \ u_{YY} \sim (-x)^{a-3b}, \qquad (7.32)$$

and matching the three terms gives $a = 1/2$, $b = 1/3$ so that velocity in the sublayer behaves like

$$u \sim \lambda Y + (-x)^{1/6} F_1'(\eta) + \cdots . \qquad (7.33)$$

At the edge of the sublayer, assuming F_1' tends to a constant (see also Brown and Stewartson (1970)), then the perturbation in the outer part of the oncoming boundary layer was just be an inviscid displacement of the oncoming flow where in particular the displacement had to behave like $(-x)^{1/6}$ as $x \to 0^-$.

This enabled Smith to define the lower deck of a triple deck through

$$\text{Lower deck}: \ x = \lambda^{-5/4} R^{-3/8} X, \ y = \lambda^{-3/4} R^{-5/8} Z. \qquad (7.34)$$

To match the oncoming flow perturbation he expanded the stream-function

$$R^{1/2} \psi \sim R^{-1/4} \lambda^{-1/2} \Psi_0(X, Z), \qquad (7.35)$$

and the pressure

$$p \sim R^{-1/4} \lambda^{1/2} p_2(X). \qquad (7.36)$$

The stream-function Ψ_0 and the pressure p_2 satisfied a boundary layer equation

$$\Psi_{0Z}\Psi_{0XZ} - \Psi_{0X}\Psi_{0ZZ} = -p_2'(X) + \Psi_{0ZZZ}, \qquad (7.37)$$

with the usual boundary layer conditions $\Psi_0 = \Psi_{0Z} = 0$ on the wall, $\Psi_0 \sim Z$ as $X \to -\infty$ but there were also both triple deck and far field conditions to be satisfied. Let $\tilde{\kappa}_0 = \lambda^{9/8} \kappa_0$.

The match between the inner and main decks brought in a displacement function:

$$\Psi_0 \sim \frac{1}{2} Z^2 + A_1(X) Z \ \text{as} \ Z \to \infty, \qquad (7.38)$$

while upstream the pressure had to satisfy

$$p_2(X) \sim -\tilde{\kappa}_0 (-X)^{1/2} \ \text{as} \ X \to -\infty. \qquad (7.39)$$

Matching between the main deck and the outer potential region provided the final relation between p_2 and A_1,

$$A_1' + \tilde{\kappa}_0 X^{1/2} H(X) = -\frac{1}{\pi} \int_{-\infty}^{\infty} \frac{p_2(s) + \tilde{\kappa}_0 (-s)^{1/2} H(-s)}{X - s} ds, \qquad (7.40)$$

with $H(X)$ a step function.

It was by computing solutions for various values of $\tilde{\kappa}_0$ that enabled Smith to show reasonably conclusively that there was a solution satisfying the appropriate asymptotic conditions for A_1 and p_2 both upstream and downstream for

$$\tilde{\kappa}_0 \approx 0.44. \tag{7.41}$$

It was clear that the upstream shear λ played an important role in taking the fundamental solution computed by Smith back to the physical application and there appears to be some judgement about what value would be most appropriate. Since Sychev's hypothesis was the the limiting form of the flow for $R \to \infty$ was smooth separation and since the variation of κ with separation angle (α_s) just beyond smooth separation (α_{ss}) is approximately linear it should be the case that the separation angle for a cylinder will be approximated by

$$\alpha_s(R) \sim \alpha_{ss} + O(R^{-1/16}) \tag{7.42}$$

so later Smith (1979b) used the wall shear at α_{ss} from a conventional boundary layer calculation to provide a value for λ.

We have not yet discussed the solution downstream of separation. Smith showed that the solution would develop in a multi-structured set of layers. The details would take us well beyond an introductory examination of interactive boundary layers but the general outline can be set out. The main part of the boundary layer displaces away from the wall according to the displacement function $-A(X)$ and along that line the stream-function has a similarity form with a layer of thickness $O(X^{1/3})$ developing where the velocity gradient is reduced from its value in the separating layer to zero. Below that is an inviscid region of slow moving fluid and finally adjacent to the wall a backward moving layer with a similarity form depending on $X^{11/16}$.

The pressure solution calculated by Smith (1977a) across the triple deck structure was smooth. What happened was that pressure changes, of order $R^{-1/4}$ occurred over a length $R^{-3/8}$ and on the body scale the gradient became increasingly steep, of order $R^{1/8}$ while at the same time the region of the steep pressure gradient decreased (and matched to the free streamline pressure gradient $\kappa x^{-1/2} \sim R^{-1/16} R^{3/16} = R^{1/8}$). Thus separation was not the result of a slow change in the boundary layer from an adverse pressure gradient acting over a long section of the body but happened from a large adverse pressure gradient acting over a very short length scale. This represented a major move away from the traditional view of how separation came about.

7.4 Separation from a cylinder

Once the universal constant $\tilde{\kappa}_0$ had been determined by Smith much of the structure of separated flow about a cylinder could be examined in the light of Sychev's hypothesis. This was done in Smith (1979b).

The simplest prediction which comes from Sychev's hypothesis is that of separation angle. We have seen that both Brodetsky's and Woods' methods gave the same result for the dependence of κ on angle (see Figure 6.15). Taking a simple linear approximation for κ against angle beyond smooth separation gives for a unit radius cylinder and α measured in degrees,

$$\kappa(\alpha) \approx 0.012(\alpha - \alpha_{ss}). \tag{7.43}$$

so finally we arrive at a prediction of separation angle for a circular cylinder,

$$\alpha(R) \sim 55 + 37\lambda^{9/8} R^{-1/16}. \tag{7.44}$$

This is slightly unsatisfactory since the upstream shear λ remains to be specified. Smith used a value $\lambda = 0.72\sqrt{2}$ from numerical solutions of a boundary layer equation over the forward part of the cylinder. The prediction (7.44) gives separation angles which are really quite low compared to such data as is available (see, for instance, Figure 8 in Thoman and Szewczyk (1969)). We have already seen that boundary layer calculations using Heimenz's data for the external flow field, for example that shown in Figure 6.21, gave a much higher value for λ (taken as the wall shear at $\alpha = \alpha_{ss}$) perhaps as much as 1.8. On the other hand it is important to appreciate Smith's caution that at moderate Reynolds numbers, $R^{-1/16}$ represents a substantial angle so any prediction of separation angle using what is only the leading term of an asymptotic expansion may be inaccurate unless the Reynolds number is very large.

In addition to the separation angle, Brodetsky's method provided a pressure drag coefficient (6.69) in terms of the leading coefficient of his series expansion, A_1 defined in (6.34). Since A_1 is known in terms of the separation angle it is also effectively known in terms of the Reynolds number. Smith used this to provide an estimate for the pressure drag coefficient. Values of A_1 are given in Table 6.1, page 134 from which for angles close to smooth separation (angle measured in degrees)

$$A_1 \approx A_{1ss} - 0.016(\alpha - \alpha_{ss}), \tag{7.45}$$

which gives a pressure drag (neglecting terms of order $R^{-1/8}$)

$$C_p \sim 0.88(1 - 1.12\lambda^{9/8} R^{-1/16}). \tag{7.46}$$

Numerical constants in (7.44) and (7.46) are slightly different from those given in Smith (1979b) because of minor differences between Smith's implementation of Brodetsky's theory and that given earlier here. In addition to the pressure drag there will be a friction drag coefficient from the cylinder surface ahead of separation. The pressure drag coefficient neglected any pressure reduction behind the cylinder (Brodetsky's solution assumed that the free streamline pressure was the same as that far upstream) and such pressure reduction would increase the drag coefficient.

What we have not discussed yet is whether Sychev's hypothesis about separation might be incorporated into different models for the global flow field. Of course one view would be that Sychev's hypothesis provides that the limiting inviscid flow field is a Kirchoff free streamline flow with an infinite wake. The view in Smith (1979b) was that it was not necessary to focus too much on the limiting form, rather Sychev's hypothesis implied that the length of the wake at finite Reynolds numbers would be large compared to the body scale. In view of the way the Kirchoff wake grew in width with distance from the body, Smith proposed that the wake should be of length $O(R)$ and of width $O(R^{1/2})$. Further, if the wake was closed after a distance of order R from the body then by assuming the separated region had an elliptical shape, Smith proposed that the pressure behind the body might be estimated from quite coarse global considerations. Additionally he observed that in this case the pressure reduction behind the body would not be confined to the separation region but would affect the boundary layer and flow immediately outside the separated region. If in the region of the triple deck around separation the pressure were everywhere reduced then Smith argued that Sychev's hypothesis would still apply locally except with the free streamline pressure (and hence speed) reduced appropriately. This enabled the estimate for the separation angle (7.44), and for the drag coefficient (7.46) to be modified by a pressure reduction in the separation region.

While the global model proposed in Smith (1979b) had many very convincing elements, because it introduced viscous effects of order $R^{-1/2}$ through wake closure in a non-rational way (that is, into an expansion which had already neglected terms of order $R^{-1/8}$) there remained unanswered problems about the global structure of the separated flow field.

7.5 Comparison with numerical solutions

Interestingly enough, advances beyond the ideas put forward in Smith (1979b) were prompted largely as a result of numerical computations of the full Navier–Stokes equations by Fornberg (1980, 1985). Differences between calculations in these two works were themselves revealing of the great difficulty in computing separated flows at even moderate Reynolds numbers.

Fornberg's view was that there had been a number of different reasons why previous calculations had run into difficulty in computing flow around a cylinder at Reynolds numbers greater than a hundred or so. The formulation of the problem in terms of stream-function – vorticity produced some intrinsic problems because of the way boundary conditions were specified, for instance on the body the stream-function was explicitly given whilst the vorticity was implicitly defined. The conditions in the far field, often approximated by uniform flow when that approximation was poor at distances dictated by practical computational boundaries. The process of alternating between a stream-function and a vorticity equation seemed the cause of some problems. Discretisation of a general convection diffusion equation remains a major challenge in computational fluid dynamics today (see for instance Morton (1996)) and that is certainly true for the vorticity equation. Fornberg aimed to deal with these as fully as possible and in doing so illustrated both the difficulty in computing flow around a cylinder at larger Reynolds numbers and has provided a tool which has helped to shape our view of what high Reynolds number outcomes might be possible.

In Fornberg (1980) the global flow field was considered by transforming the equations onto a logarithmically stretched polar grid. One difficulty which arises from this scheme is that the wake becomes confined to one axis, $\eta = 0$. To deal with that Fornberg used a second mesh which again could be specified conformally and values transferred between the two mesh grid points by a high order interpolation scheme. To avoid the alternating application of the vorticity convection diffusion equation and the stream-function – vorticity equation, Fornberg applied a Newton method simultaneously to the two equations. Finally to deal with the region beyond the computational grid, instead of just taking the outer condition to be uniform flow a potential flow was found which matched to the solution just inside the boundary and that was then used to provide an outer boundary condition.

The results he obtained showed that the separated region increased in length linearly with Reynolds number until around $R = 145$ when there was a sudden quite dramatic shortening in the separation length. The width of the separated region increased somewhat less than linearly until the shortening occurred when the width increased dramatically. The lengths of the vorticies calculated by Fornberg were, given previous calculations remarkably long and as noted in Smith (1981), excepting the sudden decrease in length at the highest Reynolds numbers, were much in agreement with estimates in Smith (1979b) for the vortex pressure, skin friction on the cylinder ahead of separation and vortex length.

In Fornberg (1985) a single conformal transformation,

$$\xi + i\eta = (x + iy)^{1/2} + (x + iy)^{-1/2},$$

was used to map the region around the cylinder and the wake to a rectangular region. The computational region extended over 175 cylinder radii downstream of the cylinder. The boundary conditions on the transformed region were (i) $\psi = \omega = 0$ on the line of symmetry ahead of the cylinder, $\xi = 0$, (ii) $\psi = 0$, ω defined by the stream-function – vorticity equation on $\eta = 0$, $0 \leq \xi < 2$ and $\psi = \omega = 0$ for $\eta = 0$, $\xi \geq 2$, (iii) on the right boundary an Oseen approximation for the vorticity lead to $\xi^2\omega$ constant across the last two grid points whilst the stream-function came from neglecting $\psi_{\xi\xi}$ in the stream-function-vorticity equation and integrating across the wake, (iv) on the top boundary the vorticity was taken as $\omega = 0$ while the stream-function across the top two rows of the mesh were related by assuming a potential flow outside the region defined by the last but one row. The results obtained by Fornberg are the most plausible computations made so far for flow around a cylinder at moderate Reynolds numbers up to $R = 300$. They showed some remarkable features, not least of which was that the vortex continued to grow with length which was linear in Reynolds number, its width while initially increasing slowly according to $R^{1/2}$ increased linearly for Reynolds numbers between 200 and 300. These were seminal results since they led to an almost complete change of view about the likely high Reynolds number structure of the separated wake behind the cylinder.

7.6 Prandtl–Batchelor flow

In order to fully appreciate the potential significance of Fornberg's computations we need to go back a little and discuss Prandtl–Batchelor flow. As we have described, Batchelor (1956b) questioned whether an infinite wake model was an appropriate limit for vanishing viscosity or whether there might be a solution which consisted of regions of constant vorticity. Prandtl had previously noted that an inviscid rotational flow with closed streamlines would need the streamlines to enclose constant vorticity, see also Batchelor (1956b). There has been relatively little theoretical work on such flows and for instance even in his 1980 paper Fornberg felt it necessary to add *if they exist* when alluding to Prandtl–Batchelor solutions for flow behind a cylinder. It is only recently that existence results for some problems such as Acker (1998) have emerged. Early studies on Prandtl–Batchelor flows are described in Smith (1986c) and there is an interesting introduction to the area in Moore et al. (1988). There is also a chapter on two-dimensional vortex regions in Saffman (1992).

Amongst the early studies were Feynman and Lagerstrom (1956), Wood (1957) (and recently Edwards (1997), Kim (1998)) who were mainly interested in cylindrical flows. Childress (1966) used a slender body approximation to calculate some of the earliest solutions for Prandtl–Batchelor flows. As an example of his technique consider a notched wall shown in Figure 7.3 where the notch is $y = -\epsilon$ for $-1 \leq x \leq 1$ and contains a region of constant vorticity with irrotational flow above the dividing streamline. Let the pressure and velocity far upstream be $\rho U^2 p_\infty$, U and the pressure, speed in the fluid be $\rho U^2 p$, Uq. The total pressure immediately above the dividing streamline, with speed q_+ is given by

$$\rho U^2 p_\infty + \frac{1}{2}\rho U^2 = \frac{1}{2}\rho U^2 (2p + q_+^2). \tag{7.47}$$

Let the total pressure in the notch be $\rho U^2 p_c$ and the speed immediately below the dividing streamline Uq_- then

$$\rho U^2 p_c = \frac{1}{2}\rho U^2 (2p + q_-^2). \tag{7.48}$$

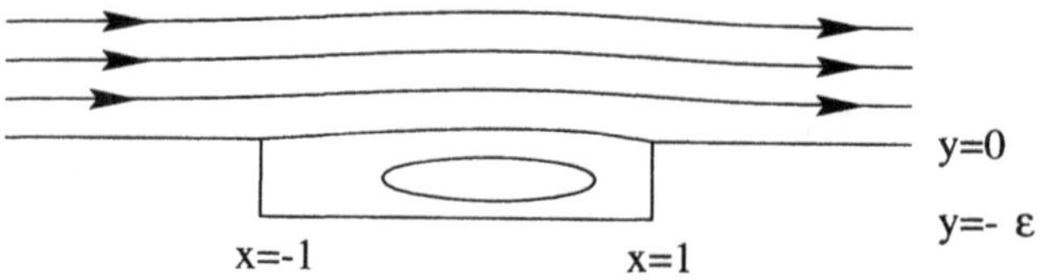

FIG. 7.3. Geometry for application of Childress' method to calculate a Prandtl–Batchelor flow.

Suppose that the constant vorticity in the notch is $\epsilon^{-1/2}\omega$ and bounded by a dividing streamline $y = \epsilon f(x)$ with the pressure continuous across the dividing streamline. In that case the fluid speed either side of the dividing streamline has to satisfy

$$q_+^2 - q_-^2 = 1 - c_p, \qquad (7.49)$$

with

$$c_p = 2(p_c - p_\infty) \qquad (7.50)$$

a pressure coefficient in the vortex satisfying $0 \leq c_p \leq 1$.

Childress sought an approximate solution for small ϵ in the region of constant vorticity by assuming derivatives across the notch were larger than those along the notch. Using a stretched coordinate $y = \epsilon Y$ the stream-function *in the notch* can be expanded

$$\psi(x,y) = \epsilon^{3/2}\Psi(x,Y) \sim \epsilon^{3/2}\Psi_0(x,Y) + o(\epsilon^{3/2}), \qquad (7.51)$$

and the leading term for the perturbation series satisfies just

$$\frac{\partial^2 \Psi_0}{\partial Y^2} = -\omega. \qquad (7.52)$$

The solution with $\Psi_0 = 0$ on $Y = -1$ and $Y = f(x)$ is

$$\Psi_0(x,Y) = -\frac{1}{2}\omega[Y^2 - Y(f(x) - 1) - f(x)], \qquad (7.53)$$

so that

$$q_-^2 \sim \frac{1}{4}\epsilon\omega^2(f-1)^2. \qquad (7.54)$$

Outside of the notch the flow is potential with boundary condition $\psi_x = -\epsilon f'(x)$ on the free streamline. Letting

$$\psi \sim y + \epsilon\psi_1(x,y) + o(\epsilon), \qquad (7.55)$$

the solution from potential theory is

$$\psi_1(x,y) = \frac{1}{\pi}\int_{-1}^{1} f'(\xi)\tan^{-1}\frac{y}{x-\xi}d\xi. \qquad (7.56)$$

Hence the speed outside of the dividing streamline is

$$q_+^2 \sim 1 + \frac{2\epsilon}{\pi}\int_{-1}^{1} \frac{f'(\xi)}{x-\xi}d\xi. \qquad (7.57)$$

If the pressure coefficient is taken as $c_p = 2\epsilon h$, then we finally arrive at a non-linear integral equation for the dividing streamline shape $f(x)$,

$$-\frac{1}{\pi}\int_{-1}^{1}\frac{f'(\xi)}{x-\xi}\mathrm{d}\xi + \frac{1}{8}\omega^2(f(x)-1)^2 = h. \qquad (7.58)$$

This equation has to be solved with boundary conditions $f(\pm 1) = f'(\pm 1) = 0$ and for h given but ω unknown or vice versa. Obviously the formulation of this example will require attention near the ends of the notch where the longitudinal and lateral length scales are no longer different.

Childress considered a number of examples, a Riabouchinsky vortex and flow behind a wedge or a back step. A Riabouchinsky flow was also considered by Pullin (1984) and a model for flow behind a backward facing step has been developed by O'Malley et al. (1991). Saffman and Tanveer (1984) have calculated a solution behind a plate with a forward facing flap at the trailing edge. Other examples are Wilmott and Fitt (1992), (axisymmetric slender body problems) Turfus (1993), (vortices in a channel and behind a flat plate) Chernyshenko (1993), (stratified flow in a channel) Chernyshenko and Castro (1993) , (flow past a row of bodies) Bunyakin et al. (1996, 1998) (airfoil with a vortex in a cavity), Pullin (1981)(vorticity layer on a wall).

While different numerical methods have been tried for calculation of Prandtl–Batchelor flows the development of suitable methods is still evolving, particularly for dealing with non-uniqueness and asymmetry. Numerical methods based on parametrisation of the dividing streamline were developed and successfully applied by Moore et al. (1988).

In the case of slender dividing streamline considered by Childress, Riley (1987) used a similar method which we shall describe shortly.

7.6.1 The Sadoviskii vortex

Whilst the model developed by by Childress used an interaction with a boundary to provide context for a Prandtl–Batchelor flow, Sadoviskii (1971) considered a model flow where there were no boundaries (other than a line of symmetry being equivalent to a boundary in the inviscid flow) but two regions of opposite signed vorticity existed in a uniform flow, see Figure 7.4.

Suppose regions of constant vorticity are bounded by $y = 0$ and $y = \pm f(x)$ where $f(x) \equiv 0$ for $|x| > L$. In the region $y > 0$ the problem was to determine

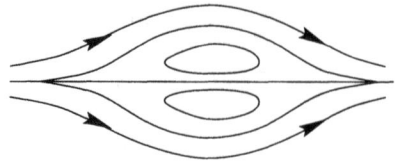

FIG. 7.4. Schematic form of a Sadoviskii vortex

a stream-function ψ, the function $f(x)$ and relations between ω, L and h such that $\psi = 0$ on $y = 0$ and $\psi = 0$ on $y = f(x)$, $-L \leq x \leq L$ with

$$\nabla^2 \psi = \begin{cases} 0 & y > f(x), \\ -\omega & -L \leq x \leq L, \ y < f(x). \end{cases} \quad (7.59)$$

There is also a condition on $y = f(x)$ that

$$|\nabla \psi|_+^2 - |\nabla \psi|_-^2 = 1 - c_p, \quad (7.60)$$

where again the subscripts refer to values above and below the dividing streamline and c_p is defined in (7.50). The far field condition was that $\psi \sim y$. The boundary shape function had to satisfy $f(\pm L) = f'(\pm L) = 0$ at the end points.

Sadoviskii approached this problem using the known solutions for point vortex singularities and writing the stream-function in terms of (a) the flow due to the uniform distribution of vorticity in the vortex, (b) a flow due to a vortex sheet of unknown magnitude on $y = f(x)$ and (c) the potential flow of the far field. This lead to two non-linear integral equations relating the vortex shape and the vortex sheet strength which were solved numerically to obtain a solution.

Later Pierrehumbert (1980) considered finite distributions of vorticity which might correspond to a steady flow far field. This problem was considered again by Saffman and Tanveer (1982) who correctly accounted for a logarithm term neglected by Pierrehumbert near the points $y = 0$, $x = \pm L$.

Of course, in the cases where $f(x)$ varies slowly there is a slender boundary approximation using the method of Childress which leads to a similar non-linear integral equation to (7.58) for $f(x)$,

$$-\frac{2}{\pi} \fint_{-L}^{L} \frac{f'(\xi)}{x - \xi} d\xi + \frac{1}{4} \omega^2 [f(x)]^2 = 1 - c_p. \quad (7.61)$$

This is an integrated form of a steady Benjamin–Ono equation (one form of the Benjamin–Ono equation for a function $u(x,t)$ is $u_t + u u_x - \mathcal{H}[u_{xx}] = 0$ where $\mathcal{H}$ is a Hilbert transform, see for instance Thomee and Vasudeva-Murthy (1998) or original work in Benjamin (1967) and Ono (1975)).

Extensive solutions for the full problem were given in Smith (1986a) and Moore et al. (1988) using very different numerical methods. Smith adopted an iterative procedure based around direct numerical solution of the Poisson equation inside the vortex for a given shape function $f(x)$ by using a stretched coordinate $Y = y/f(x)$ and transforming coordinates to (x, Y). A correction was applied to $f(x)$ proportional to the difference between the local pressure inside the vortex and that immediately outside and iterated using under-relaxation.

Moore et al. (1988) adopted the parametrisation $x = -L \cos \theta$, $0 \leq \theta \leq \pi$ and represented the shape function by a Fourier series,

$$f(\theta) = \sin^3 \theta \sum_{n=0}^{\infty} a_n \cos n\theta, \quad (7.62)$$

where the $\sin^3\theta$ ensures that $f = f' = 0$ at the endpoints. The condition (7.60) across the dividing streamline could also be written in terms of a vortex sheet along the dividing streamline where the vortex sheet strength was defined by

$$\gamma = |\nabla\psi|_+ - |\nabla\psi|_-. \tag{7.63}$$

This in turn was represented as a Fourier series

$$\gamma(\theta) = \sum_{n=0}^{\infty} b_n \cos\theta. \tag{7.64}$$

The solution was developed by a similar method to Sadoviskii. If the two Fourier series for the shape and vortex sheet strength were truncated and collocation applied to the non-linear integral equations at suitable number of points then the result was a set of non-linear equations for the coefficients of the Fourier series and ω. The non-linear equations were solved using Newton's method. Moore et al. (1988) also discussed a pressure iteration scheme but again from an analytic point of view rather than involving finite difference solutions as had Smith.

The general principles can be simply illustrated by applying the method of Riley (1987) to the slender form of the problem, (7.61). The length L can be taken as $L = 1$ through rescaling ω and $1 - c_p$, so using Riley's method let

$$f'(x) = \sum_{n=2}^{\infty} A_n \sin n\theta, \tag{7.65}$$

and integrating (remembering that $\dfrac{\mathrm{d}f}{\mathrm{d}\theta} = \sin\theta \dfrac{\mathrm{d}f}{\mathrm{d}x}$)

$$f(x) = \sum_{n=2}^{\infty} A_n \Big[\frac{\sin(n-1)\theta}{n-1} - \frac{\sin(n+1)\theta}{n+1}\Big], \tag{7.66}$$

where starting the series at $n = 2$ ensures that $f = 0$ at the ends. The Hilbert integral in the non-linear integral equation can be evaluated exactly giving for (7.61),

$$2\sum_{n=2}^{\infty} A_n \cos n\theta + \frac{1}{4}\omega^2 \Big\{\sum_{n=2}^{\infty} A_n \Big[\frac{\sin(n-1)\theta}{n-1} - \frac{\sin(n+1)\theta}{n+1}\Big]\Big\}^2 = 1 - c_p. \tag{7.67}$$

If the Fourier series are truncated at $n = N$ and collocation applied at $\theta_j = j\pi/(N+1)$, $j = 1, \ldots, N$ then a set of N non-linear equations are obtained for ω, $A_2, \ldots, A_N$. An example of the shape of the Sadoviskii vortex obtained by solving the set of equations using Newton iteration is shown in Figure 7.5 for $c_p = 0.9$. Even the simplified set of non-linear equations which come from assuming a slender shape seem very dependent on the starting solution for convergence. As noted by Moore et al. (1988) the slender theory becomes inaccurate quickly as

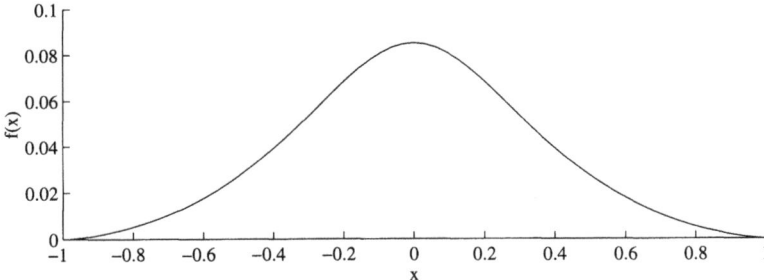

FIG. 7.5. Upper boundary of a Sadoviskii vortex for $c_p = 0.9$ calculated using Riley's method for a slender shape.

c_p decreases from one but the general nature of the numerical problem using a Fourier representation is clear from (7.67).

The results of Smith (1986a) and Moore et al. (1988) for the vortex strength $\omega(c_p)$ and the width of the vortex have established the properties of Sadoviskii vortices with fore and aft symmetry. Smith conjectured that there might be asymmetric solutions, Moore et al. (1988) evaluated the eigenvalues of the Jacobian matrix and found that the symmetric solutions are either unique or if there are other solutions they are not connected to the symmetric solutions.

7.6.2 Rotational corner flow

An extremely interesting flow from a theoretical viewpoint is the corner flow introduced by Leal (1973) as part of a study on separation. The general flow and geometry is shown in Figure 7.6. Two streams directed in opposite directions meet where there is a plate parallel to the two incoming streams. The outer potential flow corresponding to this situation is simple with stream-function $\psi = -xy$. If the plate is given by $|x| \leq 1$, $y = 0$ then the inviscid flow would correspond to an outer velocity $u_e = -x$ so this provided a decelerating outer flow and under some circumstances separation should be expected on the plate. The really substantial advantage this had as a model separation problem was that the size of the separation region was not expected to become large as the Reynolds number becomes large.

The question of Prandtl–Batchelor flows associated with the corner vortex has been described most extensively in Moore et al. (1988). In particular they note that a transformation of one quadrant of the flow to a half plane changes the flow to a type of Sadoviskii vortex but with non-uniform vorticity.

Solutions of the full Navier–Stokes equations for this geometry have been given in McLachlan (1990, 1991b) and some encouraging results have emerged at large Reynolds number such as the formation of regions of almost constant vorticity. However, McLachlan also conjectured that the occurence of secondary separation may lead to a number of regions of constant vorticity, each bounded by a vortex sheet. Another interesting feature of this flow is that the reattachment

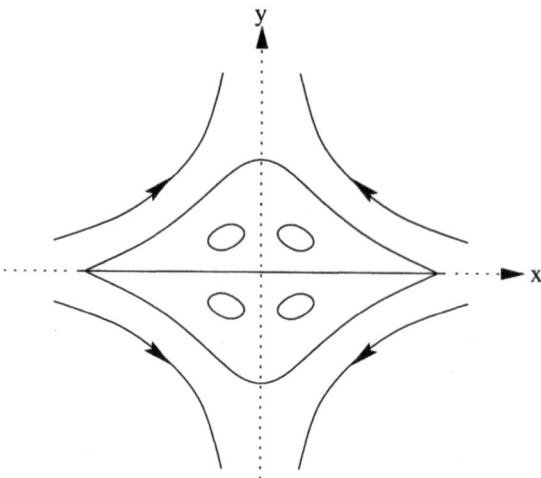

FIG. 7.6. Rotational corner flow introduced in Leal (1973).

point, which has to become cusped in a Prandt–Batchelor flow. This may prove an example of the reattachment model proposed in Peregrine (1985) but more amenable to computational verification than that in flow behind a cylinder.

7.7 Summary

This brings us back to Fornberg's calculations. The calculations in Fornberg (1980) seemed if anything to support the model of Smith (1979b) although there were some details awry, see Smith (1981). The corrected large Reynolds number results in Fornberg (1985) were much closer to the form of a Sadoviskii vortex than anything previously computed. Shortly after Fornberg's first paper, Peregrine (1985) (submitted in 1980, it was accepted for publication in 1984) had proposed a serious objection to the idea of a vortex whose wake was $O(R)$ long but only $O(R^{1/2})$ wide. He argued that the wake might be approximated by two vortices a distance W apart. Since the velocity at the top of the wake would be U and the velocity on the centreline zero the vorticity across the eddy would be of order U/W and as the area of each eddy would be of order LW the two vortices would induce a velocity in the other of order

$$\frac{\text{vorticity} \times \text{eddy area}}{\text{distance apart}} \sim \frac{\frac{U}{W}LW}{W} = \frac{UL}{W} \sim U, \tag{7.68}$$

so that the length and width should be asymptotically the same order ar $R \to \infty$.

Around the same time Smith (1985) was able to propose a model for very large eddies which had the main eddy structure built upon a Sadoviskii type vortex but was able to retain the significant features near the body which had been developed in Smith (1979b). Near the cylinder and near the tail of the vortex he proposed 'buffer' zones of asymptotic size $R^{1/2}$ which would enable

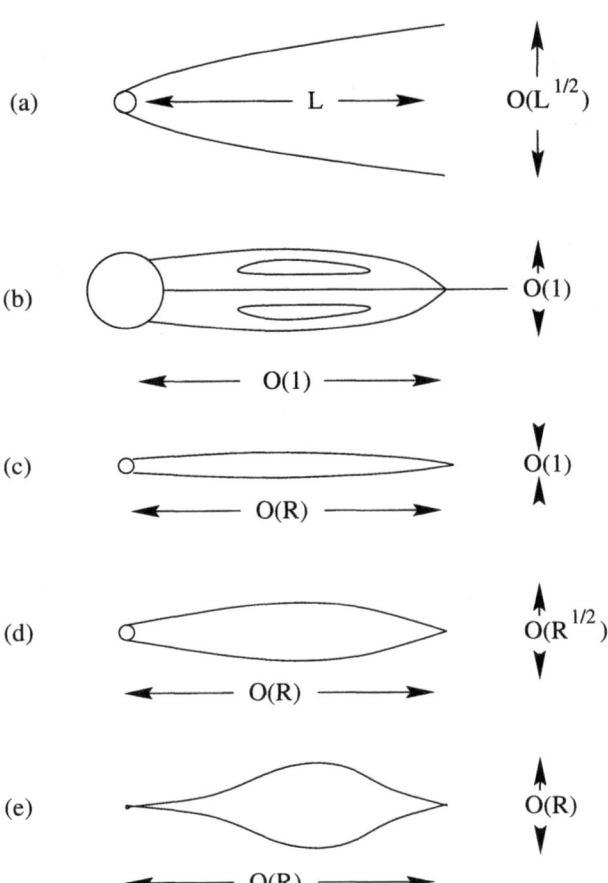

FIG. 7.7. Possible $R \to \infty$ inviscid limits for flow around a cylinder (a) Kirchoff free streamline, width $\sim O(L^{1/2})$, (b) Batchelor (1956b), length and width $\sim O(1)$, (c) Acrivos et al. (1965), length $\sim R$, width $\sim O(1)$, (d) Lagerstrom (1975), Smith (1979b), length $\sim R$, width $\sim R^{1/2}$, (e) Peregrine (1985), Smith (1985, 1986b), length $\sim R$, width $\sim R$.

matching to the Sadoviskii vortex. Incorporation of a Sadoviskii vortex has also been carried out in Chernyshenko (1988) although he proposed some different features to those in Smith (1985).

Thus it appears that the leading order asymptotic structure and characteristics for laminar high Reynolds number flow about a cylinder (and by implication, other bluff bodies) have been determined. Whether they are useful at practical Reynolds numbers is still emerging, certainly the calculations of McLachlan for a flat plate suggest that the Reynolds number has to be so large that the theory may be difficult to apply quantitatively at moderate Reynolds numbers. It is also interesting that the asymptotic models generally lead to non-linear computational problems and further advances may rest on our ability to compute accurate solutions to some extremely difficult problems.

In order to illustrate the sequence of models which have been proposed we show in Figure 7.7 possible separation shapes, starting with the Kirchoff limit with a wake that grows indefinitely. The idea that there might be two regions of constant vorticity of similar size to the body has not stood up to comparison with computed solutions of the full Navier–Stokes equations. Similarly the suggestion that the wake would be long and thin has not been successful. The model with a wake of length $O(R)$ but of width $O(R^{1/2})$ has proven to be a precursor for a model of a large eddy of length and width of order R. However, there remains uncertainty about the nature of any large eddy which forms because of the different predictions of the asymptotic drag coefficient which comes from the different models. For further reading see Chapter 6 in Sychev et al. (1998).

Part III

Channel Flow

8

INTRODUCTION TO CHANNEL FLOW

8.1 Introduction

Two-dimensional channel flow is important when flow occurs through a duct which is much wider than it is high. In such situations the effect of the sides of the duct is relatively unimportant and the flow through much of the duct is essentially two-dimensional. Of course, it is much easier to organise manifolding for pipes than channels, so the greater number of flow devices used for heat and mass transfer are based on flow though or around cylinders but there are a significant number of situations where heat or mass transfer devices are made using plates and the flow between the plates is best described as channel flow.

Consider a channel of height H through which there is a steady flux of fluid Q per unit width. If the $\hat{x}$-axis is aligned with the channel and the $\hat{y}$-axis is transverse to the channel, we might suppose that the channel boundaries are $\hat{y} = hf(\hat{x}/LH)$ and $\hat{y} = H + hg(\hat{x}/LH)$, where h and LH are specified lengths and f and g are given functions both of which vanish far upstream and downstream. Our interest is in channel geometries where the longitudinal disturbance of the boundaries is large compared to the channel height, so that $L \gg 1$ and where the indentation of the boundaries into the channel is small, so that $h \ll H$. Far upstream we expect the flow to be Poisseuille, so that depending on how the flux through the channel is non-dimensionalised, the velocity in the $\hat{x}$ direction is

$$\hat{u} \sim \frac{1}{2}\frac{U_s}{H^2}\hat{y}(H - \hat{y}), \tag{8.1}$$

where U_s is a velocity scale. As the flux is specified, the velocity scale is easily evaluated to be $U_s = 12Q/H$. The pressure far upstream will satisfy

$$\hat{p} \sim -\frac{\mu U_s}{H^2}\hat{x}. \tag{8.2}$$

Non-dimensionalise distances with respect to H, velocities with respect to U_s and pressure with respect to $\mu U_s / H$. Then using a Reynolds number $R = U_s H/\nu = 12Q/\nu$, the Navier–Stokes equations become

$$R\mathbf{u}.\nabla\mathbf{u} = -\nabla p + \nabla^2 \mathbf{u}, \tag{8.3}$$

which together with the continuity equation

$$\nabla.\mathbf{u} = 0, \tag{8.4}$$

and the boundary conditions will determine the flow.

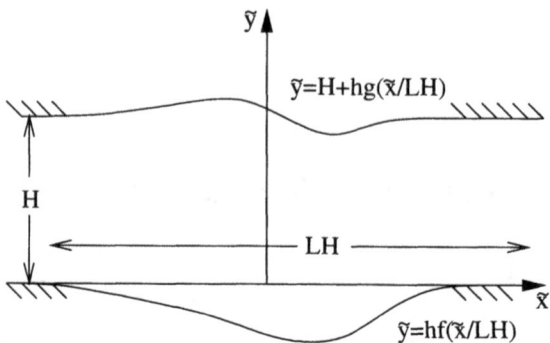

FIG. 8.1. Channel definition.

Note that using a definition of the velocity scale U_s such that the undisturbed Poisseuille flow is $\frac{1}{2}y(1-y)$ results in Reynolds numbers which are much larger than sometimes used in the literature. It is more common to see Reynolds numbers defined by Q/ν or $Q/2\nu$, that is either one twelfth or one twenty-fourth the numerical value of R as defined here. See also page 289 ff for a comparison of different published definitions of Reynolds number. As indicated, it is important to keep in mind that using a Reynolds number based on $Re = Q/2\nu$ with value say 50 would correspond to a value $R = 1200$ here and so computations are described for values of the Reynolds number R which are much higher than is perhaps conventional in the literature.

In order to prepare for an asymptotic expansion of the Navier–Stokes equations define parameters ϵ and σ by

$$\epsilon = \frac{1}{L}, \quad \sigma = \frac{h}{H}, \tag{8.5}$$

so that the boundaries of the channel are

$$y = \sigma f(\epsilon x), \quad \text{and} \quad y = 1 + \sigma g(\epsilon x). \tag{8.6}$$

Also write the Poisseuille flow far upstream as

$$U_0(y) = \frac{1}{2}y(1-y), \tag{8.7}$$

and there the pressure will satisfy $p \sim -x$.

Our aim is then to determine an asymptotic expansion of equations (8.4), (8.3) for flow between the boundaries (8.6) as $R \to \infty$ and with ϵ and σ suitably small. Since there is no external flow region the triple deck formulation cannot simply be carried over to this problem, but Smith (1976a) realised that a model based on an inviscid rotational core flow region together with viscous boundary layer type regions near the walls could be applied to this problem.

Before going into details of the solution, this problem shows there to be a critical difference between asymmetric channel indentations and symmetric

Introduction

indentations (where $g = -f$). In order to demonstrate this before we come to the fine detail of the solution we will first take a heuristic look at the expansion problem. Since the wall perturbation occurs over a long length scale, we will use a scaled longitudinal variable,

$$X = \epsilon x, \tag{8.8}$$

in the solution.

Consider the core of the flow, where we might expect the flow to be a perturbation of the oncoming Poisseuille flow. Suppose we try a perturbation of the form

$$u \sim U_0(y) + \delta U_1(X, y),$$
$$v \sim \epsilon \delta V_1(X, y),$$
$$p \sim -x + P_s P_1(X, y), \tag{8.9}$$

where δ and P_s are as yet undetermined and the difference of ϵ between u and v is to satisfy the continuity equation using the scaled X coordinate. It is attractive to say that the perturbation to the velocities should be of the same order as the wall perturbation (so that $\delta = \sigma$) but for the moment leave these distinct.

If (8.9) is substituted into the Navier–Stokes equations (8.3), then the x-momentum equation is,

$$R\epsilon\delta(U_0 U_{1X} + V_1 U_0') + R\epsilon\delta^2(U_1 U_{1X} + V_1 U_{1y}) = -\epsilon P_s P_{1X} + \delta U_{1yy} + \epsilon^2 \delta U_{1XX}. \tag{8.10}$$

What might be expected from an analogy with the triple deck is that the term $U_0 U_{1X} + V_1 U_0'$ will vanish so that

$$U_1(X, y) = A(X) U_0'(y), \quad V_1(X, y) = -A'(X) U_0(y), \tag{8.11}$$

where $-A(X)$ is an unknown displacement function.

The pressure perturbation would then be determined by a balance with the second inertial term, resulting in

$$P_{s<\text{asymmetric}>} = R\delta^2. \tag{8.12}$$

However, the solution (8.11) is always asymmetric about the centreline of the channel, so if the **flow** is to be symmetric about the centreline, it is not possible to have $U_0 U_{1X} + V_1 U_0' = 0$ and the pressure gradient has to balance with this term. Consequently in the symmetric case,

$$P_{s<\text{symmetric}>} = R\delta. \tag{8.13}$$

To relate δ and σ the flow has to be examined in one of the wall regions. Consider the lower wall region and define

$$y = \sigma Y, \tag{8.14}$$

where since the oncoming Poisseuille velocity is linear in y near the wall, the form of the perturbation expansion is

$$u \sim \sigma u_0(X, Y),$$
$$v \sim \sigma \epsilon v_0(X, Y)$$
$$p \sim -x + P_s p_1(X, Y), \tag{8.15}$$

where again the v scaling is determined by the continuity equation. If these are substituted into the Navier–Stokes equations, the x-momentum equation is

$$R\epsilon\sigma^2(u_0 u_{0X} + v_0 u_{0Y}) = 1 - \epsilon P_s p_{1X} + \sigma^{-1} u_{0YY} + \epsilon^2 \sigma u_{0XX}. \tag{8.16}$$

In order to extract a boundary layer type equation it is necessary that

$$R\epsilon\sigma^2 \sim \sigma^{-1}, \quad \text{and} \quad \epsilon P_s \sim \sigma^{-1}. \tag{8.17}$$

If the pressure scale is eliminated using (8.12) and (8.13) then two cases have to be considered separately.

In the case of an **asymmetric** disturbance to the flow $\delta = \sigma$, while if the disturbance is **symmetric** then $\delta = \sigma^2$.

In the two cases it is the **core velocity** perturbation scale which changes, the pressure perturbation scale for each channel is the same,

$$P_s = R\sigma^2. \tag{8.18}$$

As a result, if the flow is asymmetric, the core expansion for the u-velocity should begin

$$u \sim U_0(y) + \sigma U_1(X, y) + \cdots . \tag{8.19}$$

If the flow is symmetric about the channel centreline, the expansion should begin

$$u \sim U_0(y) + \sigma^2 U_1(X, y) + \cdots . \tag{8.20}$$

Before leaving this heuristic introduction to channel flow, it is worth examining the y-momentum equation.

In the core region

$$R\epsilon^2 \delta U_0 V_{1X} + R\epsilon^2 \delta^2(U_1 V_{1X} + V_1 V_{1y}) = -P_s P_{1y} + \epsilon \delta V_{1yy} + \epsilon^3 \delta V_{1XX}, \tag{8.21}$$

so that the pressure can only be constant across the core if $R\epsilon^2 \delta \ll P_s$.

Introduction

In the case of an asymmetric disturbance to the flow, the critical case where transverse pressure variation must be considered is $\sigma = \epsilon^2$ which gives $\epsilon = R^{-1/7}$. Hence for an asymmetric disturbance,

$$P_1 = P_1(X), \text{ only if } \epsilon << R^{-1/7}. \tag{8.22}$$

In the case of a symmetric disturbance then the critical case is $R\epsilon^2\sigma^2 \sim R\sigma^2$, or $\epsilon \sim 1$ and $\sigma \sim R^{-1/3}$. Hence for symmetric disturbances we expect that

$$P_1 = P_1(X), \text{ only if } \sigma << R^{-1/3}, \ \epsilon << 1. \tag{8.23}$$

An other important point comes from the x-momentum equation in the core. If ϵ were as small as R^{-1} then the first inertia term on the left of (8.10) would balance with a viscous term on the right and we could not neglect viscous effects across the core region, so we must also require that $\epsilon >> R^{-1}$.

In order to fix ideas a little more, consider the special case where $\sigma = \epsilon^a$ so that

$$\epsilon = R^{-1/(3a+1)}. \tag{8.24}$$

For an asymmetric disturbance the critical case where transverse pressure variation is important is $a = 2$ and provided $0 < a < 2$ we will be able to use $P_1 = P_1(X)$.

The result that $R^{-1} << \epsilon << R^{-1/7}$, $(0 < a < 2)$, whilst seeming straightforward might appear problematical with further thought. What intuition would predict is that as the indentation increases, the Reynolds number where cross channel pressure variation is important, would decrease. What we seem to have is that the disturbance magnitude $\epsilon^a = R^{-a/(1+3a)}$ is larger for smaller a. Of course the length of the disturbance region is also changing with R and indeed if we look at the slope of the wall perturbation (assuming f' and g' are bounded) it will be of order ϵ^{a+1}, and $R^{-(a+1)/(3a+1)}$ decreases as a decreases between two and zero so there is no contradiction in taking $R^{-1} << \epsilon << R^{-1/7}$ for an asymmetric disturbance.

A further way to regard the region of parameter space where this theory is applicable is to rewrite (8.24) in terms of the length of the perturbation region,

$$R^{1/7} << L << R, \text{ as } R \to \infty. \tag{8.25}$$

Then if the indentation region is very long (order R) viscous terms cannot be neglected in the core and some form of lubrication theory would be appropriate, if the indentation length decreases to order $R^{1/7}$ then pressure variation across the core is important while between these two limits we will be able to apply this theory, all of which is in accord with our intuition.

In his original development of this theory, Smith (1976a) formulated his solution in terms of the slope of any wall perturbations. If we denote a typical wall slope by $\alpha = \sigma\epsilon$ then then the conditions for $P_1 = P_1(X)$ are

$$\text{Asymmetric disturbance}: \quad R^{-1} \ll \alpha \ll R^{-3/7},$$
$$\text{Symmetric disturbance}: \quad R^{-1} \ll \alpha \ll R^{-1/3}, \qquad (8.26)$$

conditions first given in Smith (1976a).

8.2 Asymmetric channels: $R^{-1} \ll \epsilon \ll R^{-1/7}$

8.2.1 Non-linear formulation

Having seen some of the ideas which influence the formulation of an interactive perturbation analysis of channel flow, we shall now consider the simplest case, that of an asymmetric perturbation to the channel walls with $\sigma = \epsilon^a$, where the height of any indentation to the walls is sufficiently small for the pressure perturbation to be uniform across the channel; that is where $0 < a < 2$ and $\epsilon = R^{-1/(3a+1)}$. We shall also need to use the Poisseuille flow stream-function $\psi_0(y) = y^2/4 - y^3/6$ where $U_0(y) = \psi_0'(y)$.

Core Region

In the core of the channel, where $\epsilon^a \ll y$, $\epsilon^a \ll 1 - y$, expand the stream-function by

$$\psi \sim \psi_0(y) + \epsilon^a \psi_1(X, y) + \cdots, \qquad (8.27)$$

and the pressure by

$$p \sim -x + \epsilon^{2a} R P_1(X) + \cdots, \qquad (8.28)$$

where the independence of P_1 with respect to y was shown in the introduction. We have also seen that this stream-function expansion will lead to the solution

$$\boxed{\psi_1(X, y) = A(X) U_0(y),} \qquad (8.29)$$

in the core, where $A(X)$ is an undetermined function as is $P_1(X)$. The function $-A$ gives the displacement of streamlines in the core.

Lower wall layer

If we turn to the lower wall region and define a near wall coordinate, Y, by

$$y = \epsilon^a Y, \qquad (8.30)$$

expand the stream-function as

$$\psi \sim \epsilon^{2a} \Psi_0(X, Y) + \cdots, \qquad (8.31)$$

and retain the same expansion for the pressure as in the core, then

$$\Psi_{0Y} \Psi_{0XY} - \Psi_{0X} \Psi_{0YY} = -P_1'(X) + \Psi_{0YYY}, \qquad (8.32)$$

with boundary conditions on the plate $\Psi_{0X}(X, f(X)) = \Psi_{0Y}(X, f(X)) = 0$. Since the core flow has a non-zero velocity perturbation as $y \to 0$ (recall $U'(0) \neq$

0), the core solution will provide $\psi \sim \frac{1}{2}\epsilon^{2a}[Y^2/2 + A(X)Y] + o(Y)$ as $y \to 0$ which means that we must match

$$\Psi_0(X, Y) \sim \frac{1}{4}Y^2 + \frac{1}{2}A(X)Y + o(Y), \quad \text{as } Y \to \infty. \tag{8.33}$$

We must also require that the expansion matches far upstream, $\Psi \sim Y^2/4$ as $X \to -\infty$.

The non-linear wall condition can be simplified by a transformation due to Prandtl (Smith (1976a)), let

$$\eta = Y - f(X), \tag{8.34}$$

then the equation for $\Psi_0(X, \eta)$ is the same as (8.32),

$$\boxed{\Psi_{0\eta}\Psi_{0X\eta} - \Psi_{0X}\Psi_{0\eta\eta} = -P_1'(X) + \Psi_{0\eta\eta\eta},} \tag{8.35}$$

with boundary conditions $\Psi_{0X}(X, 0) = \Psi_{0\eta}(X, 0) = 0$, $\Psi_0 \sim \eta^2/4$ as $X \to -\infty$ and

$$\boxed{\Psi_0 \sim \eta^2/4 + [A(X) + f(X)]\eta/2 + o(\eta) \text{ as } \eta \to \infty.} \tag{8.36}$$

The system of equations for the lower wall layer cannot be solved yet because $A(X)$ remains unknown.

Upper wall layer

Next consider the region near the upper wall, and define a near wall coordinate Z by

$$1 - y = \epsilon^a Z, \tag{8.37}$$

expand the stream-function as

$$\psi \sim \frac{1}{12} - \epsilon^{2a}\tilde{\Psi}_0(X, Z) + \cdots, \tag{8.38}$$

and again retaining the pressure expansion from the core, we obtain a similar equation to (8.32),

$$\tilde{\Psi}_{0Z}\tilde{\Psi}_{0XZ} - \tilde{\Psi}_{0X}\tilde{\Psi}_{0ZZ} = -P_1'(X) + \tilde{\Psi}_{0ZZZ}, \tag{8.39}$$

with boundary conditions that $\tilde{\Psi}_{0X}(X, -g(X)) = \tilde{\Psi}_{0Z}(X, -g(X)) = 0$, $\tilde{\Psi}_0 \sim Z^2/4$ as $X \to -\infty$ but now to match with the core, there has been a sign change (corresponding with U_0' changing sign between the two walls), so that

$$\tilde{\Psi}_0 \sim \frac{1}{4}Z^2 - \frac{1}{2}A(X)Z + o(Z) \text{ as } Z \to \infty. \tag{8.40}$$

If we apply Prandtl's transformation to this system,

$$\xi = Z + g(X), \tag{8.41}$$

we will have

$$\tilde{\Psi}_{0\xi}\tilde{\Psi}_{0X\xi} - \tilde{\Psi}_{0X}\tilde{\Psi}_{0\xi\xi} = -P_1'(X) + \tilde{\Psi}_{0\xi\xi\xi}, \tag{8.42}$$

with boundary conditions $\tilde{\Psi}_{0X}(X,0) = \tilde{\Psi}_{0\xi}(X,) = 0$, $\tilde{\Psi}_0 \sim \xi^2/4$ as $X \to -\infty$ but the condition for matching with the core is now

$$\tilde{\Psi}_0 \sim \xi^2/4 - [A(X) + g(X)]\xi/2 + o(\xi), \quad \text{as } \xi \to \infty. \tag{8.43}$$

Determination of displacement function $-A(x)$

The equations for the two boundary layer like perturbations, (8.35) and (8.42) are identical, they have the same conditions at the walls and behave the same upstream. Smith (1976a) deduced that their solution far from their respective walls should also be the same, so that

$$A + f = -(A + g), \tag{8.44}$$

and hence

$$A(X) = -\frac{1}{2}(f(X) + g(X)). \tag{8.45}$$

Thus the displacement of the core is just the average of the wall displacements and with the displacement function known, the wall layer problems are fully posed and can be solved numerically.

This leads us to what might be called the fundamental problem for a wall layer: given a function $B(X) = (f(X) - g(X))/2$, determine functions $\Psi_0(X, \eta)$ and $P_1(X)$ satisfying

$$\Psi_{0Y}\Psi_{0X\eta} - \Psi_{0X}\Psi_{0\eta\eta} = -P_1'(X) + \Psi_{0\eta\eta\eta}, \tag{8.46}$$

which also satisfy the boundary conditions,

$$\Psi_0(X,0) = \frac{\partial \Psi_0}{\partial \eta}(X,0) = 0, \tag{8.47}$$

$$\Psi_0(X,\eta) \sim \frac{1}{4}\eta^2 + \frac{1}{2}B(X)\eta, \quad \text{as } \eta \to \infty, \tag{8.48}$$

and far upstream the shear tends to a constant value,

$$\frac{\partial^2 \Psi_0}{\partial \eta^2}(X,\eta) \sim \frac{1}{2} \quad \text{as } X \to -\infty.$$

8.2.2 Linearised theory

We have seen that the core flow-wall region interaction leads to a non-linear boundary layer problem. If we suppose that the wall functions, f and g are small (compared to 1) we could linearise the non-linear boundary layer problem (or at any rate the leading order term in the asymptotic expansion we have developed for the wall layer). There is a reservation in doing this; an assumption that the longitudinal velocity can be approximated in, for instance, the lower boundary layer region by a velocity

$$u \sim Y + u', \tag{8.49}$$

where $u' \ll Y$ is to some extent self contradictory since if we were considering a wall perturbation of height $f = O(\mu)$ where $\mu \ll 1$ then the appropriate scale which defines Y would also be multiplied by μ and we should still have a non-linear boundary layer problem for the velocities. However, we could for instance imagine of a wall perturbation which further downstream has f (or g) being $O(1)$ while we work in a region where $f \ll 1$. Additionally we should be concerned that very small indentations will fail to fit into the asymptotic framework developed (as evinced by the condition $\epsilon \gg R^{-1}$). Despite these remarks, there is some physical sense in considering an expansion with $f = \mu F$ and $g = \mu G$, where $\mu \ll 1$ as $R \to \infty$. We shall refer to this case as the linearised problem. The advantage of linearisation will be that we can apply Fourier transforms to the resulting equations and hence obtain transform 'solutions' in Fourier space, transform 'solutions' which if not directly invertable can nevertheless be examined asymptotically and which reveal some of the structure in the solution of the non-linear equation.

We examine only the lower wall layer after the Prandtl transformation has been made, similar conclusions follow for the upper wall layer. Assume that the leading order term in the expansion of the stream-function is given by

$$\Psi_0(X,\xi) \sim \frac{1}{4}\xi^2 + \mu\Psi_\mu(X,\xi), \tag{8.50}$$

and that the pressure is $P_1(X) = \mu p_1(X)$, so (8.42) gives the leading order equation,

$$\frac{1}{2}\xi\frac{\partial^2\Psi_\mu}{\partial\xi\partial X} - \frac{1}{2}\frac{\partial\Psi_\mu}{\partial X} = -p_1'(X) + \frac{\partial^3\Psi_\mu}{\partial\xi^3}, \tag{8.51}$$

and where we require that $\Psi_\mu \sim \frac{1}{4}\xi(F-G)$ as $\xi \to \infty$. If we let $w = \Psi_{\mu\xi\xi}$ and as before, use an overbar to represent a complex Fourier transform defined by

$$\overline{F}(k) = \frac{1}{\sqrt{2\pi}}\int_{-\infty}^{\infty} e^{-ikX} F(X) dX, \tag{8.52}$$

then

$$\overline{w}_{\xi\xi} = \frac{ik}{2}\xi\overline{w}. \tag{8.53}$$

Hence the function $\overline{w}$ satisfies a form of Airy equation. We need to be a little careful in defining roots of ik; if we take

$$\left(\frac{ik}{2}\right)^{1/3} = 2^{-1/3}|k|^{1/3}e^{i\pi\operatorname{sgn}k/6}, \tag{8.54}$$

then the solution of (8.53) is

$$\overline{w}(k,\xi) = \overline{W}(k)\operatorname{Ai}\left(\left(\frac{ik}{2}\right)^{1/3}\xi\right), \tag{8.55}$$

where $\overline{W}(k)$ is yet to be determined. This can be done by writing

$$\overline{\Psi}_{\mu\xi} = \overline{W}(k)\int_0^\xi \operatorname{Ai}\left(\left(\frac{ik}{2}\right)^{1/3}s\right)ds, \tag{8.56}$$

so that matching to the core gives

$$\overline{W}(k) = \frac{3}{4}\left(\frac{ik}{2}\right)^{1/3}(\overline{f}(k) - \overline{g}(k)). \tag{8.57}$$

Then using (8.51) at the wall, the transform of the pressure is

$$ik\overline{p}_1(k) = \left(\frac{ik}{2}\right)^{1/3}\operatorname{Ai}'(0)\overline{W}(k), \tag{8.58}$$

which simplifies to

$$\overline{p}_1(k) = \frac{3}{8}\left(\frac{ik}{2}\right)^{-1/3}\operatorname{Ai}'(0)(\overline{f}(k) - \overline{g}(k)). \tag{8.59}$$

This is a very important result since the inverse Fourier transform of $(ik)^{-1/3}$ is proportional to $X^{-2/3}H(X)$ where $H(X)$ is the Heavyside step function. Consequently when we invert (8.59) in terms of a convolution integral the pressure perturbation will always vanish *ahead* of the point where f or g is first non-zero. It was already clear that the displacement function $-A(X)$ had this property. This shows the lack of upstream influence for this range of ϵ in the boundary layer regions.

To complete the details of the solution, it is straight forward to show (see for instance Lighthill (1958)) that the inverse Fourier transform of $(ik)^{-1/3}$ is

$$\frac{\sqrt{2\pi}}{\Gamma(\frac{1}{3})}X^{-2/3}H(X),$$

and using $\operatorname{Ai}'(0) = -1/(\Gamma(\frac{1}{3})3^{1/3})$, the pressure perturbation will be given by

$$p_1(X) = -\frac{3^{2/3}\sqrt{\pi}}{2^{11/3}\Gamma^2(\frac{1}{3})}\int_0^\infty (f(X-s) - g(X-s))s^{-2/3}ds. \tag{8.60}$$

The pressure perturbation will be negative through a channel constriction and positive through a channel dilatation, all in accordance with our intuition from Bernouilli's equation.

8.3 Symmetric channels: $R^{-1} \ll \epsilon \ll 1$

In a symmetric channel, the boundary functions satisfy $f(X) = -g(X)$ and thus $A(X) = 0$ and $B(X) = f(X)$. The computational problem is identical in form to that for an asymmetric channel but has to be solved with $B(X)$ having double the value for an asymmetric channel with one flat wall. Thus while the core flow perturbation is asymptotically smaller than in the asymmetric channel case, the problem to be solved for the pressure–wall layer interaction is essentially unchanged. Effects of velocity alterations near the boundaries will if anything, be larger, as might be expected since the overall deceleration in a symmetric channel expansion (or acceleration in a constriction) will be larger than if say only one wall was perturbed.

The asymptotic theory is only slightly modified from that for an asymmetric channel, the major differences are that the core displacement function vanishes and the leading order equation for the core has to include the pressure gradient. In order to contrast this solution with that for an asymmetric channel we examine the solution in the same fashion as before.

Core region

In the core of the channel with $\sigma \ll y$ and $\sigma \ll 1 - y$ the stream-function has to be expanded

$$\psi \sim \psi_0(y) + \sigma^2 \psi_1(X, y) + \cdots, \tag{8.61}$$

and the pressure

$$p \sim -x + \sigma^2 R P_1(X), \tag{8.62}$$

and now the stream-function perturbation has to satisfy

$$\psi_0'(y) \frac{\partial^2 \psi_1}{\partial X \partial y} - \psi_0''(y) \frac{\partial \psi_1}{\partial X} = -P_1'(X). \tag{8.63}$$

In order to have symmetry the perturbation needs to satisfy $\psi_1(X, \tfrac{1}{2}) = 0$ and to match the flow upstream, $\psi_1 \to 0$ as $X \to -\infty$.

If the pressure is eliminated the stream-function perturbation must satisfy

$$U_0(y) \frac{\partial^2 \psi_1}{\partial y^2} - U_0''(y) \psi_1 = 0, \tag{8.64}$$

and at the wall

$$\psi_1(X, 0) = 2 P_1(X). \tag{8.65}$$

The solution of (8.64) for Poiseuille flow was given in Tillett (1968),

$$\psi_1 = 2P_1(X)[1 - 2y - 2y(1-y)\log\frac{y}{1-y}]. \tag{8.66}$$

The response in the core is a purely local one, given P_1 at any point X the core perturbation is completely determined without explicit reference to upstream or downstream conditions. There may of course be an implicit linkage through the determination of the pressure perturbation, P_1. This solution does have a problem in that the longitudinal velocity will have a logarithmic singularity near each wall so that further steps in the expansion for the stream-function should include terms involving $\log\sigma$.

Wall layers

Using the scaling

$$y = \sigma(\eta + f(X)), \tag{8.67}$$

and the expansion

$$\psi \sim \Psi_1(X, \eta), \tag{8.68}$$

then the stream-function still satisfies equation (8.35) and its boundary conditions with $A \equiv 0$. The upper layer satisfies the same system.

Thus the computational problem is only slightly modified, the same boundary layer system has to be solved but instead of the core perturbation being given in terms of a displacement function it is determined by the pressure perturbation in the wall layers.

8.4 Free streamline theory

Having considered the theory of channel flow from the view that there is a perturbation to Poiseuille flow as the Reynolds number becomes large, we should also consider the view according to traditional free streamline theory. Of course, to do that we need to determine free streamline flows which occur in channels. In this section we consider the application free streamline theory with the assumption that the flow far upstream is uniform irrotational flow. In the next chapter we will see that having uniform irrotational flow is probably not the correct high Reynolds number limit for Poiseuille flow but nevertheless what might be denoted 'conventional' free streamline theory for channels is of interest in its own right and may eventually provide some methods for dealing with more complicated inviscid *rotational* channel flows.

We have already seen some use of Woods (1954a, 1956) method for calculating free streamline flow around a cylinder. In this section we will apply his ideas for free streamline flow through a channel. The mappings used are slightly different from those for a cylinder but his underlying idea was the same: the flow region is mapped to a strip where on one side the flow angle θ is assumed given and on the other side the fluid speed q on a free streamline is known. Then the solution for the complex function

$$\Omega(w) = \log \frac{dz}{dw} = -\log q + i\theta,$$

could be written down in terms of integrals of the known values of q, θ on the boundaries of the strip and once Ω was known the flow field could be determined by integration as in (6.23).

In the case of channel flow there are some important differences from an external flow. The maps which will be used are shown in Figure 8.2. The physical $z = x + iy$ plane is taken to have $x = 0$ at the point of separation. The flow field illustrated could be either separation form the upper wall with the lower wall flat or separation in a symmetric channel where we only consider the upper half channel. If the flux through the channel $0 \leq y \leq 1$ is take to have unit value then the flow field in the complex potential plane, $w = \phi + i\psi$ will occur in the strip $0 \leq \Im w \leq 1$. Woods' method relies on mapping this region of the w plane to a strip in another plane, ζ, such that the free streamline becomes the line $\zeta = \pi i/2$ while the walls in contact with the fluid become the line $\zeta = 0$, see Figure 8.2, (d). The simplest way to determine the transformation from the w to the ζ plane is to introduce an auxiliary plane t where the strip in the w plane is mapped to the upper half of the t plane and then a Schwarz–Christoffel transformation is used to map the t plane to the ζ plane before eliminating the t plane.

In this case the transformation

$$t = e^{\pi w}, \tag{8.69}$$

provides the required map from the w plane to the t plane with the separation point S mapping to $t = -1$. Using a Schwarz–Christoffel transformation with angle change π at $S : t = -1$,

$$\frac{d\zeta}{dt} = \frac{k}{t+1}, \tag{8.70}$$

where k is an unknown constant. Integrating,

$$\zeta = k \log(t+1) + c, \tag{8.71}$$

with c another constant. The two constants k, c are determined by requiring that

$$t = 0 \mapsto \zeta = 0, \quad t = -2 \mapsto \zeta = \frac{\pi i}{2}, \tag{8.72}$$

so that

$$\zeta = \frac{1}{2} \log(t+1), \tag{8.73}$$

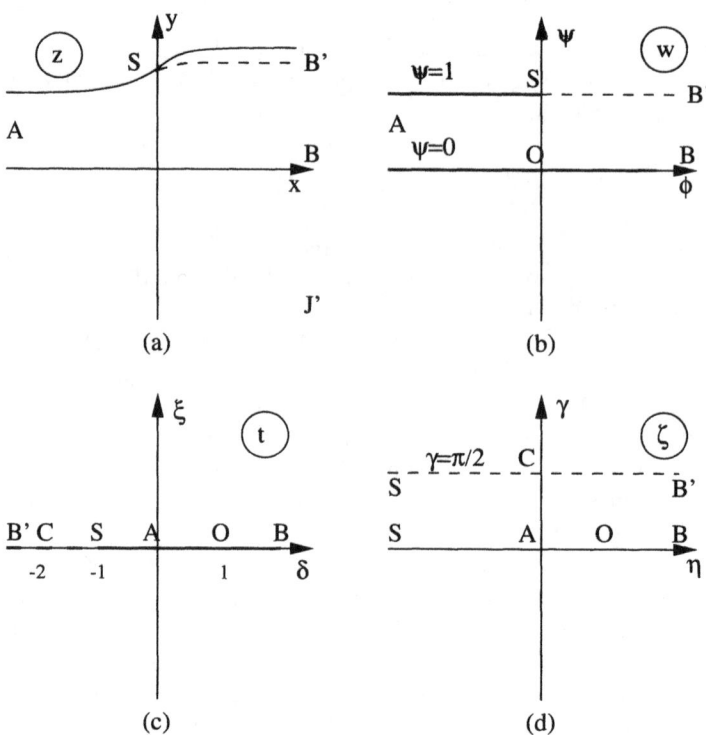

FIG. 8.2. Planes used in Woods' method of calculating free streamlines in a channel (a) physical plane (b) complex potential (c) auxiliary t-plane (d) ζ plane.

and hence
$$\zeta(w) = \frac{1}{2}\log(e^{\pi w} + 1). \tag{8.74}$$

On the walls of the channel, S-A-B in Figure 8.2, (d), the angle θ is known while on the free streamline, since the pressure will be constant, the speed will be known, denoted $q = V$. Upstream the fluid speed is assumed to be one. It should be clear that one difference from flow around a cylinder which arises here is that unless the fluid were to continue on along the line $y = 1$, the speed V must be different from one to satisfy continuity. If the downstream width of the flow B-B' is b then continuity gives $Vb = 1$. Determining V for a given separation position has to be one of the outcomes of the solution procedure.

If in the ζ plane we denote the angle θ on $\zeta = \eta + 0i$ by $\theta_0(\eta)$ and the real part of Ω on $\zeta = \eta + \pi i/2$ by $\Omega_1(\eta)$ then Woods' solution for the harmonic function $\Omega(\zeta)$ is
$$\Omega(\zeta) = \frac{1}{\pi}\int_{-\infty}^{\infty}[\theta_0(\eta)\operatorname{cosech}(\eta - \zeta) + \Omega_1(\eta)\operatorname{sech}(\eta - \zeta)]d\eta. \tag{8.75}$$

Since $\zeta = \zeta(w)$ we will without confusion use the notation $\theta_0(\eta)$ and $\theta_0(\phi) = \theta_0(\eta(\phi))$ to represent the angle of the flow direction on the wall.

As in the case of flow around a cylinder, (8.75) turns out not to be the best representation of the solution since it is derived by contour integration around a strip in the ζ-plane containing the point ζ in (8.75) and so it is not easy to consider $\zeta \to \pm\infty$ as the integral and the limit cannot be interchanged. It is better to integrate (8.75) by parts, and Woods gave that solution

$$\Omega(\zeta) = \Omega_\infty + \frac{2}{\pi}\int_{-\infty}^{\infty} \tanh^{-1} e^{\eta-\zeta} d\theta_0(\eta) - \frac{2}{\pi}\int_{-\infty}^{\infty} \tan^{-1} e^{\eta-\zeta} d\Omega_1(\eta), \quad (8.76)$$

where Ω_∞ is the limit as $\zeta \to \infty + \pi i/2$. In flow about a cylinder and in many channel flows the angle θ_0 vanishes as $\eta \to \infty$ so that then

$$\Omega_\infty = -\log V;$$

however, in some channel flows where we consider the free streamline to reattach to a wall at a point where the potential is ϕ_0 then

$$\Omega_\infty = -\log V + i\theta_0(\phi_0).$$

To apply this formula to the problem being considered it is necessary to note that the interval $(-\infty, \infty)$ for η corresponds to both the intervals $[0, -\infty)$ and $(-\infty, \infty)$ for the potential on the upper and lower walls (that is S-A and A-B in Figure 8.2, (d) and (b)). On the lower wall, A-B, $d\theta_0 = 0$ and on the free streamline $d\Omega_1 = 0$. Hence using the map $\zeta(w)$ to rewrite the integral in terms of w instead of ζ we obtain

$$\Omega(w) = -\log V - \frac{2}{\pi}\int_{-\infty}^{0} \tanh^{-1}\sqrt{\frac{1-e^{\pi\tilde\phi}}{1+e^{\pi w}}} d\theta_0(\tilde\phi). \quad (8.77)$$

If now $w \to -\infty + i$ where the speed has unit value and the angle is zero, giving $\Omega = 0$, then the free stream velocity is obtained as

$$\log V = -\frac{2}{\pi}\int_{-\infty}^{0} \tanh^{-1}\sqrt{1-e^{\pi\tilde\phi}} d\theta_0(\tilde\phi). \quad (8.78)$$

It is this result which is only possible after integration of (8.75) by parts.

Of course θ_0 is known in terms of x and not in terms of ϕ so to complete the theory it is necessary to determine the map between ϕ and x; that is, the speed along the upper wall before separation occurs. If we put $w = \phi_w + i$ in the integral and q for the speed along the wall,

$$-\log q = -\log V - \frac{2}{\pi}\int_{-\infty}^{0} \tanh^{-1}\sqrt{\frac{1-e^{\pi\tilde\phi}}{1-e^{\pi\phi_w}}} d\theta_0(\tilde\phi). \quad (8.79)$$

If the upper wall is $y = 1 + G(x)$ with distance s along the wall measured from $s = 0$ at $\phi_w = 0$, then the speed q is related to the potential on the wall $\phi_w(x)$ by

$$q = \frac{d\phi_w}{ds} = \left[1 + \left(\frac{dG}{dx}\right)^2\right]^{-1/2} \frac{d\phi_w}{dx}. \tag{8.80}$$

Finally we reach an integral equation for $\phi_w(x)$, $-\infty < x \leq 0$,

$$-\log\left\{[1 + \left(\frac{dG}{dx}\right)^2]\frac{d\phi_w}{dx}\right\} = -\log V - \frac{2}{\pi}\int_{-\infty}^{0} \tanh^{-1}\sqrt{\frac{1 - e^{\pi\phi_w(\tilde{x})}}{1 - e^{\pi\phi_w(x)}}}\frac{d\theta_0}{dx}(\tilde{x})d\tilde{x}. \tag{8.81}$$

The numerical solution of this integral equation can be found iteratively in the same way as for Woods' method for flow around a cylinder. Suppose the upstream wall is straight for $x < -L$ and let $h = L/M$, $x_j = jh$, $j = -M, \ldots, 0$ with M an integer. If the wall angle is $\theta_j = \theta_0(x_j) = \tan^{-1} G'(x_j)$, $\phi_j = \phi_w(x_j)$ and the integral evaluated by quadrature on the mid-points of intervals, then the right-hand side of (8.81) becomes

$$-\log V - \frac{2}{\pi}\sum_{k=-M}^{k=-1} \tanh^{-1}\sqrt{\frac{1 - e^{\pi(\phi_k+\phi_{k+1})/2}}{1 - e^{\pi\phi_j}}}(\theta_{k+1} - \theta_k). \tag{8.82}$$

Equating this to the left-hand side of (8.81) provides an estimate of $\frac{d\phi_w}{dx}$ which can then be integrated from $\phi_w = 0$ at $x = 0$ to obtain a new iterate for ϕ_j and a discrete version of (8.78) gives a new estimate of V. This process seems to converge in a small number of iterations.

Once the wall potential values ϕ_j have been calculated, streamlines can be determined by first calculating the (x, y) location of the equipotential $\phi = 0$ and then streamlines determined for ψ constant as ϕ varies from $\phi = 0$, in each case using (6.23),

$$z = \int e^{\Omega(w)} dw.$$

As an example, we use a channel whose upper wall changes from $y = 1$ to $y = 2$ via a cosine expansion. In order to vary the separation point, suppose separation occurs a distance $x_s < 1$ from the start of the channel expansion so that

$$G(x) = \begin{cases} 0 & x \leq -x_s, \\ \frac{1}{2}[1 - \cos(\pi(x + x_s))] & -x_s < x \leq 1 - x_s, \\ 1 & x > 1 - x_s. \end{cases} \tag{8.83}$$

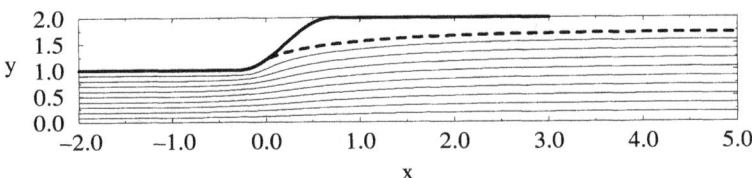

FIG. 8.3. Separated solution using Woods' method for a channel with cosine expansion and separation at $x_s = 0.3$. Streamlines shown at interval $\Delta\psi = 0.1$.

Calculated streamlines for the case $x_s = 0.3$ is shown in Figures 8.3 while the wall speed, q is shown in Figure 8.4. The free stream velocity for this case is $V = 0.582$ and the free streamline eventually expands to $y = 1.72$. The wall speed increases sharply as the expansion is approached before decreasing to the free streamline value. The singularity in the gradient of the speed at the separation point will be reflected in a singularity in the pressure gradient at separation. Numerical experiments show that for this expansion, separation must occur before $x_s \approx 0.378$ if the flow is not to intersect the wall at $y = 2$.

The type of pressure gradient singularity can be found by expanding (8.77) just upstream of the separation point. If we let $w = i - \epsilon$ with ϵ small but positive then

$$\Omega(i - \epsilon) \sim -\log V + i\theta_0(\phi_0) - \frac{2}{\pi} \int_{-\infty}^{0} \tanh^{-1}\sqrt{\frac{1 - e^{\pi\tilde{\phi}}}{\pi\epsilon}} \, d\theta_0(\tilde{\phi}), \quad (8.84)$$

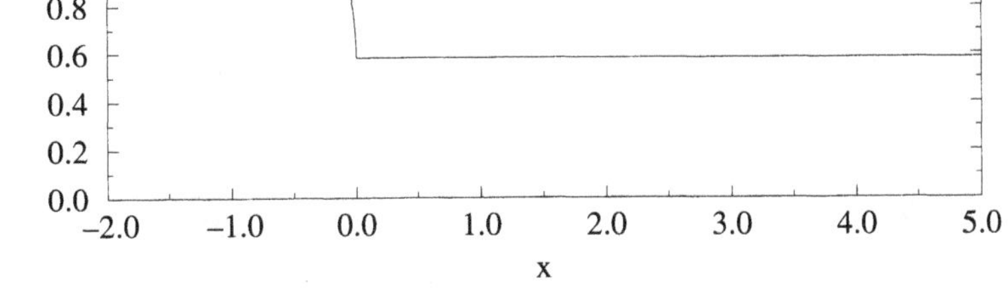

FIG. 8.4. Speed on the upper wall ($x < 0$) and on the free streamline ($x¿0$) from Woods' method for cosine expansion and separation determined by $x_s = 0.3$.

and using $\tanh^{-1} a \sim -\pi i/2 + a^{-1}$ for large a,

$$\Omega(i - \epsilon) = -\log q + i\theta \sim -\log V - \frac{2}{\sqrt{\pi}} \int_{-\infty}^{0} \frac{1}{\sqrt{1 - e^{\pi\tilde\phi}}} \, d\theta_0(\tilde\phi)\sqrt{\epsilon} + i\theta_0(0). \tag{8.85}$$

The pressure gradient on the wall will be given by

$$\frac{dp}{ds} = -q\frac{dq}{ds} = q^2 \frac{dq}{d\phi}.$$

The coordinate s along the wall, with $s = 0$ at separation, satisfies $\phi \sim s$ as $s \to 0$, so defining

$$\kappa = \frac{V^{5/2}}{\sqrt{\pi}} \int_{-\infty}^{0} \frac{1}{\sqrt{1 - e^{\pi\tilde\phi}}} \, d\theta_0(\tilde\phi), \tag{8.86}$$

then the pressure gradient upstream of separation behaves as

$$\frac{dp}{ds} \sim \kappa(-s)^{-1/2} \text{ as } s \to 0^-. \tag{8.87}$$

As with external flows, the pressure gradient has a square root singularity as separation is approached.

The singularity strength κ is shown in Figure 8.5 for the cosine expansion (8.83). Surprisingly it rises and falls as the separation position is moved along

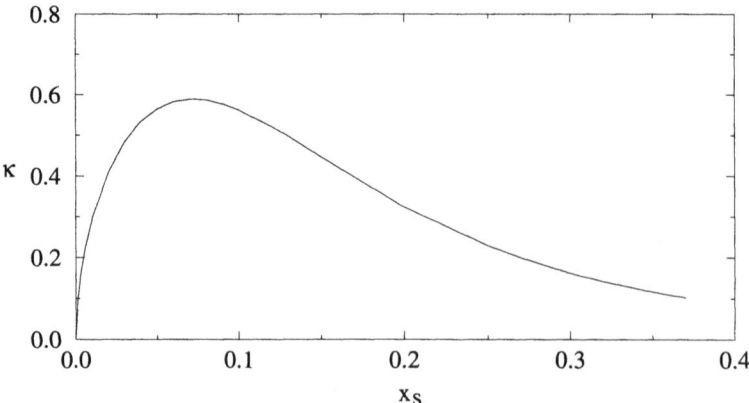

FIG. 8.5. Pressure singularity coefficient κ as separation point x_s is varied for a cosine expansion.

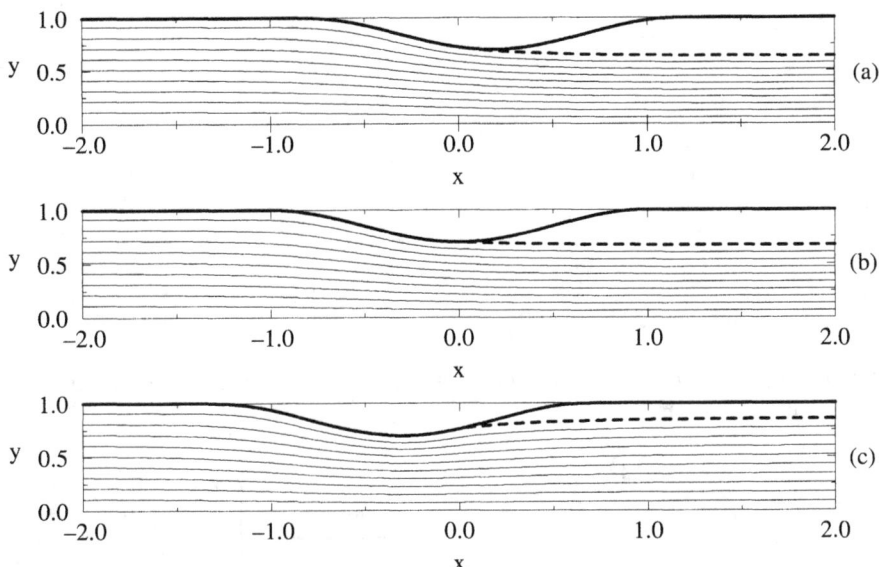

FIG. 8.6. Free streamlines for channel indentation with indentation magnitude $a = 0.3$. (a) $x_s = 0.84$, (b) $x_s = 1.0$ (c) $x_s = 1.3$. In each case the indentation is aligned so that separation is at $x = 0$

the expansion suggesting that were we to try to combine this inviscid flow structure with Sychev's proposal relating the singularity strength to Reynolds number then there would be a critical Reynolds number in order to achieve separation and that there might be two limits. In one the separation point would start at the point where κ is a maximum and move upstream, in the other the separation point would move downstream from where it first occurred, presumably leading to a flow with no free streamline in the limit $R \to \infty$. Neither of these corresponds to calculations from the Navier–Stokes equations where separation usually occurs first near the location of maximum slope and the separation point moves upstream as the Reynolds number is increased. This does suggest that the free streamline theory which will be relevant to Poisseuille flow as $R \to \infty$ will be more complicated than that which occurs with irrotational flow far upstream.

A second channel with a separation region which extends to infinity is illustrated in Figure 8.6. In this case the upper wall gives a channel contraction, defined by

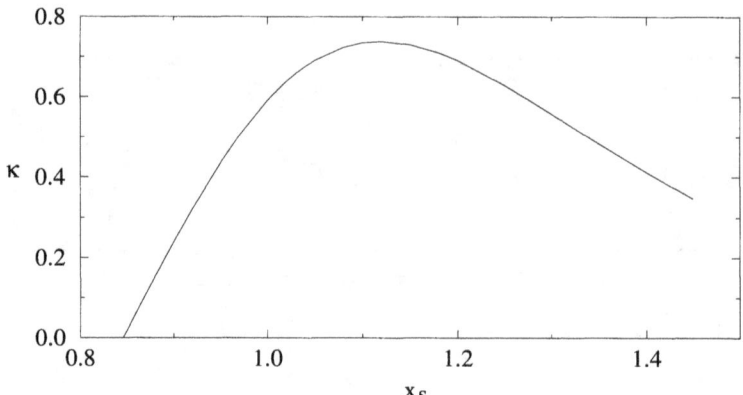

FIG. 8.7. Pressure singularity coefficient κ versus separation point for a cosine indentation with $a = 0.3$.

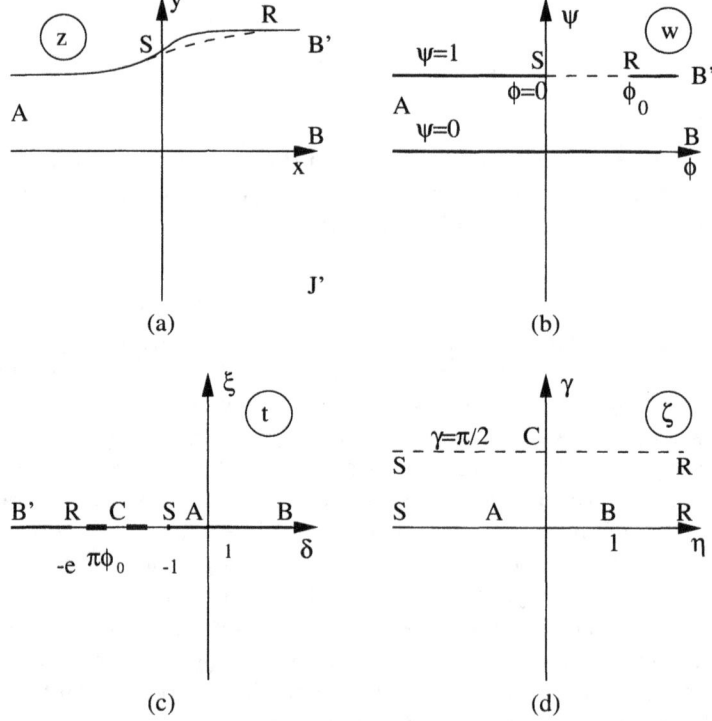

FIG. 8.8. Planes used in Woods' method of calculating free streamlines in a channel with separation-reattachment. (a) Physical plane (b) Complex potential (c) Auxiliary t-plane (d) ζ plane.

$$G(x) = \begin{cases} 0 & x \leq -x_s, \\ -\frac{\mu}{2}[1 - \cos(\pi(x + x_s))] & -x_s < x \leq 2 - x_s, \\ 0 & x > 2 - x_s, \end{cases} \quad (8.88)$$

with $\mu < 1$ determining the extent of the contraction.

The utility of Woods' method is not restricted to channel flows where the free streamline extends to infinity. An example we shall examine now using Woods' method are flows which separate from a wall and then reattach further downstream. A schematic picture of the flow region is given by Figure 8.8 (a). In the w-plane suppose separation occurs at S:$\phi = 0$ and reattachment at R:$\phi = \phi_0$.

The transformation from the w-plane to the ζ-plane can be found in the same way as before. If we use an auxiliary t-plane where

$$t = e^{\pi w},$$

then S is mapped again to $t = -1$ while R is mapped to $t = -\exp(\pi \phi_0)$. The t-plane can in turn be mapped to the ζ plane using a Schwarz–Christoffel map with angle π at R and S, so that

$$\frac{d\zeta}{dt} = \frac{k}{(t+1)(t+e^{\pi\phi_0})}, \quad (8.89)$$

where k is a constant and on integration

$$\zeta = k_1 \log \frac{t+1}{t+e^{\pi\phi_0}} + k_2, \quad (8.90)$$

with k_1, k_2 further constants. Now we shall choose to map

$$t \to \infty + 0\mathrm{i} \mapsto \zeta = 1, \text{ and } t = -1 - \lambda(e^{\pi\phi_0}) \mapsto \zeta = \frac{\pi \mathrm{i}}{2}, \quad (8.91)$$

where $\lambda < 1$ is a positive constant still to be decided. Mapping the first point gives $k_2 = 1$ while the second point gives $k_1 = 1/2$ and requires $\lambda = (e^2 - 1)^{-1}$ so that

$$\zeta(w) = \frac{1}{2} \log \frac{e^{\pi w} + 1}{e^{\pi w} + e^{\pi \phi_0}} + 1. \quad (8.92)$$

The various planes are illustrated in Figure 8.8.

With the transformation $\zeta(w)$ determined, the solution for $\Omega(w)$, (8.76) is

$$\Omega(w) = \Omega_\infty + \frac{2}{\pi} \int_W \tanh^{-1} \sqrt{\frac{(1 - e^{\pi \tilde{\phi}})(e^{\pi \phi_0} + e^{\pi w})}{(1 + e^{\pi w})(e^{\pi \phi_0} - e^{\pi \tilde{\phi}})}} \, d\theta_0(\tilde{\phi}), \quad (8.93)$$

where the integral is over the wetted wall and $\Omega_{infty} = -\log V + \mathrm{i}\theta_0(\phi_0)$. Assuming that the lower wall is straight with $d\theta_0 = 0$ then the integral is

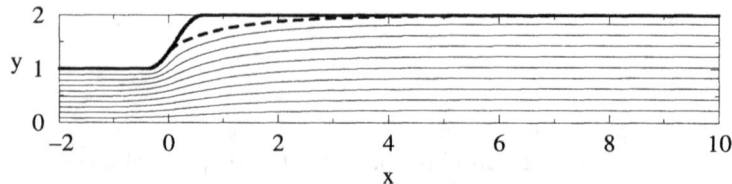

FIG. 8.9. Streamlines for separated-reattached potential flow past a cosine expansion.

$$\int_W d\theta_0(\tilde{\phi}) \equiv \int_0^{-\infty} d\theta_0(\tilde{\phi}) + \int_\infty^{\phi_0} d\theta_0(\tilde{\phi}), \qquad (8.94)$$

and finally

$$\Omega(w) = \Omega_\infty - \frac{2}{\pi}(\int_{-\infty}^0 + \int_{\phi_0}^\infty)\tanh^{-1}\sqrt{\frac{(1-e^{\pi\tilde{\phi}})(e^{\pi\phi_0}+e^{\pi w})}{(1+e^{\pi w})(e^{\pi\phi_0}-e^{\pi\tilde{\phi}})}}\, d\theta_0(\tilde{\phi}). \qquad (8.95)$$

The free stream velocity V is again determined from $w \to -\infty + 0i$,

$$\log V = -\frac{2}{\pi}(\int_{-\infty}^0 + \int_{\phi_0}^\infty)\tanh^{-1}\sqrt{\frac{1-e^{\pi\tilde{\phi}}}{1-e^{\pi\tilde{\phi}-\pi\phi_0}}}\, d\theta_0(\tilde{\phi}). \qquad (8.96)$$

There are two special cases of note: if $\phi_0 \to \infty$ with $\theta_0(\phi_0) \to 0$ then the solution (8.77) is recovered, if reattachment occurs on a downstream part of the wall which is straight with $\theta_0 = 0$ then the second integral in (8.95) vanishes and

$$\Omega(w) = -\log V - \frac{2}{\pi}\int_{-\infty}^0 \tanh^{-1}\sqrt{\frac{(1-e^{\pi\tilde{\phi}})(e^{\pi\phi_0}+e^{\pi w})}{(1+e^{\pi w})(e^{\pi\phi_0}-e^{\pi\tilde{\phi}})}}\, d\theta_0(\tilde{\phi}). \qquad (8.97)$$

In the case of a cosine expansion from unit width to a channel of width two, numerical experiments show that there is only one separation point consistent with reattachment to the downstream wall, $x_s = 0.37855\ldots$ and $\phi_0 = 4.99\ldots$. The flow is illustrated in Figure 8.9.

In the more general case where the free streamline reattaches on a section of curved wall it is necessary to slightly modify Woods' procedure presented above. The solution procedure is still iterative from an initial guess but now the two integral parts of (8.95) have to be approximated by quadrature. Assume separation and reattachment occur at x_s and x_r respectively and as an initial guess use $\phi_w(x) \approx x$ and $\phi_0 \approx x_r - x_s$. Then the free streamline speed V can be estimated from the integral (8.96) and the free streamline trajectory calculated from (8.95). Together with calculation of q on the wall this provides an estimate for ϕ'_w on the wall and on the free streamline which enables an updated estimate

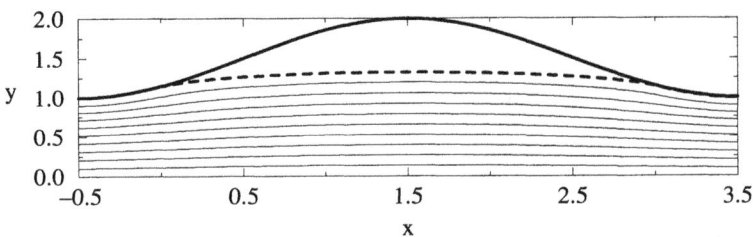

FIG. 8.10. Streamlines for separated–reattached potential flow past a cosine expansion–contraction of length 4.

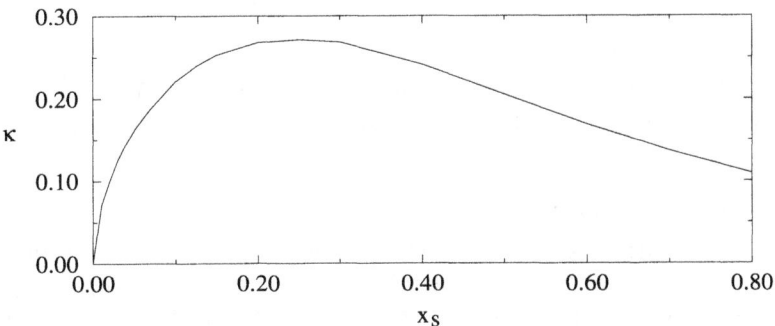

FIG. 8.11. Pressure singularity coefficient κ versus separation point for separated-reattached flow past a cosine expansion-contraction of length 4.

of ϕ_w and ϕ_0 to be calculated and the iteration repeated. Convergence appeared to be very rapid, usually within a few iterations.

An example which can be easily used for viscous flow calculations is a cosine expansion-contraction,

$$G(x) = \begin{cases} 0 & x \leq -x_s, \\ \frac{\mu}{2}[1 - \cos(\pi(x+x_s)/L)] & -x_s < x \leq 2L - x_s, \\ 1 & x > 1 - x_s, \end{cases} \quad (8.98)$$

where the parameter μ determines the magnitude of the expansion-contraction.

In the case $\mu = 1$ typical streamlines and free surface are shown in Figure 8.10 for an expansion-contraction of length 4 and with separation at $x_s = 0.5$. The free streamline solution also results in a singularity at separation which increases in strength as x_s increases from zero before decreasing for values of x_s greater than around 0.25, see Figure 8.11.

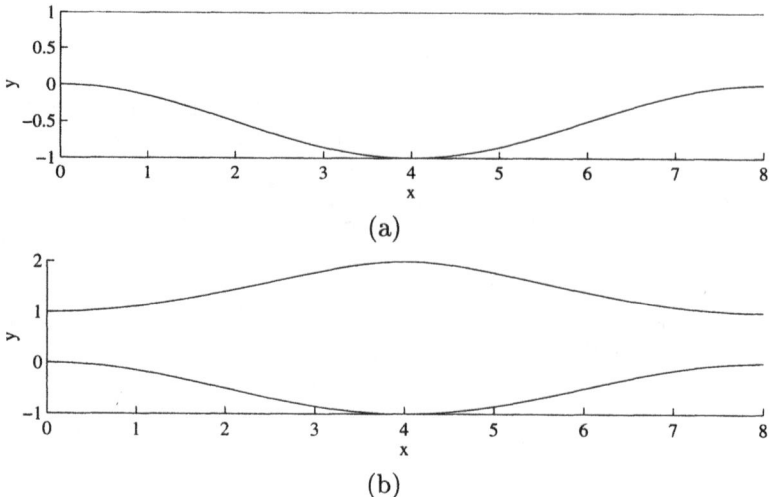

FIG. 8.12. Example channel sections for periodic geometry (a) asymmetric channel (b) symmetric channel. In each case the channel has periodicity $L = 8$.

8.5 Computed examples

In computing solutions to the interactive equations we have to consider calculations using the canonical problem and also how they are related to a physical problem. In some ways it is easiest to start with a well defined physical problem. Since it is connected to applications in mass transfer devices we shall use the following problem. Let the channel lower wall have a sinusoidal variation of non-dimensional length L and depth D (relative to the channel width) so that we need to consider only $0 \leq x \leq L$ and expect the any solution to be L-periodic in x. We shall not pursue the question of conditions on the wall shape far upstream or downstream, it is sufficient to take the channel to be one where the depth of each sinusoidal variation is the limit $\gamma \to 0^+$ of a nominal depth $De^{-\gamma|n|}$ with n the number of periods from $x = 0$.

We will look at two channels, one is asymmetric with

$$g(x) = 0, \qquad (8.99)$$

and one is symmetric, with

$$g(x) = -f(x), \qquad (8.100)$$

Given the form of the boundary again take $X = \epsilon x$ where,

$$\epsilon = L^{-1}. \qquad (8.101)$$

and introduce a new parameter, μ, such that the depth of the furrow satisfies

$$D = \epsilon^a \mu, \qquad (8.102)$$

and the wall displacement function f is given by

$$f(x) = \frac{\mu}{2}(1 - \cos\frac{2\pi x}{L}). \tag{8.103}$$

If $\mu > 0$ then this wall of the channel consists of a series of expansions, or furrows from $y = 0$, if $\mu < 0$ then the wall is a series of constrictions from $y = 0$.

Substituting for ϵ in (8.24), then as $R = L^{1+3a}$,

$$a = \frac{1}{3}(\frac{\log R}{\log L} - 1), \tag{8.104}$$

and hence (8.102) can be rewritten

$$\mu = DL^{-1/3}R^{1/3}. \tag{8.105}$$

Of course, as we have seen, the restriction that $a < 2$ for this theory to hold means we should only consider wall disturbances L asymptotically longer than $R^{1/7}H$ with $L \to \infty$ and $R \to \infty$. Shorter length wall variation will result in a cross channel pressure gradient, to be considered in the next chapter. The relation (8.105) was given in Sobey (1980). The significance of (8.105) is that there will be a critical value of μ for which separation occurs and there will be only one value once the channel shape is specified, denoted μ_c, so that the flow will be separated whenever $\mu \geq \mu_c$ and thus given the furrow length and depth, (8.105) gives an estimate of the critical Reynolds number for separation, R_c, as

$$R_c = \mu_c^3 L D^{-3}. \tag{8.106}$$

This result is in accord with our intuition, the larger L or the smaller D, then the larger the critical Reynolds number for separation. If μ_c can be calculated, this provides a prediction which can be tested against computed solutions of the Navier–Stokes equations.

Calculating a solution to the fundamental interactive wall layer problem is of course only part of providing a solution to test against computed solutions of the Navier–Stokes equations. It is also necessary to compute a composite expansion for comparison with other computations. The matching between the lower wall layer and the core flow is for $Y \to \infty$ in the wall layer and $y \to 0$ in the core. However, if $\mu > 0$ then the wall layer occupies the physical region $y < 0$. The normal method of obtaining a composite expansion, adding the individual expansions (represented in a suitable variable, usually the outer variable and in this case y) and subtracting either the expansion of the inner solution in the outer region or the outer solution in the inner region (and in this case if we use the outer variable, it would be the first of these) breaks down unless great care is used in determining the 'outer' solution for $y < 0$. If there is a channel expansion on the lower wall, the actual wall position is located at $y < 0$ for ϵ fixed and

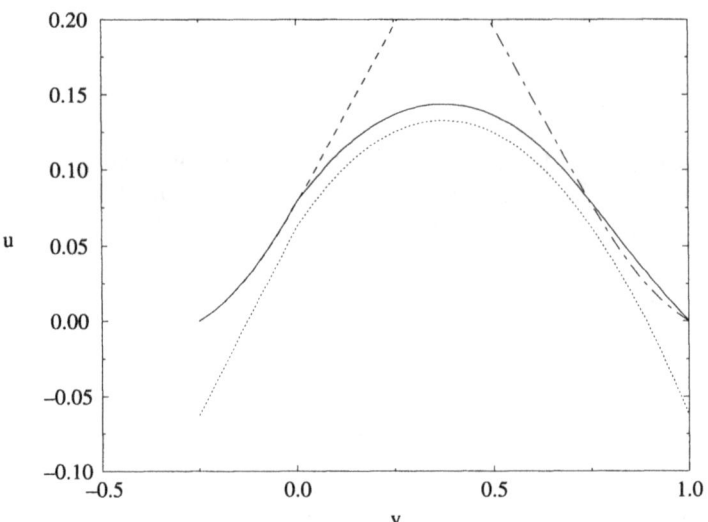

FIG. 8.13. Example calculation of 2-2 term composite expansion for velocity u. (a) ($\cdots$) core velocity, (b) (- - - -) lower wall layer velocity, (c) (- $\cdot$ - $\cdot$ -) upper wall layer velocity, (d) (———) composite profile. See text for further details.

finite but the matching still has to be carried out for $y \to 0^+$. The core velocity is given by

$$u \sim \frac{1}{2}y - \frac{1}{2}y^2 + \frac{1}{2}\epsilon^a A(X) - \epsilon^a A(X)y + o(\epsilon^a), \tag{8.107}$$

and when $y = O(\epsilon^a) > 0$ this can be reordered

$$u \sim \frac{1}{2}y + \frac{1}{2}\epsilon^a A(X) + o(\epsilon^a), \tag{8.108}$$

so to the order we are considering, the extension of the core flow to cover the region $y < 0$ is assumed to be

$$u_{\text{core}}(X, y) = \begin{cases} \frac{1}{2}y(1-y) + \epsilon^a A(X)(\frac{1}{2} - y) & : \ 0 \leq y \leq 1, \\ \frac{1}{2}y + \frac{1}{2}\epsilon^a A(X) & : \ y < 0. \end{cases} \tag{8.109}$$

If the u velocity in the lower and upper wall layers is denoted $u_L(X, Y)$ and $u_U(X, Z)$, the 2-term inner, 2-term outer composite expansion is

$$u_{\text{composite2-2}}(X, y) = u_{\text{core}}(X, y) + u_L(X, \epsilon^{-a} y) - \frac{1}{2}y + \frac{1}{2}\epsilon^a A(X)$$
$$+ u_U(X, \epsilon^{-a}(1-y)) - \frac{1}{2}(1-y) - \frac{1}{2}\epsilon^a A(X). \tag{8.110}$$

It should be clear that assuming the upper velocity contribution minus its far field expansion has gone to zero near the lower wall, these definitions give

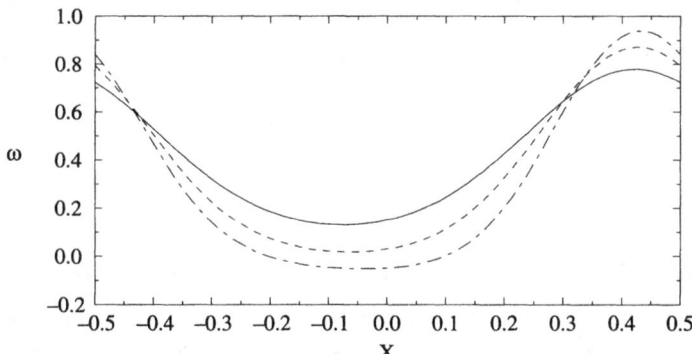

FIG. 8.14. Wall vorticity (ω) calculated for three channel indentations, (—) $\mu = 2.$, ($\cdots$) $\mu = 3.$, ($-\cdot-$) $\mu = 4$.

$u_{\text{composite2}-2} = u_L$ for $y < 0$ and that there will be a change in velocity gradient of order ϵ^a at $y = 0$.

An example of calculating this composite expansion for an asymmetric channel is shown in Figure 8.13. The calculation is for $L = 8$, $D = 0.5$ and the velocity profile is that across the widest part of the channel, $x = 4$. The wall layer solutions grow linearly with distance far from the walls, the core flow becomes negative near either wall. The composite profile satisfies $u = 0$ at either wall but the overall flux is in error by an amount of order ϵ^a. In this example $\epsilon = 0.125$ and $a = 1$. Overall, composite expansions in two-space dimensions have not been widely used to generate solutions and much further work is needed to establish the best form of composite expansion and its comparison with solutions of the full Navier–Stokes equations.

Asymmetric example

The first example is an asymmetric case where $g = 0$ and f is defined by (8.103). In this channel, with one flat wall and an expansion on the lower wall the flow separates on the lower wall at quite low Reynolds numbers, depending on the expansion depth. The wall vorticity is shown in Figure 8.14 for three cases, $\mu = 2., 3., 4.$ and where separation first occurs when $\mu \approx 3.25$ at $X \approx 0.44$ a little upstream of the maximum point in the wall expansion which occurs at $X = 0.5$.

Applying the criterion (8.105) this gives a prediction of separation for this particular wall geometry of Reynolds numbers greater than R_c where

$$R_c \approx 34LD^{-3}. \qquad (8.111)$$

Symmetric example

Results for the symmetric channel shown in Figure 8.12 (b) can be deduced from the asymmetric case by considering a double value of μ in an asymmetric channel with one flat wall. This provides an estimate of the critical Reynolds number for separation in this type of channel as

$$R_c \approx 4.25 L D^{-3}; \qquad (8.112)$$

that is $\mu_c = 1.625$ for the particular sinusoidal wall variation in Figure 8.12 (b).

8.6 Numerical solution of the Navier–Stokes equations

Numerical solution of the Navier-Stokes equations in two-dimensional channels is relatively straightforward for modest Reynolds numbers. The solutions we shall discuss here have been obtained in two ways. One is from the unsteady stream-function – vorticity formulation using time marching but steady boundary conditions. The other is by solving the steady equations iteratively. Each of these methods has an extensive literature.

The convection diffusion equation for the vorticity can be discretised in a number of ways. The results here are from a central difference discretisation of the vorticity equation. The stream-function has been determined from the vorticity using a multigrid method. Simple iterative methods such as Jacobi, Gauss–Seidel or relaxation on their own are much less efficient than either multigrid or conjugate gradient iteration. In channels with smoothly varying walls, such as a cosine expansion or contraction, it is straightforward to transform the channel to a rectangular computational region and then to discretise the convection–diffusion equation for vorticity and the Poisson equation for the stream-function in transformed coordinates. Further details of the numerical methods we have used and example programs can be found through material in appendix A. An example conformally mapped mesh is shown in Figure 8.15 for a 256×32 mesh although practical meshes are usually finer, for instance 512×64. The suggestion to use smooth walls to study separation in channels was made to me by Lighthill around 1977. General experience since then indicates the wisdom of that advice since separation at moderate Reynolds number in channels does seem much more affected by global considerations such as length and depth of an expansion region rather than fine detail of the wall geometry and by using a smooth wall questions about local flow near a corner can be side-stepped.

As an example of a separated channel flow consider an asymmetric channel in Figure 8.12 (a) where $L = 8$ and the depth $D = 1$. This represents a severe test of an asymptotic theory since the parameter $\epsilon = 0.125$ has a large value. The theory being considered here will be valid for $R \ll \epsilon^{-7} = 2097152$ (that is, cross channel pressure variation should not be significant for these Reynolds numbers) and in Figure 8.16 the streamlines are shown for $R = 1800$. There is a reasonably large vortex against the lower wall and the dividing streamline is convex towards the flow.

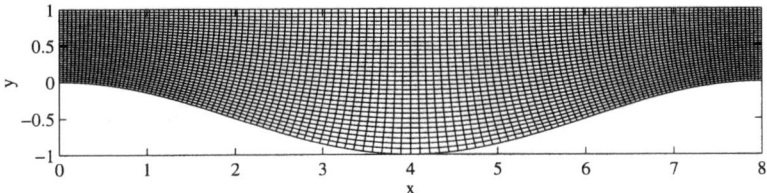

FIG. 8.15. Example conformally mapped mesh for solution of Navier–Stokes equations in a channel.

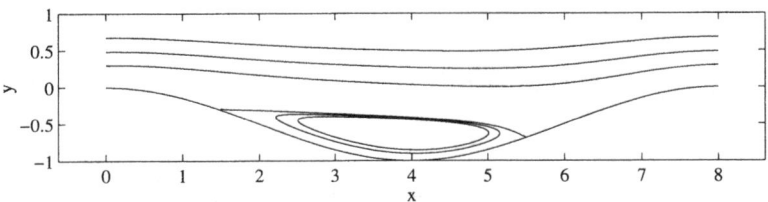

FIG. 8.16. Computed solution from the Navier–Stokes equations for the stream-function at $R = 1800$.

If the asymptotic theory is used this channel and Reynolds number correspond to $\mu = 6.08$. The flow near the lower wall scaled to physical units is shown in Figure 8.17. Some features are reproduced quite well. The length of the vortex is given reasonably accurately and the wall velocity gradient, shown in Figure 8.18 is also predicted reasonably well (for purposes such as estimating the average value or applying Reynolds analogy to estimate heat or mass transfer). The vortex itself is not predicted with such resolution, there being a significant part of the dividing streamline which is concave towards the flow. Further the theory predicts that the same feature would be found on the opposite wall which is not the case in the computed flow. Given the rather large value of ϵ and the low value of R it is perhaps not surprising that agreement is only partial.

An additional prediction from the asymptotic theory is the Reynolds number for separation (8.111) and that is difficult to test conclusively because of

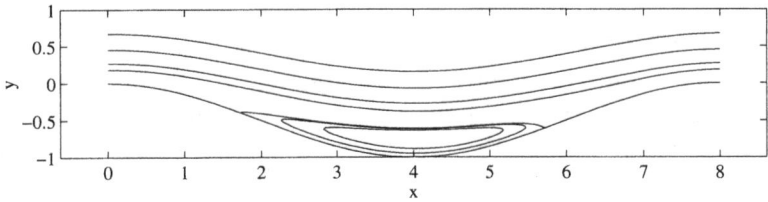

FIG. 8.17. Triple deck solution for lower wall layer rescaled to physical coordinates for $R = 1800$.

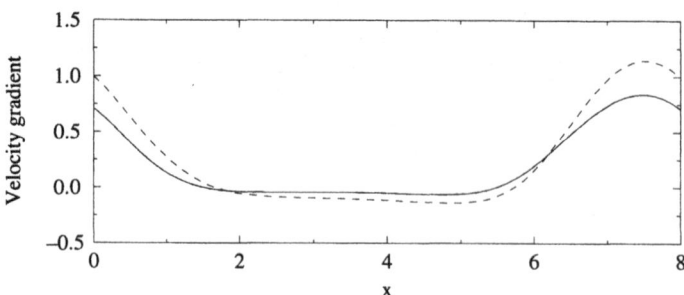

FIG. 8.18. Comparison of Navier–Stokes (——) and triple deck (- - -) calculation of wall shear ar $R = 1800$.

the difficulty in computing solutions for the laminar Navier–Stokes equations at very large Reynolds numbers. However, some comparison can be made and in Figure 8.19 we show estimates of the critical Reynolds number for separation from computed solution of the Navier–Stokes equations at three channel furrow depths, $D = 0.5, 0.35, 0.3$ which seem reasonably in agreement with asymptotic theory.

It needs to be remembered that the prediction of this theory is not just for separation on the lower wall but also on the upper flat wall. It is the case that there are neither observations nor calculations which show such upper wall separation at moderate Reynolds numbers. Separation on the upper wall can be observed in unsteady flow and in unsteady calculations and one has to assume that were we able to compute sufficiently large Reynolds number solutions of the Navier–Stokes equations then separation on the upper flat wall would occur.

Of course, one has to distinguish the essential correctness of an asymptotic theory with its utility and what these examples attempt to illustrate is both the correctness of the asymptotic theory and the utility for prediction of some gross quantities such as a critical Reynolds number for separation and wall stress even though fine detail of the flow may not be accurate at practical Reynolds numbers.

8.7 Flow near a corner

In many channel flows the boundary varies smoothly but some of the most interesting applications arise when there is steady flow through a sharp channel expansion, for instance one important example is a right-angled expansion. One misconception which once existed within fluid mechanics and particularly within computational fluid mechanics concerned the nature of flow near a corner of angle greater than π. There was a general uncertainty about the flow because at low Reynolds number the flow had a weak singularity in the vorticity at such a corner and at infinite Reynolds number the flow also has a singularity at the

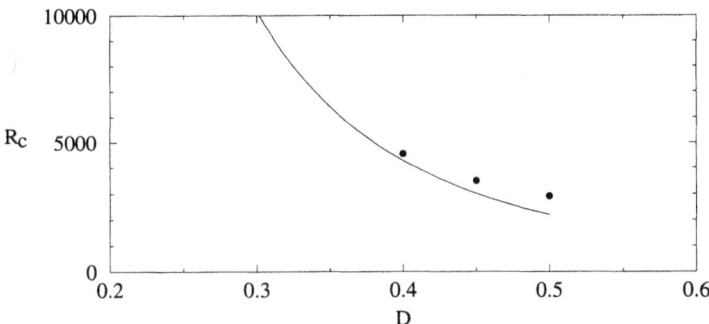

FIG. 8.19. Critical Reynolds number for separation in an asymmetric channel of length $L = 8$ versus expansion depth D: (—) interactive boundary layer prediction, filled circles show computed solutions of Navier–Stokes equations.

corner This resulted in concern that flow at finite Reynolds numbers must be accompanied by separation **from** the corner.

These ideas were based on the well known singularities which occur at $R = 0$ (Stokes flow) and at $1/R = 0$ (potential flow). However. the singularities at $R = 0$ and at $1/R = 0$ are fundamentally different and in the case $R = 0$, Weinbaum (1968) has shown that the dividing streamline should not in general, emanate from the corner since the vorticity will have the same sign on either side of the corner. In the case where the Reynolds number is large but finite, interactive boundary layer theory indicates the same conclusion and it is only in the limit $R \to \infty$ that the corner singularity will be accompanied by separation. In this section we shall draw together these ideas to examine the behaviour of the flow near a corner and the location of a separation point using both low Reynolds number theory and interactive boundary layer theory and consider some practical consequences for computation of flow near corners.

Whilst this section does not deal directly with a central aspect of interactive boundary layer theory, it is a problem where interactive boundary layer theory supports and is consistent with a suggestion that at finite Reynolds number, separation does not occur at a corner but a little beyond the corner.

The study of flow near a sharp corner can be traced back at least to Rayleigh (1911) who considered the special case of Stokes flow near a trailing or leading edge but who did not obtain singular solutions for more general angles. Later Dean and Montagnon (1949) established that there could be solutions for other angles and that these would have a weak singularity in the vorticity at the corner. Moffatt (1964) looked particularly at solutions where an infinite set of regions of rotating fluid existed in an acute angled corner. That work was closely followed by a comprehensive study from Lugt and Schwiderski (1965a,b,c).

Consider a corner of half angle α, Figure 8.20. In the case of potential flow suppose the complex potential is

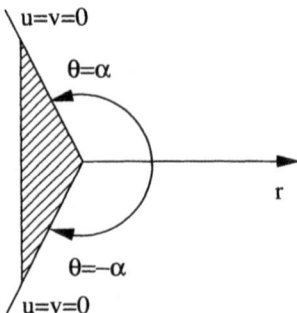

FIG. 8.20. Geometry for flow around a corner of angle 2α.

$$w = Az^\beta, \tag{8.113}$$

and for the lines $\theta = \pm\alpha$ to have $\psi = 0$ write $A = |A|e^{i\alpha\beta}$ and choose β such that

$$2\alpha\beta = \pi, \tag{8.114}$$

with consequence that the velocity

$$u - iv = \frac{dw}{dz} = \beta Az^{\beta-1}, \tag{8.115}$$

would have a singularity at $r = 0$ if $\beta < 1$ or $\alpha > \pi/2$. When the velocity is singular, so too is the pressure and hence for $1/R = 0$ any flow might be expected to adjust so as to have smooth separation from the corner to avoid such a singularity. Since the flow is given by a complex potential (and so u and v are harmonic) the vorticity must vanish regardless of any singularity in the velocity.

If there is Stokes flow, then solutions for that problem can be found by looking for a stream-function

$$\psi \sim r^m f_m(\theta), \tag{8.116}$$

and showing that the functions f_m would have form

$$f_m(\theta) = A_m \cos m\theta + B_m \cos(m-2)\theta, \tag{8.117}$$

for antisymmetric flow about $\theta = 0$ and

$$f_m(\theta) = C_m \sin m\theta + D_m \sin(m-2)\theta, \tag{8.118}$$

for symmetric flow about $\theta = 0$, where A_m, B_m, C_m, D_m are constants. It is the antisymmetric case which might appear to be of most interest here where

$$B_m = -\frac{\cos m\alpha}{\cos(m-2)\alpha} A_m, \tag{8.119}$$

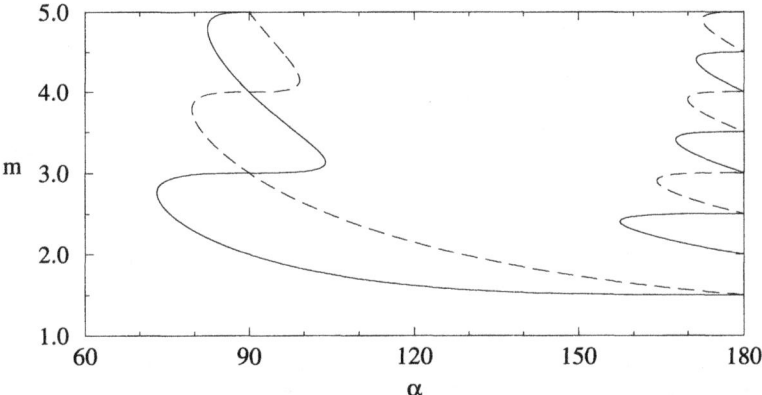

FIG. 8.21. Real roots for $m(\alpha)$ for the symmetric (- - -) and antisymmetric (—) Stokes flow about a corner of half angle α (shown in degrees).

is required for the stream-function to vanish on the wall and

$$m \sin 2\alpha + 2 \cos m\alpha \sin(m-2)\alpha = 0, \qquad (8.120)$$

for the normal velocity to vanish on the wall. Solutions of this equation were given by Moffatt and further solutions, including the symmetric case where

$$m \sin 2\alpha - 2 \sin m\alpha \cos(m-2)\alpha = 0, \qquad (8.121)$$

were given by Weinbaum (1968). For each problem there are a sequence of values for m leading to a series solution for flow about the corner. The real solutions are illustrated in Figure 8.21, there are also complex roots for acute angles, where α is small, leading to an infinite series of vortices in such a corner, see Moffatt for further details.

In the case $\alpha = \pi/2$ there is no corner and $\psi \sim r^2$ but for

$$\alpha = \frac{\pi}{2} + \epsilon, \qquad (8.122)$$

then it is straightforward to show that there is a solution

$$m \sim 2 - \frac{\epsilon}{\pi}, \qquad (8.123)$$

with consequence that the wall vorticity ω will satisfy

$$\omega \sim r^{-\epsilon/\pi}, \text{ as } r \to 0, \ \epsilon > 0, \qquad (8.124)$$

and so be singular for any angle greater than $\pi/2$, although the vorticity will be integrable so that, for instance, force or moment on the corner will remain

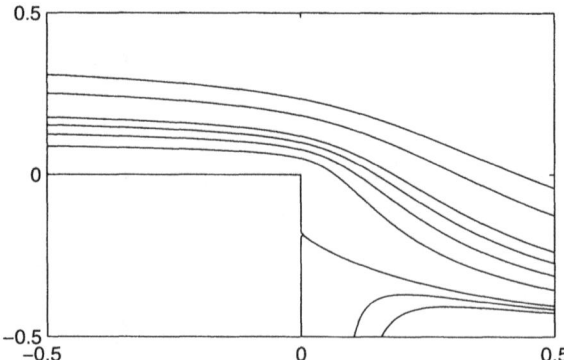

FIG. 8.22. Solution using Weinbaum's method for Stokes flow around a corner with equal symmetric and antisymmetric parts.

bounded. In the case of a right angled corner, Moffatt showed that the smallest value of m was $m = 1.544...$ implying a singularity in the vorticity of order $r^{-0.455...}$.

Thus the two limits for flow around a corner are that at $R = 0$ the velocity is zero and the vorticity has a weak singularity, and that at $1/R = 0$, the velocity is singular but the vorticity is zero. Since for all finite Reynolds numbers we impose as a boundary condition that the velocity is zero on the wall what remains is how any weak vorticity singularity interacts with a dividing streamline.

In the case of Stokes flow Weinbaum considered in detail the case of a right-angled expansion corner with $\alpha = 3\pi/4$. It is clear from Figure 8.21 that there are only two real solutions for $m(\alpha)$, $m_1 = 1.54448...$ for the antisymmetric case and $m_2 = 1.90852...$ for the symmetric case. Weinbaum argued that close to the corner the solution must be a linear combination of the functions for these two values so that the stream function might be written

$$\psi \sim a r^{m_1} f_{m_1}(\theta) + b r^{m_2} f_{m_2}(\theta), \text{ as } r \to 0, \qquad (8.125)$$

where

$$f_{m_1}(\theta) = \cos m_1 \theta + \cot \frac{3m_1 \pi}{4} \cos(m_1 - 2)\theta, \qquad (8.126)$$

and

$$f_{m_2}(\theta) = \sin m_2 \theta - \tan \frac{3m_1 \pi}{4} \sin(m_1 - 2)\theta, \qquad (8.127)$$

with a and b being undetermined constants whose ratio would reflect conditions further from the corner. In addition, if the inertial terms of the Navier–Stokes equations were included then this expansion would still give valid leading order terms for $Rr \to 0$. Weinbaum also gave the solution for the pressure near the corner and that too has a weak singularity of order $m_1 - 2 \approx 0.4554....$

Weinbaum then showed that although (8.125) gave a weak singularity in the vorticity, the vorticity did not go through zero at the corner, rather the vorticity vanished someway down the rear step and it was from that point that a dividing streamline must emanate. Contours of ψ from (8.125) are illustrated in Figure 8.22 for $\psi = -0.01, -0.05, 0.\ 0.025, 0.05, 0.075, 0.1, 0.2$ for the case $a = b = 1$. The angle the dividing streamline made with the wall was independent of the values of the constants a and b and depended only on the geometry, a result which Weinbaum acknowledges was independently discovered by H.E. Topakoglu.

Of course, since we might expect that for any Reynolds number the Stokes equations will hold close enough to a corner (on the scale $r = O(R^{-1})$) then Weinbaum's results implied that the dividing streamline should approach the corner as $R \to \infty$.

Our next step is to examine solutions using interactive boundary layer theory near a corner even though we expect the Stokes solution to be embedded within such a solution. The reason for this further solution is that when we come to consider numerical methods, the Stokes solution will often only be valid only on length scales which are much smaller than practical computational meshes whereas the interactive boundary layer solution may be valid on such practical mesh scales. It is also the case that we should only expect to obtain interactive boundary layer solutions for corner angles which are only small perturbations from a straight wall and thus not very large corner angles. Nevertheless since both inviscid and Stokes flow models predict a singularity at a corner even when the angle is small it is possible to obtain insight from application of interactive theory to corners with only small deflection.

The question of what might happen near a corner had been addressed by Stewartson (1970b) for a compressible exterior flow over an airfoil with a sharp corner. His results showed that the wall shear, while having a maximum at the corner, was nevertheless finite at the corner. The theory for a corner in the wall of a channel was set out clearly in Smith (1976a) who considered a linear dilatation and a linear expansion followed by an exponentially decreasing contraction. His results also clearly showed that there was no singularity at a sharp expansion corner on the interactive theory length scale. It is also interesting to see Figure 4 in Bogolepov (1985) for compressible flow behind a thick plate.

This can be seen from the linearised theory already considered. If a linear dilatation occurs on the lower wall with X shifted to the corner,

$$f(X) = \begin{cases} 0 & X < 0, \\ -\mu X & X \geq 0, \end{cases} \quad (8.128)$$

then (8.57) gives the transform of the wall shear perturbation as

$$\overline{W}(k) = -\frac{3\mu \mathrm{Ai}(0)}{2^{7/3}\sqrt{2\pi}} (ik)^{-5/3} \quad (8.129)$$

so that the wall shear perturbation should be given by

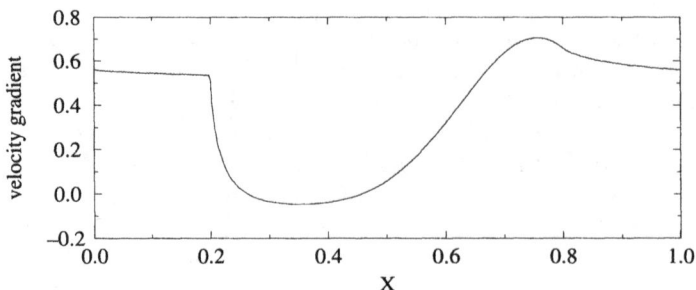

FIG. 8.23. Wall velocity gradient for channel with a corner and $\mu = 30$.

$$W(X) = \begin{cases} 0 & X < 0, \\ -\dfrac{3\mu\mathrm{Ai}(0)}{2^{7/3}\Gamma(5/3)} X^{2/3}, & X \geq 0. \end{cases} \qquad (8.130)$$

Consequently the linearised theory predicts a wall shear

$$\frac{\partial u}{\partial y} \sim \begin{cases} \frac{1}{2}, & X < 0, \\ \frac{1}{2} - 0.2341...\mu X^{2/3}, & X \geq 0, \end{cases} \qquad (8.131)$$

and separation should occur some distance downstream from the corner.

The same picture emerges from solution of the interactive equations. As an example in a periodic channel consider

$$f(X) = \begin{cases} 0, & 0 \leq X < 0.2, \\ -\frac{1}{45}\mu(5X-1)(5X-4)^2, & 0.2 \leq X < 0.8, \\ 0, & 0.8 \leq X \leq 1. \end{cases} \qquad (8.132)$$

chosen so that the wall variation is zero at $X = 0.2$, 0.8 and has zero slope at $X = 0.8$ but slope $-\mu$ at $X = 0.2$. In this case separation occurs at a somewhat greater distance from the corner than predicted by linearised theory but of course the wall slope is varying significantly from its value at the downstream side of the corner and a reduced wall slope should delay the onset of separation. An example of wall velocity gradient is shown in Figure 8.23 along with streamlines in Figure 8.24 for the case $\mu = 30$. Since the channel is taken to be periodic there is some development of the wall shear ahead of the corner but the streamlines are apparently straight there and change direction sharply above the corner. This is obviously an undesirable physical feature but inescapably part of the leading order interactive solution and would be removed (at least to a higher derivative) were a second order theory to exist.

While interactive boundary layer theory takes separation to be at a corner in the limit of infinite Reynolds number, at finite Reynolds number it is expected that separation will occur on a length scale at least order $R^{-1/7}$ from the corner.

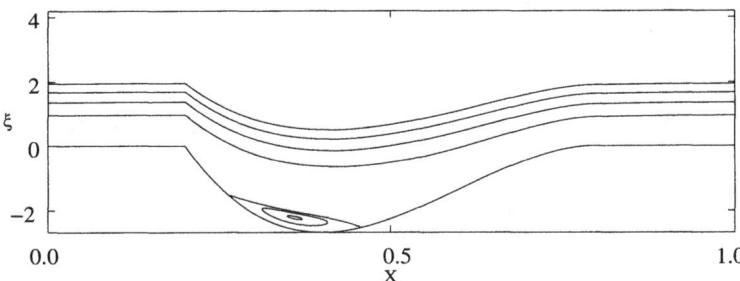

FIG. 8.24. Streamlines for flow through a channel with a corner and $\mu = 30$.

If Sychev's hypothesis can be applied to an internal flow the length scale will be of order $R^{-1/16}$. Hence the fine-ness of a mesh needed to resolve separation near a corner should be judged by these scales while any integrable viscous singularity affecting say the vorticity will be restricted to much smaller length scales and so probably not resolvable with practical meshes. What we argue next is that in practice, there is no real need to attempt to resolve these small viscous length scales and any integrable viscous singularity in the vorticity can be properly accounted for on longer length scale meshes.

So we now turn to the application of these ideas to computation of vorticity near the most common corner, a right angled corner, particularly for stream-function vorticity schemes but the ideas are also of interest for computations in two and three dimensions using primitive variables where instead of a singularity in the vorticity, a pressure singularity is relevant. The use of point values at mesh points has to some extent been replaced by interpretation of mesh values as finite volume averages and under that assumption many concerns about corner vorticity recede, provided any singularity in vorticity is integrable. A very early attempt to account for a corner singularity in potential flow was made by Motz (1946) who located the singular point away from any mesh points and used a local analytic expansion to treat mesh points near the corner singularity.

The development of finite difference and finite element approximations has centred around nodal values which approximate point values of a function, for instance in calculating a function $u(x, y)$ at a nodal point (x_i, y_j) the value $U_{i,j}$ is regarded as a direct approximation to $u(x_i, y_j)$. In finite differences equations for the values $U_{i,j}$ would usually come from local Taylor series expansions using point values. In finite elements the nodal values are used again as point values to generate local approximations from given local basis functions and the local approximation is then substituted into some integral form of the differential equation to obtain equations relating the nodal values.

The finite volume view of approximation is that the value $U_{i,j}$ is an approximation to the average value of $u(x, y)$ over some element of area and relations between the values $U_{i,j}$ are derived from integrals of the governing equations over area elements, usually applying integral theorems to relate area or volume inte-

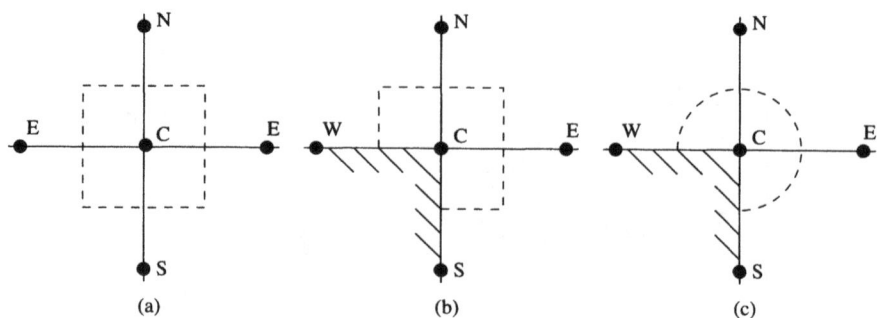

FIG. 8.25. Finite volume region for (a) a normal vertex and (b),(c) for a corner vertex.

grals to line or surface integrals of the flux of a quantity through the boundary of the element. Clearly this should work well for linear divergence type conservation laws but less well for non-linear ones, a feature which is common with finite difference and finite element discretisations.

Consider this for the case of the Poisson equation for vorticity,

$$\omega = -\nabla^2 \psi, \qquad (8.133)$$

in the context of Figure 8.25 where we are concerned with the node C whose neighbours are for convenience denoted W, S, E, N. The exact nodal value w_C is defined on a uniform mesh of size h by the average

$$\omega_C = \frac{1}{h^2} \int_{x_C-h/2}^{x_C+h/2} \int_{y_C-h/2}^{y_C+h/2} w(x,y) \mathrm{d}x \mathrm{d}y. \qquad (8.134)$$

Integrating equation (8.133) over the region enclosed by the dotted line in Figure 8.25 (a) and applying an integral theorem,

$$h^2 \omega_C = -\oint \nabla \psi . \mathbf{n} \mathrm{d}s, \qquad (8.135)$$

where $\mathbf{n}$ is the outward normal and s is distance around the dotted boundary. The integral on the right-hand side can then be related to differences between the nodal values leading to the usual five-point operator (even though nodal values represent averages derivatives still have leading order representations in terms of usual differences) for the *approximate* value W_C in terms of approximate values for the average stream-function values held at the nodes C, W, S, E, N.

It is when this is applied to a corner that the result is different to a usual finite difference formula. The region at the corner over which the finite volume average is defined is that between the wall and the dotted line in Figure 8.25 (b) so that the finite volume representation of (8.133) becomes

$$W_C = -\frac{4}{3h^2}[\psi_N + \psi_E - 2\psi_C]. \qquad (8.136)$$

Since the vorticity is integrable at the corner there is no concern that W_C could represent an indefinite quantity.

Since the leading order analytic solution for Stokes flow near such a corner has been found, what change would occur if the average vorticity were to be found from the analytic solution? This question has been considered by Floryan and Czechowski (1995) for calculation of Stokes flow but they stopped short of incorporating their work into a finite volume formulation.

If the average value at a node is again used as the leading order estimate for the point value at a node, then (8.125) can be used to estimate the constants a and b as

$$a = \frac{\psi_N + \psi_E - 2\psi_C}{h^{m_1} f_{m_1}(\pi/4)}, \quad b = \frac{\psi_N - \psi_E}{h^{m_2} f_{m_2}(\pi/4)}. \tag{8.137}$$

The application of (8.135) with the solution (8.125) is relatively complicated, little might be lost by applying the finite volume method to a sector of a circle as shown in Figure 8.25 (c). In that case

$$\frac{3\pi}{16} h^2 \omega_c = -a m_1 \left(\frac{h}{2}\right)^{m_1} \int_{-\alpha}^{\alpha} f_{m_1}(\theta) d\theta, \tag{8.138}$$

the symmetric solution (via b) giving no contribution to the corner vorticity.

Combining (8.137) and (8.138) then

$$W_C = -\frac{16 m_1 \int_{-\alpha}^{\alpha} f_{m_1}(\theta) d\theta}{3\pi 2^{m_1} f_{m_1}(\pi/4)} \frac{\psi_N + \psi_E - 2\psi_C}{h^2} = -\frac{1.752...}{h^2}[\psi_N + \psi_E - 2\psi_C], \tag{8.139}$$

giving an average vorticity at the corner which is a little higher than that estimated in (8.136). The rectangular volume from Figure 8.25, (b), could be dealt with by numerical quadrature of the analytic solution.

Of course, there are a number of other possible approaches to incorporating the local Stokes flow solution into a numerical scheme, for instance the vorticity at N and E could be used to determine the constants a and b instead of the stream-function, for further reading see Floryan and Czechowski (1995). It should also be apparent that a local analytic description could be incorporated into a finite element method to provide special local element relations near a corner.

8.8 Summary

In this introduction to channel flow is should be clear that an asymptotic analysis of a channel flow revolves around assumptions about the length and magnitude of any wall perturbation as well as whether the flow is symmetric. The division of the channel into an inviscid rotational core and two wall layers gives an asymptotic base which is quite robust even for wall variations which are much larger than those considered here and which form the subject of the next channel. Comparison with solutions of the full Navier–Stokes equations is difficult because the Reynolds number can easily become far too large for current computational techniques to cope but such comparisons as are possible are mostly favourable. Perhaps the least acceptable part of the theory is that separation will occur simultaneously on both walls, even when there is a flat wall opposite a channel expansion–contraction.

9

UPSTREAM INFLUENCE

9.1 Introduction

We saw in the last chapter that if a channel indentation was sufficiently small then the leading order term for the pressure was constant *across* the channel and further, the flow would be undisturbed until the channel indentation first occurred. It also emerged that the structure which was found for small channel indentations, an inviscid rotational core perturbation and two non-linear wall layers would still carry over to greater wall changes provided the pressure variation across the channel could be properly accounted for. The development of this theory lead to one of the important predictions of interactive boundary layer theory, that of the asymptotic distance upstream for which a channel indentation causes a significant perturbation to the oncoming Poisseuille flow.

Since we also saw that the scalings for a symmetric and an asymmetric perturbation (about the channel centreline) were different we have to consider details for the two cases separately. The theory for asymmetric disturbances is due to Smith (1976b, 1977b) while the symmetric case was developed in Smith (1979b).

In the case of an asymmetric disturbance there is a further important difference from the smaller wall indentations: the displacement function $-A(X)$ is no longer known but has to be determined as part of the numerical solution and as we shall see $A(X)$ being unknown is central to the numerical solution of the problem.

9.2 Asymmetric channels: $\epsilon \sim R^{-1/7}$

In the case of an asymmetric disturbance we saw that the important scale for which the pressure varied across the channel was a wall perturbation of size $\sigma = \epsilon^2$ and where $\epsilon \sim R^{-1/7}$. In that case the transverse momentum equation (8.21) in the core region had a leading order expansion,

$$P_{1y} = -U_0 V_{1X}, \tag{9.1}$$

and since the core disturbance will still be given by

$$V_1(X, y) = -A'(X) U_0(y), \tag{9.2}$$

the pressure in the core is determined by

$$P_1(X, y) = P_L(X) + A''(X) \int_0^y U_0^2(s) \, ds, \tag{9.3}$$

where $P_L(X)$ is the pressure for $y = 0$. If the pressure at $y = 1$ is denoted $P_U(X)$ then the pressures on either side of the core flow are related by

$$P_U(X) = P_L(X) + qA''(X), \qquad (9.4)$$

where for oncoming Poiseuille flow, the constant has the value $q = 1/120$. Thus we have the following expansions for the three regions.

Core region

In the core of the channel, with $\epsilon^2 \ll y$, $\epsilon^2 \ll 1 - y$, expand the stream-function by

$$\psi \sim \psi_0(y) + \epsilon^2 \psi_1(X, y) + \cdots, \qquad (9.5)$$

and the pressure by

$$p \sim -x + \epsilon^4 RP_1(X, y) + \cdots. \qquad (9.6)$$

As before these expansions lead to a solution in the core

$$\psi_1(X, y) = A(X)U_0(y), \qquad (9.7)$$

and

$$P_1(X, y) = P_L(X) + A''(X) \int_0^y U_0^2(s) \, ds, \qquad (9.8)$$

where the displacement function $-A(X)$ is an undetermined function as is the pressure in the lower layer, $P_L(X)$.

Lower wall layer

In lower wall region define a near wall coordinate, Y, by

$$y = \epsilon^2 Y, \qquad (9.9)$$

expand the stream-function as

$$\psi \sim \epsilon^4 \Psi_0(X, Y) + \cdots, \qquad (9.10)$$

and retain the same expansion for the pressure as in the core, then

$$\Psi_{0Y}\Psi_{0XY} - \Psi_{0X}\Psi_{0YY} = -P'_L(X) + \Psi_{0YYY}, \qquad (9.11)$$

with boundary conditions on the plate $\Psi_{0X}(X, f(X)) = \Psi_{0Y}(X, f(X)) = 0$. Again, matching the core solution provides

$$\Psi_0(X, Y) \sim \frac{1}{4}Y^2 + \frac{1}{2}A(X)Y + o(Y), \quad \text{as } Y \to \infty. \qquad (9.12)$$

and matching far upstream: $\Psi \sim Y^2/4$ as $X \to -\infty$.

Using Prandtl's transformation
$$\eta = Y - f(X), \tag{9.13}$$
then the equation for $\Psi_0(X, \eta)$ is the same as (9.11),
$$\boxed{\Psi_{0\eta}\Psi_{0X\eta} - \Psi_{0X}\Psi_{0\eta\eta} = -P'_L(X) + \Psi_{0\eta\eta\eta},} \tag{9.14}$$
with boundary conditions $\Psi_{0X}(X,0) = \Psi_{0\eta}(X,0) = 0$, $\Psi_0 \sim \eta^2/4$ as $X \to -\infty$ and
$$\boxed{\Psi_0 \sim \eta^2/4 + [A(X) + f(X)]\eta/2 + o(\eta) \text{ as } \eta \to \infty.} \tag{9.15}$$

Upper wall layer

In the region near the upper wall, and define a near wall coordinate Z by
$$1 - y = \epsilon^2 Z, \tag{9.16}$$
expand the stream-function as
$$\psi \sim \frac{1}{12} - \epsilon^4 \tilde{\Psi}_0(X, Z) + \cdots, \tag{9.17}$$
and using the pressure expansion from the core, we obtain
$$\tilde{\Psi}_{0Z}\tilde{\Psi}_{0XZ} - \tilde{\Psi}_{0X}\tilde{\Psi}_{0ZZ} = -P'_U(X) + \tilde{\Psi}_{0ZZZ}, \tag{9.18}$$
with boundary conditions that $\tilde{\Psi}_{0X}(X, -g(X)) = \tilde{\Psi}_{0Z}(X, -g(X)) = 0$, $\tilde{\Psi}_0 \sim Z^2/4$ as $X \to -\infty$ but now to match with the core, there has been a sign change (corresponding with U'_0 changing sign between the two walls), so that
$$\tilde{\Psi}_0 \sim \frac{1}{4}Z^2 - \frac{1}{2}A(X)Z + o(Z) \text{ as } Z \to \infty. \tag{9.19}$$
If we apply Prandtl's transformation to this system,
$$\xi = Z + g(X), \tag{9.20}$$
we will have
$$\boxed{\tilde{\Psi}_{0\xi}\tilde{\Psi}_{0X\xi} - \tilde{\Psi}_{0X}\tilde{\Psi}_{0\xi\xi} = -P'_U(X) + \tilde{\Psi}_{0\xi\xi\xi},} \tag{9.21}$$
with boundary conditions $\tilde{\Psi}_{0X}(X,0) = \tilde{\Psi}_{0\xi}(X,) = 0$, $\tilde{\Psi}_0 \sim \xi^2/4$ as $X \to -\infty$ but the condition for matching with the core is now
$$\boxed{\tilde{\Psi}_0 \sim \xi^2/4 - [A(X) + g(X)]\xi/2 + o(\xi), \text{ as } \xi \to \infty.} \tag{9.22}$$

The boundary region equations, (9.14) and (9.21) parallel those considered in the last chapter except for the different pressure gradients, P'_L and P'_U which are related by
$$P'_U(X) = P'_L(X) + qA'''(X), \tag{9.23}$$
and as already observed, the displacement function $-A$ will emerge as part of the numerical solution process.

9.3 Upstream influence

This section is concerned with the most fundamental difference between disturbances of order $\epsilon = O(R^{-1/7})$ and ones where ϵ is smaller, a distinction developed in Smith (1977b) where he first demonstrated the upstream extent to which a disturbance to Poiseuille flow would be significant. He had developed much of the theory in an earlier work, Smith (1976b), where he considered small unsteady perturbations to Poiseuille flow but the significance for upstream interaction in steady flow was mostly explored in the later paper in the context of a free interaction problem.

In order to draw out the theory it is easiest to consider the following functions in the two wall layers,

$$\frac{\partial^2 \Psi_0}{\partial \eta^2} = \frac{1}{2} + w_L(X, \eta), \quad \frac{\partial^2 \tilde{\Psi}_0}{\partial \xi^2} = \frac{1}{2} + w_U(X, \xi), \tag{9.24}$$

under the assumption of linearised theory where in some sense the wall perturbation functions are small (see too the discussion in the last chapter about linearised theory). In the linearised case we have the following equations for the two wall layers:

First, in the lower layer,

$$\frac{\partial^2 w_L}{\partial \eta^2} = \frac{1}{2}\eta \frac{\partial w_L}{\partial X}, \tag{9.25}$$

with

$$\frac{\partial w_L}{\partial \eta}(X, 0) = P'_L(X), \tag{9.26}$$

and

$$\int_0^\infty w_L(X, s) \, ds = \frac{1}{2}(A(X) + f(X)), \tag{9.27}$$

while in the upper layer,

$$\frac{\partial^2 w_U}{\partial \xi^2} = \frac{1}{2}\xi \frac{\partial w_U}{\partial X}, \tag{9.28}$$

with

$$\frac{\partial w_U}{\partial \xi}(X, 0) = P'_U(X), \tag{9.29}$$

and

$$\int_0^\infty w_U(X, s) \, ds = -\frac{1}{2}(A(X) + g(X)). \tag{9.30}$$

These equations are coupled by the pressure relation across the core region which was written in the form

$$P'_U(X) = P'_L(X) + qA'''(X), \qquad (9.31)$$

with $q = 1/120$ in the $\epsilon \sim R^{-1/7}$ case and $q = 0$ for smaller indentations. Allowing q to vary restores the case considered in the last chapter where the pressure was the same in each wall layer. In each layer the velocity and pressure disturbance must vanish far upstream.

As before the linearised equations can be solved using Fourier transforms where an overbar will denote the transform with respect to X and k the transform variable. The solution follows much of section 8.2.2 closely. Letting $p = ik/2$; in the lower layer

$$\overline{w}_L(k,\eta) = \overline{W}_L(k)\mathrm{Ai}(p^{1/3}\eta), \qquad (9.32)$$

and the boundary condition (9.27) gives

$$\overline{W}_L(k) = \frac{3}{2}p^{1/3}(\overline{A} + \overline{f}), \qquad (9.33)$$

while in the upper layer

$$\overline{w}_U(k,\eta) = \overline{W}_U(k)\mathrm{Ai}(p^{1/3}\xi), \qquad (9.34)$$

and the boundary condition (9.30) becomes

$$\overline{W}_U(k) = -\frac{3}{2}p^{1/3}(\overline{A} + \overline{g}). \qquad (9.35)$$

Using the pressure relation (9.31) and the conditions on the walls, (9.26) and (9.29), the transform of the displacement function $-A$ is determined by

$$\overline{A}(k) = -\frac{\overline{f}(k) + \overline{g}(k)}{2[1 - \tilde{q}(ik)^{7/3}]}, \qquad (9.36)$$

where

$$\tilde{q} = -\frac{2^{2/3}q \int_0^\infty \mathrm{Ai}(s)\mathrm{d}s}{\mathrm{Ai}'(0)} = -\frac{2^{2/3}q}{3\mathrm{Ai}'(0)} > 0.$$

The distinction between the cases $\epsilon \ll R^{-1/7}$ ($q = 0$) and $\epsilon \sim R^{-1/7}$ ($q \neq 0$) becomes immediately apparent: in the former case the denominator of (9.36) is just a constant and the displacement function is the average wall displacement which vanishes ahead of any change in channel geometry. In the

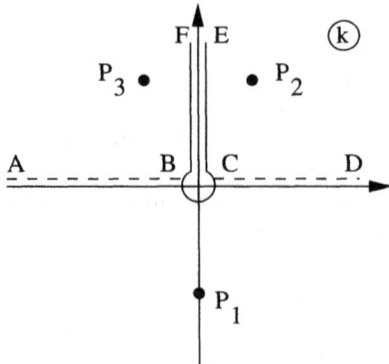

FIG. 9.1. Poles (P_1, P_2, P_3) and branch cut (ECBF) in the complex k-plane.

later case the displacement function will be a convolution integral of the average wall displacement and the inverse of

$$\overline{B}(k) = \frac{1}{1 - \tilde{q}(ik)^{7/3}} \qquad (9.37)$$

and so not zero ahead of any change in channel geometry. This then is the origin of upstream influence.

The inversion of this transform is routine provided we observe the requirement for

$$\arg\{(ik)^{1/3}\} \leq \pi/3$$

which is necessary to use the Airy function solution. As a result $-\frac{3}{2}\pi \leq \arg k \leq \frac{1}{2}\pi$ and for $\tilde{q} > 0$ there are three poles of importance, at

$$k = \tilde{q}^{-3/7} e^{-\pi i/2}, \quad k = \tilde{q}^{-3/7} e^{5\pi i/14}, \quad k = \tilde{q}^{-3/7} e^{-19\pi i/14},$$

and there is a branch cut on the positive imaginary axis, as shown in Figure 9.1.

In the upstream region, where $X < 0$, the inverse Fourier transform will require $\Im k \leq 0$ so that the pole at

$$k = \tilde{q}^{-3/7} e^{-\pi i/2}$$

is appropriate to determine the behaviour there while in the downstream section with $X > 0$ the inverse Fourier transform requires the path ABCD to be deformed around the poles and onto the branch cut ECBF. To simplify algebra, let $c = \tilde{q}^{-3/7}$. As a consequence of the different paths for different sign of X the function $B(X)$ is given by

$$B(X) = \begin{cases} \frac{3}{7}\sqrt{2\pi}ce^{cX}, & X < 0 \\ -\dfrac{\sqrt{3}\tilde{q}X^{-10/3}}{\sqrt{2\pi}}b(X,\tilde{q}) + O(e^{-cX}), & X > 0 \end{cases} \qquad (9.38)$$

with

$$b(X,\tilde{q}) = \int_0^\infty \frac{r^{7/3}}{1 - \tilde{q}X^{-7/3}r^{7/3} + \tilde{q}^2 X^{-14/3} r^{14/3}} e^{-r} dr.$$

Once the function $B(X)$ had been calculated, the displacement function was given by

$$A(X) = -\int_{-\infty}^\infty B(X-s)\frac{f(s)+g(s)}{2}ds, \qquad (9.39)$$

and similarly the pressure could be determined in terms of a convolution integral.

Thus the main result from linearised theory was that in the case $\epsilon = O(R^{-1/7})$ where the pressure varied across the channel, there would be an influence of any wall perturbation which extended a distance upstream of order $R^{1/7}$ times the channel width. If R was asymptotically large, Smith considered that any distance would be so far upstream that the local details of the wall perturbation would become irrelevant and any perturbation to the main flow would have to develop as a free interaction in a parallel channel. To answer whether this was possible it was necessary to solve the non-linear equations governing the two wall layers in a parallel channel but assuming that far upstream, the displacement function satisfied

$$A(X) \sim \beta_1 e^{cX}, \text{ as } X \to -\infty, \qquad (9.40)$$

where β_1 was unknown and c as before. Smith's view was that the value of β_1 would only emerge from requiring that the flow return to Poisseuille flow far downstream.

The question of whether the interactive equations together with pressure variation across the channel could support separation upstream of any channel wall variation was answered affirmatively in Smith (1977b). Although the constant β_1 may be unknown, if an arbitrary origin shift $-X_{start}$ were introduced in X, then the function A would behave as

$$A(X) \sim \beta_1 e^{cX_{start}} e^{c(X - X_{start})} \equiv de^{c(X - X_{start})}, \qquad (9.41)$$

where d might be made arbitrarily small with the effect of only varying the origin of the solution. So the interactive equations can be perturbed by an arbitrarily small amount and a numerical solution sought for the subsequent flow development. This enabled Smith to show that the coupled interactive boundary layer

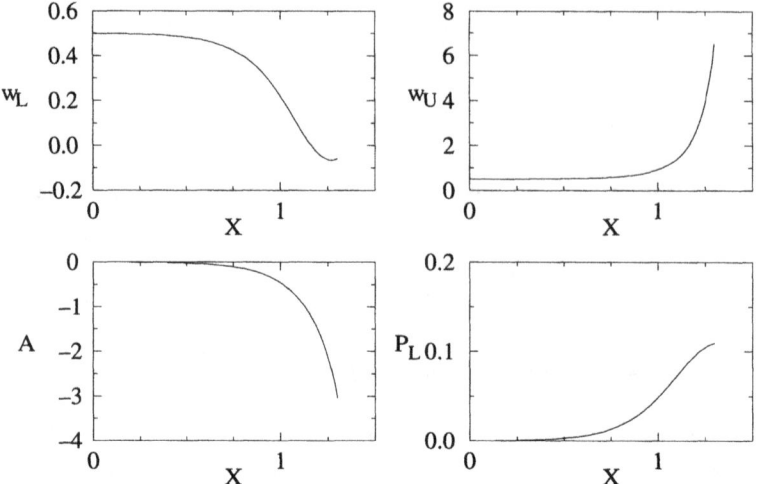

FIG. 9.2. Free interaction solution in a channel: Velocity gradient normal to the lower and upper walls, w_L and w_U; displacement function $-A$ and pressure in lower layer; lower layer pressure, P_L, versus distance from start of calculation.

equations did indeed support the development of a free interaction solution with separation.

As an example of this development we have perturbed the function $A(X)$ to have the values

$$A(X_{start}) = 0.001, \ A(X_{start} - h) = 0.001e^{-ch}, \ A(X_{start} - 2h) = 0.001e^{-2ch},$$

with the stream-function and velocity having their upstream values and then computed the subsequent development of the solution with X. The A values provide the start of a free interaction and the solution in a *parallel channel* separates a short distance downstream of the start of the computation. The results of such a calculation are shown in Figure 9.2 where the velocity gradient on the lower wall falls to zero near $X = 1.165$. The precise value is of course irrelevant as changing the starting perturbation d produces an origin shift. The function A changes dramatically in size as separation is approached and the calculation could not be continued much further downstream than that shown, around $X = 1.3$ for the starting value of $d = 0.001$. The pressure exhibited a gradual increase throughout the free interaction development.

In addition the streamlines, Figure 9.3, show that for most of the interaction region the flow remains parallel to the wall before separation occurs relatively

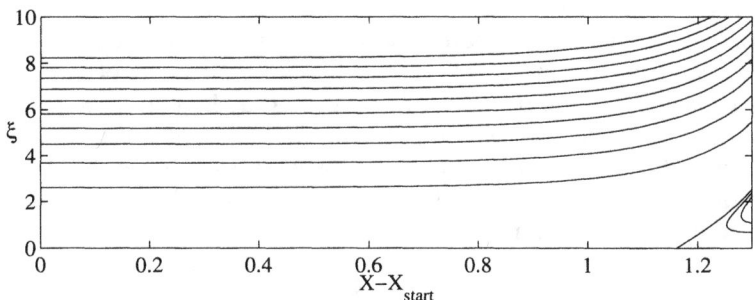

FIG. 9.3. Streamlines of free interaction solution for lower wall layer showing start of separation. The X-coordinate has been offset by the starting value, X_{start}, for the computation.

abruptly. The curvature of the separation streamline is away from the wall. Calculations such as this demonstrate that the flow upstream of any wall variation can develop as a free interaction but they also suggest that the calculation cannot be continued indefinitely in the X direction. The curves for the displacement function $-A$ and the upper wall velocity gradient looks alarmingly as if a singularity in the solution develops some distance downstream of separation.

Perhaps surprisingly, the singularity which develops is not primarily connected with the separated layer and any failure to account for inflow which must occur downstream of separation. Smith realised that the difficulty originated in the upper layer. A glance at the streamlines in the upper layer which are shown in Figure 9.4 reveals some of the difficulty. The streamlines are all curving sharply down towards the wall with a consequence that the flow must be accelerated near the wall (and so the rapid increase calculated for w_U). Without some mechanism to change the curvature of the streamlines, it would appear that a singularity must ensue just from continuity.

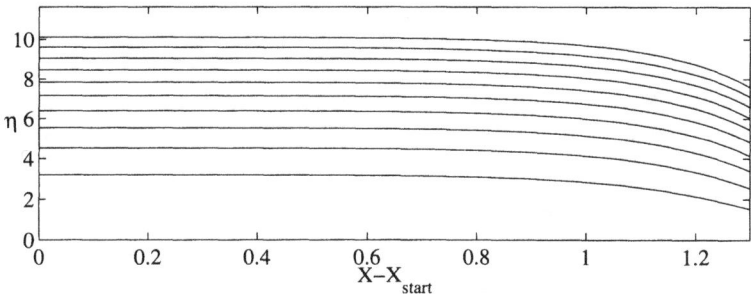

FIG. 9.4. Streamlines of free interaction solution for upper wall layer. The X-coordinate has been offset by the starting value, X_{start}, for the computation.

Smith was able to demonstrate that a singularity must occur downstream of the separation point by using expansions within subregions of the wall layers. Supposed that separation occurs at $X = X_s$ and a singularity structure occurs at $X = X_0$. The effect of the compression of the upper layer is to induce an increase in velocity in a layer which will become thinner as $X \to X_0$. Hence Smith introduced a sublayer within the upper layer which would terminate at $X = X_0$. Defining a new similarity variable

$$s = \frac{\xi}{(X_0 - X)^{3/2}}, \qquad (9.42)$$

then the stream-function near the upper wall could be expanded

$$\tilde{\Psi}(X, \xi) \sim (X_0 - X)^{-1/2} f_0(s) + o((X_0 - X)^{-1/2}), \qquad (9.43)$$

and the displacement function $-A$ had to have an expansion

$$A(X) \sim -\frac{2}{5(X_0 - X)^2} \qquad (9.44)$$

upstream of the point $X = X_0$ and the function f_0 satisfied a Falkner–Skan equation.

While the sublayer in the upper wall layer was relatively simple to analyse, the structure in the lower layer was far more intricate. Smith found it necessary to use three sublayer expansions (although one expansion could itself be thought of in terms of two sublayers). If the detaching layer were defined around $\eta \sim -A(X)$ then defining

$$\bar{\eta} = \eta + A(X), \qquad (9.45)$$

the outermost expansion within the lower layer would be for $\bar{\eta}$ of order one where the stream-function would have an expansion

$$\Psi(X, \eta) \sim F_0(\bar{\eta}) + (X_0 - X) f_1(\bar{\eta}) + \cdots, \qquad (9.46)$$

and that expansion would have to match with the core flow when $\bar{\eta} \to \infty$. Near $\bar{\eta} = 0$ it was necessary to introduce new layer dependent on the similarity variable

$$\bar{s} = \frac{\bar{\eta}}{(X_0 - X)^{1/3}} \qquad (9.47)$$

according to

$$\Psi \sim (X_0 - X)^{2/3} G_0(\bar{s}) + \cdots \qquad (9.48)$$

with matching in (9.48) as $\bar{s} \to \infty$ against $\bar{\eta} \to 0$ in (9.46). Below this second layer was another similarity layer, this time set against the wall with

$$\tilde{s} = \eta(X_0 - X)^{5/6} \tag{9.49}$$

so that the layer is increasing in thickness as the singularity is approached. The stream-function was then

$$\Psi \sim (X_0 - X)^{11/6} g_0(\tilde{s}) + \cdots \tag{9.50}$$

where the matching was between $\bar{s} \to -\infty$ in (9.48) against $\tilde{s} \to \infty$ in (9.50). As the lower wall layer was only of thickness $O((X_0 - X)^{-5/6})$ whereas the limit of $\bar{s} \to -\infty$ was on a scale $\eta \sim (X_0 - X)^{-2}$ (the same scale as the displacement function $-A$) Smith divided the lowest region into a wall layer of thickness $O((X_0 - X)^{-5/6})$ and an inviscid layer which joined between the lower wall layer and the similarity layer around $\bar{\eta} = 0$. The structure of the approach to the singularity is illustrated in Figure 9.5

While the function g_0 satisfied a Falkner–Skan equation and the function G_0 was already known from Stewartson and Williams (1973) but the function F_0 was indeterminate, and Smith argued that this indeterminacy was both needed for and would be resolved by matching to the oncoming flow profile.

Although numerical calculations cannot be continued to the singular point X_0 (indeed they have to stop a some distance upstream of the singular point), by extrapolating data for the region after separation but still ahead of the singular point (for instance by plotting $(-A)^{-1/2}$ versus X), Smith estimated that distance between the separation point and the singular point would satisfy

$$X_0 - X_s \approx 0.49. \tag{9.51}$$

Perhaps one of the remarkable results in Smith (1977b) is that the intricate asymptotic structure which leads up to the singularity at $X = X_0$ provides a quite rational interpretation when considered in terms of the natural coordinates, x and y with x far upstream of the singularity providing as $x \to -\infty$ it is still the case that $X_0 - xR^{-1/7} \ll 1$.

In the core the velocity was given by

$$u \sim U_0(y) + \epsilon^2 A(X) U_0'(y), \tag{9.52}$$

and with $X = \epsilon x$ and $A(X) \sim (X_0 - X)^{-2}$ then the velocity satisfies

$$u \sim U_0(y) - \frac{2}{5} \frac{U_0'(y)}{x^2}, \tag{9.53}$$

for x large and negative.

Note that on the scale of $-x$ being considered the decay is algebraic rather than the exponential change we started with on the X scale because of

$$e^{cX} = e^{cxR^{-1/7}} \sim 1 + cxR^{-1/7} + \cdots.$$

The adjustment of the core velocity far ahead of any change in channel geometry led Smith to propose a radically new structure for flow ahead of a contraction

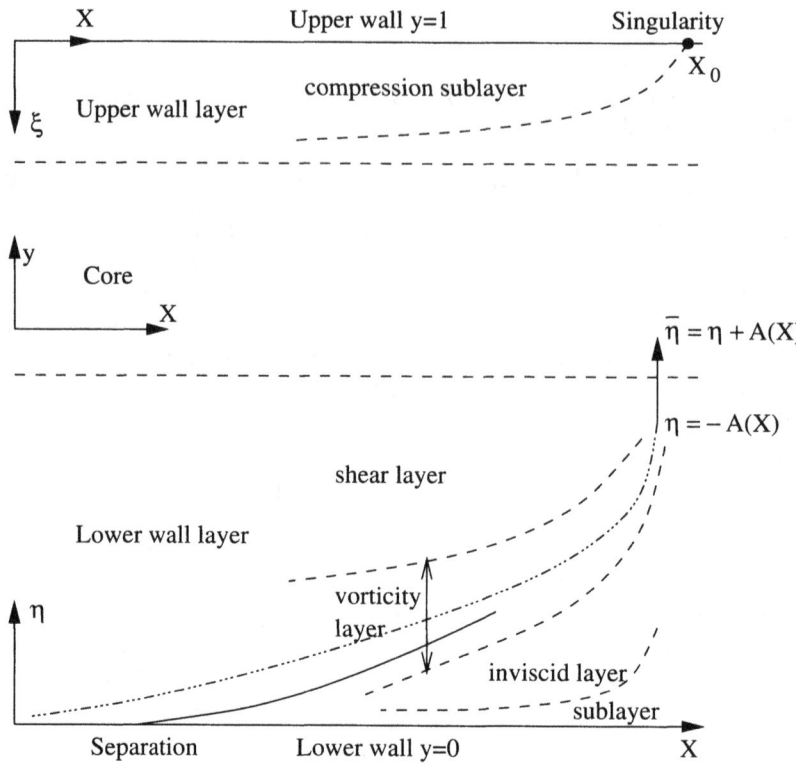

FIG. 9.5. Schematic diagram of the development of free interaction solution and singularity at $X = X_0$. (a) The upper layer, (X, ξ), develops a compression sublayer of thickness $O((X_0 - X)^{2/3})$ (b) The lower layer, (X, η) has a vorticity layer of thickness $O((X_0 - X)^{1/3})$ about $\eta = -A(X)$, a layer above the vorticity layer where the shear is approximately constant and a layer below which can be subdivided in to an inviscid layer and a wall layer of thickness $O((X_0 - X)^{-5/6})$.

in a channel. If separation occurs ahead of a contraction through a free interaction then he believed that the uncertainty in the function F_0 in (9.46) could be explained by an uncertainty in the location of the singular point X_0 relative to the contraction. That is, the long x length scale represented by order one values of X would give way to a shorter length scale where x is order one but there would be an uncertainty in that value which in turn would allow the contraction to occur ahead of the singularity. This provided an asymptotic description for $R \to \infty$ whereby separation occurred at

$$x_s = -0.49 R^{1/7} + D,$$

and D was an order one constant which ensured that the change in channel geometry occurred before the singularity would be reached.

Upstream influence 275

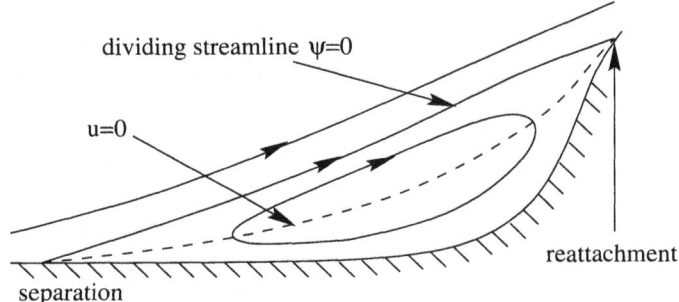

FIG. 9.6. Streamlines in the vicinity of an upstream separation region showing the dividing streamline and the line where $u = 0$.

Even more radical was Smith's proposed structure for the upstream separated flow region. This is best illustrated by reference to what might be considered a typical diagram of an upstream separated region at moderate Reynolds number, Figure 9.6. The usual way this would be drawn would show the separation streamline from the separation point to the reattachment point with a recirculation region between the separation streamline and the wall. The expansion for the vorticity layer found by Smith was not centred around the dividing streamline, rather the layer was about the line where the u-velocity was zero, shown as a dashed line in Figure 9.6. What happens when the Reynolds number becomes infinite? The implications in Smith (1977b), Smith and Duck (1980) and later work is that the two lines become identical in the limit $R \to \infty$ and the infinite Reynolds number flow would be given by an inviscid deformation of the oncoming vorticity. The fluid between the dividing streamline and the wall would then be at rest. At large but finite Reynolds numbers, any slow motion under the dividing streamline would come in as a higher order effect.

Thus the line where $u = 0$ was the important feature representing an effective boundary for the development of the core region (rather than the physical boundary at the wall). If this were given by a curve

$$y = S(x), \qquad (9.54)$$

where on the x scale $S(x)$ would be asymptotically $S \sim x^{-2}$ as $R \to \infty$, $x \to -\infty$ then the determination of this curve would be critical to achieve a description of the flow structure ahead of a change in channel geometry and it implied that the flow approaching a contraction would be very different from Poisseuille flow.

In one sense the prediction that separation occurs a distance of order $R^{1/7}$ upstream is surprising, because of its implications for the limiting infinite Reynolds number flow, but in another sense it is a remarkably short distance upstream. Even if the Reynolds number is of order $R \sim 10^5$ then the upstream scaling parameter has $R^{1/7} \sim 5$ so that any disturbance *from* a channel contraction is unlikely to be significant more than five or six channel widths upstream. Hence in computing laminar solutions of the full Navier–Stokes equations at Reynolds numbers which are much smaller than 10^5 it is unlikely that a calculation needs to consider more than a handful of channel widths upstream of most channel contractions.

This concludes what might be considered introductory material on the theory of upstream influence in an asymmetric channel. There have been a number of further studies which take the theory forward. The description of separation in flow through an asymmetric channel was continued in Smith and Duck (1980) using a Kirchoff free streamline model in combination with an interactive boundary layer model. This followed a study of an axi-symmetric constriction by Smith (1979b) and we shall shortly briefly discuss that application to a symmetric channel by Smith (1982). Other ideas relevant to this area are also be found in Smith and Daniels (1981).

9.4 A numerical example

Computation of solutions to the full Navier–Stokes equations to compare with the asymptotic theory are quite difficult because of the very high Reynolds numbers required for close comparison. As we have seen in the previous chapter some very good qualitative comparisons can be made and indeed some quantitative comparisons. In the cases where $\epsilon \sim R^{-1/7}$ the Reynolds numbers needed for really satisfactory comparison become much too great for relatively straight forward computations. Keep in mind that the conditions in the preceding section where

$$R^{-1} \ll \epsilon \ll R^{-1/7},$$

correspond to

$$\epsilon^{-1} \ll R \ll \epsilon^{-7},$$

whereas now we are at the upper limit, $R \sim \epsilon^{-7}$. Consequently what we show here is aimed not at direct comparison with solutions of the full Navier–Stokes equations but rather at some of the differences which come out when the pressure varies across the channel.

As a model problem we use a channel with a periodic indentation on one wall, shown in Figure 9.7 which is defined in (X, η) coordinates by

$$f(X) = \begin{cases} 0, & 0 \leq X < \frac{5}{12}, \\ \mu(1 - \cos(12\pi(X - \frac{5}{12}))) & \frac{5}{12} \leq X < \frac{7}{12}, \\ 0, & \frac{7}{12} \leq X \leq 1, \end{cases} \qquad (9.55)$$

and repeating with period one in X. This provides an example of a smooth indentation on the lower wall.

Streamlines in the lower and upper layer from the interactive boundary layer solution are shown in Figure 9.7 for the case $\mu = 2$ where separation occurs

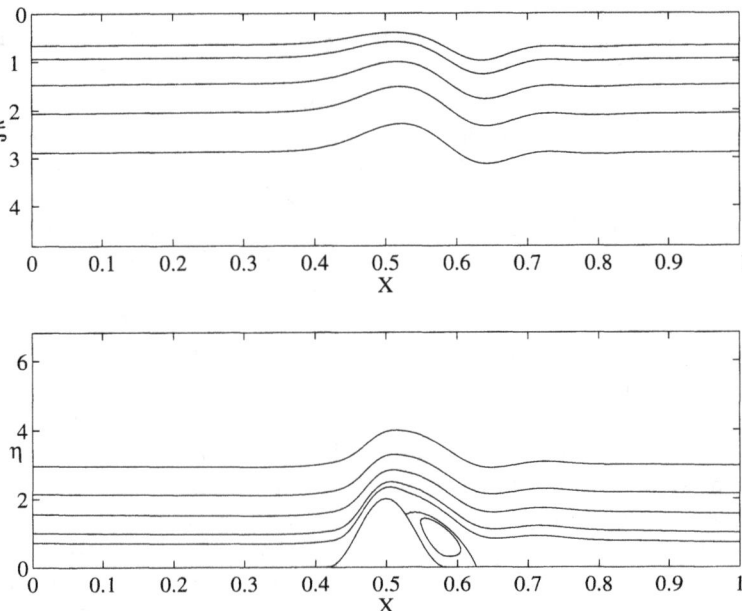

FIG. 9.7. Streamlines in lower and upper wall layers.

behind the indentation but not ahead of it. There is no separation on the opposite wall although there is a region of lowered shear. The velocity gradient on each wall and the pressure in the two layers is shown in Figure 9.8. A dip in the lower wall gradient ahead of the indentation is presumably a precursor to upstream separation on the lower wall although it is difficult to obtain stable solutions to the interactive equations as the indentation size increases using the simple method we have described. Possibly more sophisticated methods or methods using vastly greater resolution would be more successful. These results were calculated with 150 points in the X direction and 50 points across each of the two wall layers.

9.5 Symmetric channels

We turn now to symmetric channels where the variation of the two walls is both symmetric and large enough to cause a cross channel pressure variation. This will first arise when $\epsilon = O(1)$ and $\sigma = R^{-1/3}$. In that case the simple theory presented in 8.3 has to be modified because the shortened length of the wall disturbance leads firstly to an explicit axial dependence in the core perturbation equations and secondly increasing wall height leads to a significant change in upstream influence. It might seem then that flow through a symmetric channel would be a straightforward analogue of that for an asymmetric channel and that is true to some extent, but Smith (1978, 1982) showed that considerably greater development of the theory could be made leading to an asymptotic structure for flow through a symmetric indentation which was comparable with the channel width, and in doing so highlighted some differences between flow in a symmetric and an asymmetric channel. His work was mostly aimed at flow through constrictions and he proposed that the problem should be viewed in terms of three different asymptotic constriction magnitudes. That the pressure was varying across the channel implied ϵ to be order one so the length of any wall perturbation was already comparable with the channel width. The description derived in the last chapter showed that the first critical indentation height would be $\sigma = O(R^{-1/3})$ and Smith proposed to call this a **fine** indentation. Since the leading order oncoming flow is linear near the wall and the core perturbation depends on the square of a perturbation parameter, Smith realised that a second critical case should occur if the oncoming wall layer had thickness of order $R^{-1/3}$ as in the fine case (and so velocity with the same scale), but the wall perturbation could be of order $R^{-1/6}$ to provide a comparable perturbation of order $R^{-1/3}$ in the core which could then be matched to outer edge of the wall layer. This second wall perturbation was asymptotically much greater than in the fine case and Smith denoted that case a **moderate** constriction. If the constriction size were given by $\mu R^{-1/6}$ then Smith considered the case $\mu \to R^{1/6}$ as a limit within the expansion for $R \to \infty$ as a viable process thus arriving at a wall perturbation which was comparable with the channel width which he denoted a **severe** constriction.

9.5.1 Fine indentation: $\sigma \sim R^{-1/3}$

A cross channel pressure gradient needs to be taken into account when $\sigma \sim R^{-1/3}$. The asymptotic structure already seen of an inviscid rotational core and two wall layers remains.

Core flow

In the case $\epsilon = O(1)$ and a symmetric channel, the stream-function in the core should be expanded

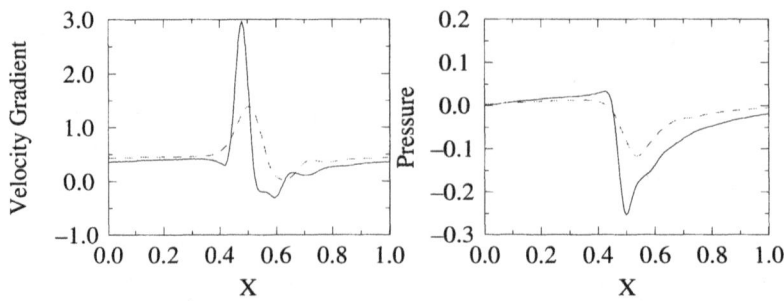

FIG. 9.8. Velocity gradient and pressure in lower (——) and upper (- - -) wall layers.

$$\psi \sim \psi_0(y) + R^{-2/3}\psi_1(X,y) + \cdots, \tag{9.56}$$

with perturbation velocities $R^{-2/3}(U_1, V_1)$ and pressure $R^{1/3}P_1(X,y)$ so that the the momentum and continuity equations give leading order equations

$$U_0 U_{1X} + V_1 U_{1y} = -P_{1X}, \tag{9.57}$$

$$U_0 V_{1X} = -P_{1y}, \tag{9.58}$$

$$U_{1X} + V_{1y} = 0. \tag{9.59}$$

These can be replaced by a single equation for the stream-function,

$$U_0(y)\{\frac{\partial^2 \psi_1}{\partial X^2} + \frac{\partial^2 \psi_1}{\partial y^2}\} = U_0''(y)\psi_1, \tag{9.60}$$

where the boundary conditions from the core are $\psi_1 = 0$ on $y = 1/2$ for symmetry, $\psi_1 \to 0$ upstream to match the oncoming Poisseuille flow and that ψ_1 is bounded downstream. In applying this last condition Smith anticipated that the flow may not return to Poisseuille form downstream which needs $\psi_1 \to 0$ as $X \to \infty$: that is still possible but so too are other outcomes.

Wall layers

The equation (9.60) for the stream-function perturbation still lacks one boundary condition on the wall. Smith obtained this by considering the wall layer where

$$y = R^{-1/3}Y, \tag{9.61}$$

and the stream-function

$$\psi \sim R^{-2/3}\Psi_0(X,Y) + \cdots. \tag{9.62}$$

If the wall pressure is $R^{1/3}P_L(X)$ then the leading order term for the stream-function satisfies a non-linear boundary layer equation

$$\Psi_{0Y}\Psi_{0XY} - \Psi_{0X}\Psi_{0YY} = -P_L'(X) + \Psi_{0YYY}, \tag{9.63}$$

with boundary conditions that $\Psi_0 \to \frac{1}{4}Y^2$ upstream, $\Psi_0 = \Psi_{0Y} = 0$ on the wall $Y = f(X)$ and $\Psi_0 \to \frac{1}{4}Y^2 + 2P_L(X)$ as $Y \to \infty$. The occurence of the pressure in the outer condition comes from considering (9.63) as $Y \to \infty$ together with the leading order expansion for the stream-function.

Matching

This provides the final condition necessary to specify ψ_1 in the core since matching between the core and the wall layer gives

$$\psi_1(X, y) \to 2P_L(X), \quad \text{as } y \to 0, \tag{9.64}$$

The solution of the equation for ψ_1 can be approached using Fourier transforms since the equation (9.60) is linear so the Fourier transform $\overline{\psi}_1(k, y)$ should satisfy

$$\frac{d^2 \overline{\psi}_1}{dy^2} - [k^2 - \frac{2}{y(1-y)}]\overline{\psi}_1 = 0, \tag{9.65}$$

with conditions $\overline{\psi}_1 = 0$ on $y = 1/2$ and $\overline{\psi}_1 = \overline{P}_1(k)$ on $y = 0$.

The equation (9.65) had arisen and been solved numerically by Tillett (1968) for a problem involving a laminar free jet issuing from a channel. He had found that the solution was a combination of the solution to 8.64) with the inhomogeneous boundary condition on $y = 0$ together with solutions of the eigenvalue problem with homogeneous boundary conditions at $y = 0$ and $y = 1/2$. There was a discrete spectrum for the eigenvalue problem leading to solutions upstream which were exponentially decaying, being of the form $e^{\gamma_n X}$, $n = 1, 2, \ldots$ and with $\gamma_1 \approx 5.175$. This decay upstream leads only to disturbances which might propagate an $O(1)$ distance upstream from the start of any wall variation. Such a disturbance may or may not include a separated region.

This completes the overall structure for the fine case developed by Smith but it then led onto the moderate constriction where the asymptotically thin wall layer essentially follows an asymptotically larger (but still small compared to the channel width) wall displacement.

9.5.2 Moderate indentation: $\sigma \sim R^{-1/6}$

The perturbation equations for the core flow can still hold for wall deflections which are larger than $O(R^{-1/3})$ and it is this case which Smith (1978) developed next. In particular suppose the wall is defined by

$$y = \mu R^{-1/6} f(X), \tag{9.66}$$

where for the present μ is an order one constant.

As indicated, the expansion in the core has to have a larger perturbation to accommodate the larger wall deflection,

$$\psi \sim \psi_0(y) + R^{-1/3} \psi_1(X, y) + \cdots, \tag{9.67}$$

so that the velocity perturbation is $R^{-1/3}(U_1, V_1)$ and the pressure $R^{1/3} P_1(X, y)$ with the stream-function perturbation ψ_1 sill satisfying (9.60). However, there is a difference. In the fine constriction a condition on the core perturbation came from the wall layer. Here a leading order condition comes from consideration of the core flow near the now much larger indentation. The stream-function at the wall can be expanded using a Taylor series:

$$\psi(X, \mu R^{-1/6} f(X)) \sim \psi(X,0) + \mu R^{-1/6} f(X) \psi_y(X,0) + \frac{1}{2}\mu^2 R^{-1/3} f^2(X) \psi_{yy}(X,0) + \cdots, \quad (9.68)$$

and then the expansion

$$\psi \sim \psi_0 + R^{-1/3} \psi_1 + \cdots \quad (9.69)$$

applied at $y = 0$ gives formally

$$\psi_1(X,0) = -\frac{1}{2}\mu^2 f^2(X) \psi_0''(0) = \frac{1}{2}\mu^2 f^2(X). \quad (9.70)$$

Thus Smith proposed that the solution of the core problem upstream of an indentation would be given in terms of the functions $\mu^2 e^{\gamma_n X}$ and considering the first and largest of these terms,

$$\mu^2 e^{\gamma_1 X} = e^{\gamma_1 X + 2\log\mu}, \quad (9.71)$$

then if in the fine constriction, separation occurred at some order one distance ahead of the constriction, it would now occur at

$$X \approx -\frac{2}{\gamma_1} \log\mu + O(1). \quad (9.72)$$

There are many further details, particularly for the case of a tubular constriction which can be found in Smith (1978, 1979b) including discussion of the flow field when there is separation both upstream and downstream of a constriction. However the major prediction of his theory follows directly from (9.72).

9.5.3 Severe indentations: $\mu \sim R^{1/6}$

If the variable μ is allowed to become large the wall perturbation will become comparable with the channel width when $\mu \sim R^{1/6}$ and Smith argued that (9.72) would still be asymptotically valid in that limit giving a constructive prediction for separation ahead of a symmetric constriction of

$$X_s \sim -\frac{1}{3\gamma_1} \log R + O(1), \quad (9.73)$$

As for an asymmetric channel, it is worth noting that while the upstream separation length is asymptotically large, for practical calculations it is quite small: particularly since the value of γ_1 is over five. Despite this comment, Dennis and Smith (1980) have managed to compute solutions of the Navier–Stokes equations ahead of a symmetric step contraction at sufficiently large Reynolds numbers to obtain very favourable comparison between the computed separation length and the prediction from asymptotic theory.

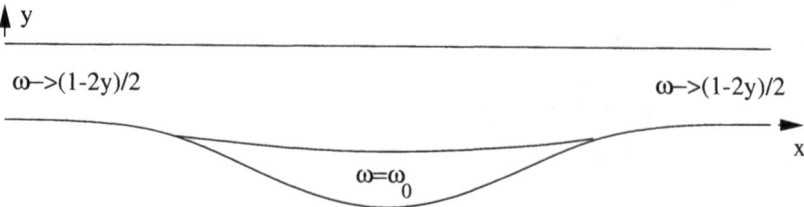

FIG. 9.9. Possible Prandtl–Batchelor flow for a channel expansion-contraction.

9.6 Prandtl–Batchelor flow in channels

One of the fascinating outcomes of the theory for channel flow is the limiting infinite Reynolds number flow structure for finite channel indentations (severe indentations in Smith's notation). We have already seen that for exterior flows there is a choice between a potential flow limit combined with free streamlines or the Prandtl–Batchelor limit where a potential flow encompasses finite regions of constant vorticity. The constant vorticity regions were proposed by Batchelor to be the result of infinitesmal viscosity acting over an infinite time. In channel flow, the structure proposed by Smith (1977b, 1978), Smith and Duck (1980) and in subsequent papers turns the external flow problem about and it is the vorticies which in the limit $R \to \infty$ have a potential character and the main channel flow which is rotational. The limiting structure is even more interesting than a simple rotational flow with a free surface since if for instance we take a flow which has upstream separation, then the conditions on the 'free streamline' are not just that the pressure is constant but that there is no slip velocity on the streamline so that the velocity vanishes there and the irrotational region which formed the vortex at finite Reynolds number is in this case stationary.

Of course one can pose a conventional Prandtl–Batchelor problem for a channel by having an irrotational main flow and finite regions near walls with constant vorticity but it seems contrary to allow vanishing viscosity to provide a constant vorticity limit within what was a vortex near a wall while the same viscosity does not affect the main channel flow. So while such a problem can be posed it would be unsatisfactory to regard it as the correct high Reynolds number limit for Poisseuille flow.

In the case of a channel expansion–contraction, such as illustrated in Figure 8.12, where there is no suggestion of upstream separation or vorticies extending far downstream it is possible to ask whether there is a solution of the Euler equations which has constant vorticity within a vortex while tending to a linear distribution of vorticity far upstream and downstream, see Figure 9.9. This would provide more of a direct analogue for channel flow of an exterior Prandtl–Batchelor flow.

9.7 Summary

The complexity introduced when the wall variation is sufficient to cause a cross channel variation in the pressure is substantial and we have not described some of the more advanced models for such flow. The extension of the influence of a channel wall variation upstream is perhaps to be expected, the influence it has on upstream separation is less expected. The limiting inviscid flow has many unexpected features but does provide a rational description of the structure of large but finite Reynolds number flows through channels.

10

COANDA EFFECT

10.1 Introduction

Amongst many interesting phenomena in fluid dynamics, one of the simplest illustrations of unexpected physical complexity in fluid flow is the Coanda effect. It is an effect named after H. Coanda (1886-1972) who observed that a jet of air flowing parallel but near a wall (in his case, the fuselage of an aircraft) would deflect so as to flow closer to the wall, appearing to 'attach' to the wall. One explanation for this comes from consideration of the pressure distribution. A jet in an unbounded fluid would entrain fluid from both sides equally and any pressure minimum would be on the centreline of the jet. However, if a wall was placed to one side of the jet then entrainment by the jet would not be equal. Whereas fluid could flow from one side towards the low pressure region of the jet that could not happen on the other side so while the low pressure could to some extent be relieved on the free side of the jet that would not be true on side near the wall. Thus a transverse pressure distribution would be set up which would 'push' the jet towards the wall resulting in the jet would appearing to attach to the wall. The wall then enables the flow to sustain a transverse pressure gradient with a pressure minimum between the jet centreline and the wall and for the jet to maintain a deflected path.

Coanda's original observations were for an external flow but the same phenomenon occurs in channel flow. In a channel where one wall provides a channel expansion or a backward step then if the flow rate is high enough, separation can occur in the lee of the expansion. If flow occurs through a symmetric expansion then separated regions can form either side of the main flow. What is observed is that the two symmetric separated regions may not remain the same size and the flow appears to attach to one or the other wall. The prediction of conditions for this effect to occur is the subject of this chapter.

10.2 Symmetry and bifurcation

Using the ideas from bifurcation theory in fluid mechanics is common nowadays but not so long ago many phenomena, for instance hysteresis, were not widely thought of in terms of such theory. As increasing numbers of computations reveal the complicated patterns associated with bifurcation and chaos, so also many fluid flows are observed to have remarkable resemblances to such patterns. The Coanda effect comes from the very beginning of a bifurcation sequence and is one of the simplest bifurcation events. There are many recent texts with suitable

Symmetry and bifurcation

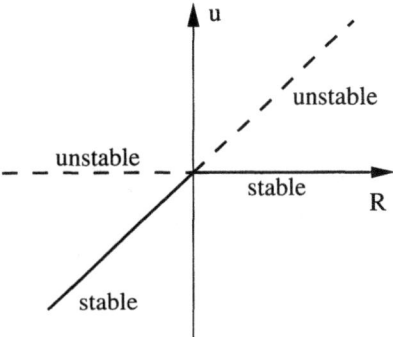

FIG. 10.1. Example bifurcation diagram for exchange of stability.

introductory material; one recommended book is Drazin (1992). Bifurcation theory does not by itself explain why any particular phenomenon occurs but it does provide a systematic method to categorise phenomena and to recognise links between changes in the state of a system.

Much of the appeal of using bifurcation theory is that to describe the response of a complex system comparison can be made to a number of simpler problems which can be applied locally to identify behaviour in the more complicated system.

Consider the simple ordinary differential equation

$$\frac{du}{dt} = F(u, R), \tag{10.1}$$

where R is some appropriate parameter. The solutions $u_0(R)$ of $F = 0$ are stationary points for which $u' = 0$. It is then sensible to ask whether a steady solution is time stable in the sense that for a fixed value of the parameter R, does the solution $u(t, R) \to u_0(R)$ as $t \to \infty$? A solution $u_0(R)$ is called stable if there is a local region about u_0 for which this does happen, otherwise the solution is unstable. It is important to realise that this is **not** implying an unstable solution $u(t, R)$ becomes unbounded for large time, only that a solution which starts arbitrarily close to u_0 at a particular R value will not approach the stationary point. It should be regarded as the normal situation in a physical system for a solution near an unstable fixed point only to move to a nearby stable fixed point. The absence of this should normally be regarded as an indicator that some important physical aspect of the problem has been neglected in developing the model.

A very simple example is an exchange of stabilities. If $F = u(u - R)$ there are two fixed points, $u_0 = 0$ and $u_0 = R$ and a solution $u = u_0 + v$, where v is in some sense small, will satisfy

$$\frac{dv}{dt} = (2u_0 - R)v,$$

with solution

$$v(t) \sim v(0)e^{(2u_0-R)t},$$

and for the two fixed points, this will only satisfy $v \to 0$ if $u_0 = R$ for $R < 0$ or $u_0 = 0$ for $R > 0$. Hence the solution $u_0 = R$ is stable for $R < 0$ but there is an exchange of stability so that the solution $u_0 = 0$ is stable for $R > 0$. This is usually represented on a bifurcation diagram by a solid line for a stable solution and a dashed line for an unstable solution; see Figure 10.1.

A more interesting class of phenomena occur when F is locally a cubic in u. In the case

$$F(u, R) = u(R - R_c - u^2), \tag{10.2}$$

then for $R > R_c$ there are three real solutions, $u_0 = 0$ and $u_0 = \pm\sqrt{R - R_c}$ whereas for $R < R_c$ there is only one real solution, $u_0 = 0$. A simple linearised stability analysis using $u = u_0 + v(t)$ would give

$$v(t) \sim v(0)e^{(R-R_c-3u_0^2)t}, \tag{10.3}$$

so that the solution $u_0 = 0$ is stable for $R < R_c$ and unstable for $R > R_c$ while the two solutions $u_0 = \pm\sqrt{R - R_c}$ are stable for $R > R_c$. A bifurcation diagram for this model problem is shown in Figure 10.2 and the bifurcation is called a supercritical pitchfork bifurcation. A characteristic of this bifurcation is that the magnitude of the stable solutions near $R = R_c^+$ varies like $\sqrt{R - R_c}$ and so this characteristic can be sought in fluid flow problems to identify the occurence of a pitchfork bifurcation (however, see also Drazin(1992), page 57, for a discussion of the effect of imperfections).

If the function F is slightly different,

$$F(u, R) = u(R - R_c + u^2), \tag{10.4}$$

then there is only one solution, $u = 0$ for $R > R_c$ while there are three, $u = 0, \pm\sqrt{R_c - R}$ for $R < R_c$ but in this case only the solution $u = 0$ for $R < R_c$ is stable. This is called a subcritical pitchfork bifurcation (see Figure 10.2(b)) and in most physical systems would be associated with a further bifurcation, for instance a turning point (see Figure 10.3) to restore the existence of at least one stable solution, giving an example of hysteresis in a system.

One other important result which comes from this simple analysis is that near the bifurcation point, $R = R_c$, disturbances have exponential decay or growth rate proportional to $-(R - Rc)$ so that if a time marching code is used to calculate a flow, the time taken for disturbances to evolve will become very large and that too can be useful or not. In some calculations it is useful because very long decay times can indicate the presence of a nearby bifurcation, but it can also be tempting to terminate calculations before a true steady solution has been reached.

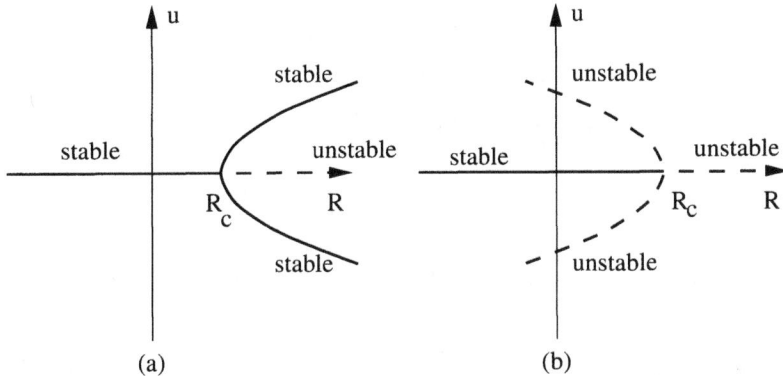

FIG. 10.2. Pitchfork bifurcation diagram showing (a) supercritical bifurcation with one stable solution for $R < R_c$ and two stable and one unstable solutions for $R > R_c$, (b) subcritical bifurcation with one unstable solution for $R > R_c$ and two unstable and one stable solutions for $R < R_c$.

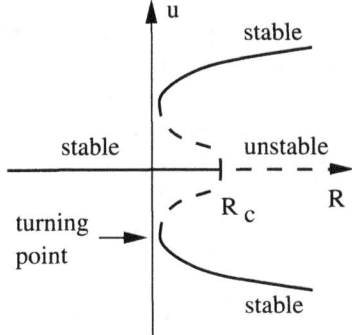

FIG. 10.3. Subcritical pitchfork bifurcation at $R = R_c$ followed by turning point bifurcation.

A last bifurcation which we need comes from a system of two variables, suppose u and v satisfy

$$\frac{du}{dt} = -v + (R - R_c - r^2)u,$$
$$\frac{dv}{dt} = u + (R - R_c - r^2)v, \qquad (10.5)$$

where $r^2 = u^2 + v^2$, then letting $\theta = \tan^{-1}\frac{u}{v}$, the polar components satisfy

$$\frac{dr}{dt} = r(R - R_c - r^2),$$
$$\frac{d\theta}{dt} = 1. \qquad (10.6)$$

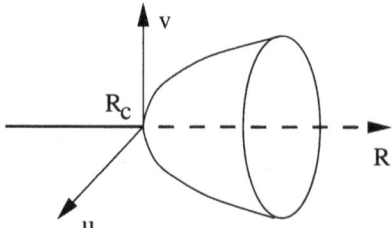

FIG. 10.4. Supercritical Hopf bifurcation at $R = R_c$.

Hence the fixed point $u = v = 0$ will bifurcate at $R = R_c$ such that r has the positive part of a supercritical pitchfork bifurcation and $\theta = t$. The stable solution will then be time periodic where for fixed $R > R_c$, the 'radius' r is constant. This bifurcation from a steady solution to a time periodic solution is called a supercritical Hopf bifurcation; see Figure 10.4. If the sign of r^2 in (10.5) is changed there will be a sub-critical Hopf bifurcation in the same way as there was a sub-critical pitchfork for (10.4).

These simple bifurcations allow us to interpret the observed sequence of events for flow through a symmetric channel expansion. At low Reynolds numbers there is a unique streaming flow with no region of reversed flow. As the Reynolds number increases, separation occurs and two symmetric counter rotating vortices form on either wall. At a slightly larger Reynolds number the flow looses symmetry with respect to the centreline via a supercritical pitchfork bifurcation and the main flow is deflected more to one side than to the other side and one vortex is larger than the other vortex. This is the Coanda effect. At larger Reynolds numbers Sobey (1985) observed in a 3:1 expansion but finite width channel that there was a loss of symmetry normal to the plane of the flow and a steady three-dimensional flow occurred. This was also observed by Sobey and Drazin (1986) for a channel with a contraction followed by an expansion. There are no other observations about the nature of this loss of symmetry but it is expected that it too should be via a supercritical pitchfork bifurcation. This observation was, however, for only a finite width channel, since an infinite width channel has no obvious "centre" about which symmetry might be lost it seems possible that a finite wavelength or time periodic disturbance would be more likely in such an idealised channel.

Returning to a finite width channel, at even higher Reynolds numbers (the exact Reynolds numbers are dependent on the expansion ratio) it has been observed that the flow becomes time periodic; see for instance Cherdron et al. (1978). Sobey and Drazin (1986) calculated that for the two-dimensional solutions (hence infinite width channel) there would be a sub-critical Hopf bifurcation followed by a turning point when the flow first became time periodic but there are no detailed observations near this transition point in finite channels. It is not known how the bifurcation sequence develops near the first onset of periodic flow in a finite width channel nor whether the time periodic solution for a two-dimensional flow calculated by Sobey and Drazin (1986) is relevant to the onset of time periodic flow in a finite width channel.

While the sequence of loosing symmetry with respect to one plane, then loosing symmetry with respect to a second plane and finally becoming time dependent which has been observed for channel flow is reasonably general, it is by no means true that all flows go through this entire sequence of events. A steady flow could jump straight to a finite amplitude oscillatory flow via a subcritical Hopf bifurcation and a turning point.

Having described one sequence of events which have been observed in channel flow, we shall now focus on the first supercritical pitchfork bifurcation when the flow remains two-dimensional but looses symmetry with respect to the centreline. In order to use the details of bifurcation theory it is necessary to identify a measure of asymmetry. Ideally the measure should be such that it is zero if and only if the flow is symmetric. Thus Sobey and Drazin (1986) used a measure based on symmetry of the stream-function about the channel centreline (assuming the channel is aligned with the x-direction),

$$\zeta^2 = \frac{1}{A} \int_A (\psi(x,y) + \psi(x,-y))^2 \, dx dy, \qquad (10.7)$$

where A represents a half of the channel on one side of the centreline. The variable ζ can have a sign associated with it depending on how symmetry is broken, that is which wall the flow deflects towards. This measure is ideal for computational purposes but impractical for observations. Fearn et al. (1990) used laser-Doppler velocimetry measurements at a single point of the transverse velocity on the centreline a little way downstream of a channel expansion to characterise the bifurcation for a right-angled channel expansion.

Using either of these measures of asymmetry gives a pitchfork bifurcation, that is the measure of asymmetry is proportional to $\sqrt{R - R_c}$ for a region $R > R_c$. In the case of a 1:3 right-angled expansion the critical Reynolds number is around $R_c \approx 656$. Note that there are various scalings for the velocity and length scale (average velocity, peak velocity, channel width and channel half width) which have been used so some conversion of Reynolds number may be necessary when comparing results of different authors.

The upstream flow has been defined so that the corresponding non-dimensional Poissueille pressure gradient has unit value, resulting in $u \sim \frac{1}{2} y(1-y)$ upstream in a channel extending over $0 \le y \le 1$ and a Reynolds number

$$R = \frac{12Q}{\nu}.$$

In Sobey and Drazin (1986) and Dennis and Smith (1980) the channel was non-dimensionalised to have width 2 and the average velocity was used as a velocity scale so that the non-dimensional problem had the stream-function varying between $\psi = \pm 1$ on the walls and

Coanda Effect

$$\text{Reynolds number } Re = \frac{Q}{2\nu} = \frac{1}{24}R.$$

In Fearn et al. (1990) the maximum velocity and channel width were used as scales so that

$$\text{Reynolds number} = \frac{3Q}{4\nu} = \frac{1}{16}R.$$

In Borgas and Pedley (1990) the Reynolds number was defined by

$$\text{Reynolds number} = \frac{Q}{\nu} = \frac{1}{12}R.$$

The different definitions merely provide a scaling difference: experiments and computations from the Navier–Stokes equations confirm that location of the bifurcation and that it is a pitchfork bifurcation.

10.3 Bifurcation solutions from Navier–Stokes equations

As an example of numerical results we show in Figure 10.5 contours of the stream-function for flow through a 1:3 sudden expansion at a Reynolds number $R = 960$ ($Re = 40$) which show that once vortex to be much larger than the other and the main flow moving to 'attach' to the wall behind the smaller vortex. A flow with alternate asymmetry can be computed for instance in an unsteady calculation by reflecting the stream-function and vorticity about the centreline and continuing the calculation. A computational code will usually calculate one branch preferentially because rounding errors in the calculation put the starting solution in the domain of attraction of one branch of the two stable branches of the pitchfork bifurcation. If a starting solution can be put into the domain of attraction of the other branch then the solution converges in time to that branch. This also means that one is always computing an imperfect bifurcation using a time marching method and it is possible to follow one branch as the Reynolds number increases, artificially swap to another branch and then follow that branch down with decreasing Reynolds number until a turning point is reached and the solution jumps back to the original branch.

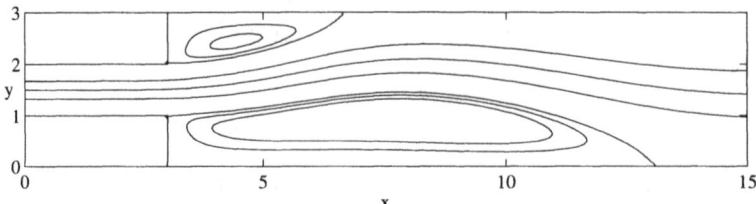

FIG. 10.5. Computed solution of Navier–Stokes equations at $R = 960$ ($Re = 40$) for a sudden 1:3 expansion.

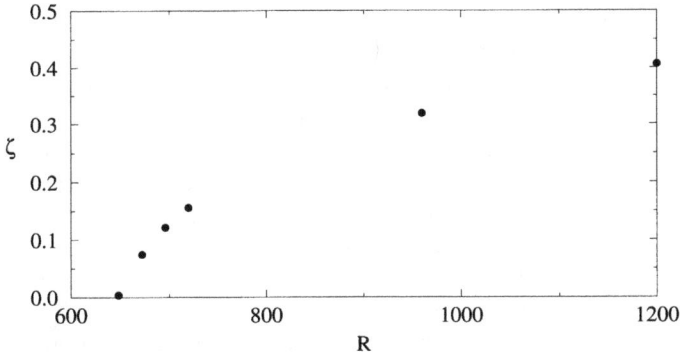

FIG. 10.6. Computed values of the asymmetry parameter ζ for a 3:1 sudden expansion. There is a negative branch which is not shown.

If the Reynolds number is varied then one branch of the bifurcation diagram emerges and is shown for a 1:3 sudden expansion in Figure 10.6. The computation of such diagrams using time marching methods can be very laborious since the convergence rate becomes very small near the bifurcation point. It is also possible for numerical methods to introduce spurious bifurcation structures; see for instance Uchibori and Sobey (1992) for problems specific to this bifurcation.

It may be tempting to think that the bifurcation structure is associated with shear layers coming from the sudden expansion and that the bifurcation is an artefact from the corners at the expansion. This is not the case; Lighthill's advice regarding sharp corners (see page 250) again proves useful here. If flow is calculated for a smooth walled symmetric expansion then a bifurcation structure is still found. An example calculation of the flow for the same Reynolds number as that in Figure 10.5 for a smooth walled 1:3 expansion is shown in Figure 10.7 for the same stream-function values. The vorticies are a little weaker than for a sudden expansion and that is the real effect of the 'sudden-ness' of the expansion. See also Cliffe and Greenfield (1982) as well as Cliffe et al. (1982) for calculation of symmetric solutions.

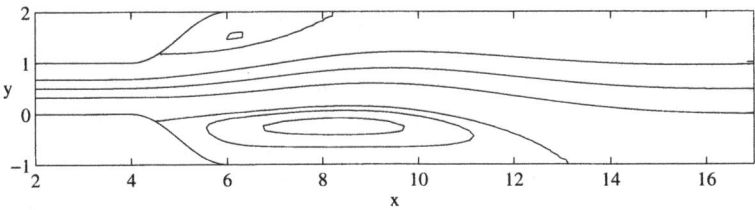

FIG. 10.7. Computed solution of Navier–Stokes equations at $R = 960$ ($Re = 40$) for a smooth walled 1:3 expansion.

Thus we come to an as yet unanswered question. If the expansion ratio (denoted D) is varied, how does the critical Reynolds number at which the flow bifurcates vary and does the structure of the bifurcation remain unaltered? Sobey and Drazin (1986) attempted to compute a solution to this question but their numerical results for steady flow suffered from spurious numerical artefacts. Some calculations exist for an expansion ratio $D = 2$, see Cliffe and Greenfield (1982). The reason this question is interesting is because when $D = 1$; that is, the expansion vanishes and flow is through a parallel channel, it is known that the flow becomes unstable via a subcritical Hopf bifurcation. It is well known from experiments that time periodic flows occur for flow through channel expansions, see for example Cherdron et al. (1978). Sobey and Drazin (1986) also calculated that steady flow through an expansion would become time dependent via a subcritical Hopf bifurcation at higher Reynolds numbers, for instance near $R = 3840$ for $D = 2$. The inference from their work was that the critical Reynolds number for the first steady pitchfork bifurcation would increase without bound as D decreased to one. What happens to the critical Reynolds number for the subcritical Hopf bifurcation is much less clear. It might not increase indefinitely or alternately it too increases indefinitely and the sub-critical Hopf for channel flow has some different origin. While this does seem remote since one might expect some structural stability to the bifurcation structure for parallel channel flow if the channel walls are only slightly perturbed, since this would have to include asymmetric wall perturbations it is possible that for D only a little greater than one there is an entirely separate sub-critical Hopf bifurcation to that which is associated with larger symmetric channel expansions. In either case there should be a very interesting point at the value of D having the same Reynolds number for a steady pitchfork bifurcation and for a sub-critical Hopf bifurcation.

This background material is intended to demonstrate that the Coanda effect is an important theoretical problem with wide implications for which we might hope that a powerful theory such as interactive boundary layer theory could be used as a tool to probe some of these questions since the Reynolds numbers involved are so great that direct calculation of solutions to the full Navier–Stokes equations seem to be some way off.

10.4 Application of interactive boundary layer theory

There has as yet only been one substantial attempt to model the Coanda effect using interactive boundary layer theory, that by Borgas and Pedley (1990). Their study concerned flow through an annular channel and in the limit of vanishing annular gap they obtained a solution for a plane channel. Their main result was for the case where the wall disturbance was sufficiently small for there to be no cross channel pressure variation. When the walls, which were undisturbed upstream, had a particular similarity form for $X > 0$ they found that there would be non-uniqueness in the solution. In the limit of a symmetric plane channel expansion the non-uniqueness resulted in a pitchfork bifurcation as the wall displacement increased in magnitude. As the expansion they considered was of

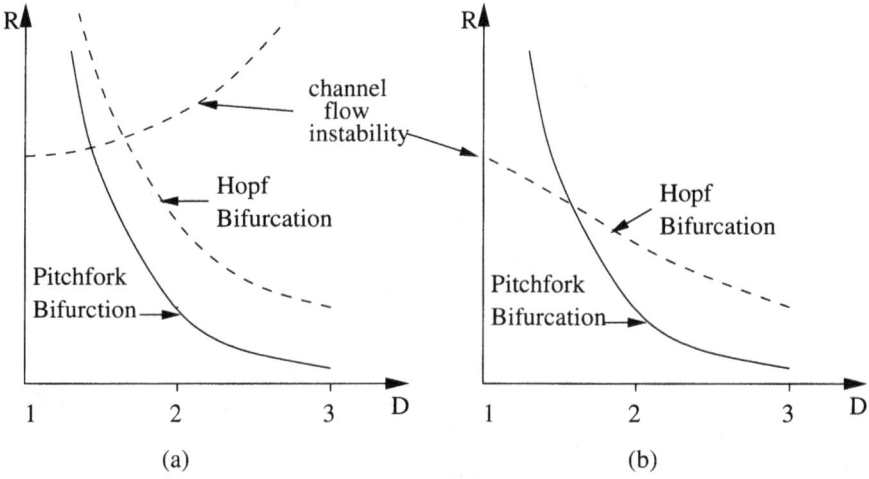

FIG. 10.8. Possible bifurcation structures for flow through a channel expansion as the expansion ratio D varies. (a) Bifurcation for parallel channel flow independent of large D bifurcations (b) Bifurcation for parallel channel flow limit of subcritical Hopf bifurcation for flow through an expansion.

a specific annular form they derived plane channel flow only as a limiting case. Since the most significant part of their work can be directly applied to a plane channel we shall consider their work in that context rather than that for flow through an annular channel.

Consider again the formulation for channel flow from section 8.2 where the channel wall variation is sufficiently small that no cross channel pressure variation occurs in the leading two terms and the flow is modelled as an inviscid rotational core with an interactive boundary layer on each wall. Each wall layer has to satisfy no slip conditions on the walls and match to the linear velocity profile of the core near the walls.

The equations for the lower and upper wall layers, (8.35) and (8.42) are identical in form to a normal boundary layer equation so they must admit similarity solutions for the stream-functions in each layer, $\Psi_0(X,\eta)$ and $\tilde{\Psi}_0(X,\xi)$ of the form

$$\Psi_0 = \sqrt{\frac{2}{m+1}} X^{(m+1)/2} M(s), \quad s = \sqrt{\frac{m+1}{2}} \frac{\eta}{X^{(1-m)/2}}, \tag{10.8}$$

and

$$\tilde{\Psi}_0 = \sqrt{\frac{2}{m+1}} X^{(m+1)/2} \tilde{M}(\tilde{s}), \quad \tilde{s} = \sqrt{\frac{m+1}{2}} \frac{\xi}{X^{(1-m)/2}}, \tag{10.9}$$

with boundary conditions $M(0) = M'(0) = 0$ and $\tilde{M}(0) = \tilde{M}'(0) = 0$ on each wall. Up to this the formulation is identical to that by Falkner and Skan (1931)

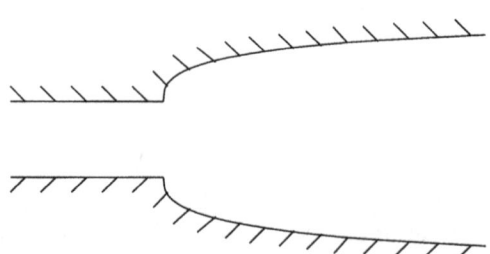

FIG. 10.9. Symmetric channel expansion for application of Borgas & Pedley's theory.

already described, now some significant differences are introduced. In the derivation by Falkner & Skan the velocity profile was constant far from the wall whereas here the velocity profile has to be linear to match with the core.

The condition on the stream-functions far from the walls,

$$\Psi_0 \sim \frac{1}{4}\eta^2 + \frac{1}{2}[A(X) + f(X)]\eta + o(\eta), \quad \text{as } \eta \to \infty, \quad (10.10)$$

and

$$\tilde{\Psi}_0 \sim \frac{1}{4}\xi^2 - \frac{1}{2}[A(X) + g(X)]\xi + o(\xi), \quad \text{as } \xi \to \infty, \quad (10.11)$$

can only be satisfied if $M(s) \sim s^2$ as $s \to \infty$ and $m = 1/3$ (with identical conditions for $\tilde{M}$). We have of course seen the need for this value of m on a number of occasions before.

Suppose then that the walls are straight for $X < 0$ and for $X \geq 0$ are given by

$$f(X) = -\mu_L X^{1/3}, \quad g(x) = \mu_U X^{1/3}. \quad (10.12)$$

The channel shape for $\mu_L = \mu_U > 0$ is shown in Figure 10.9

Let the displacement function $-A(X)$ have the same similarity form

$$A(X) = A_0 X^{1/3}. \quad (10.13)$$

We now come to the most significant difference between this model situation and the usual Falkner–Skan model. In an exterior flow the constant flow outside the boundary layer would be taken as $u_e = X^{1/3}$ so that the pressure would be $p(X) = -X^{2/3}/2$ and the pressure gradient everywhere **favourable**, $p'(X)$ being negative for $X > 0$. In the case of a channel flow the core has to accelerate or decelerate according to a fairly coarse continuity argument. If the channel narrows, the core accelerates with the pressure gradient being favourable but if the channel widens the flow has to decelerate so the pressure gradient has to be positive or **adverse**. Since this is an interactive boundary layer solution the magnitude of the pressure gradient is unknown although for a similarity solution its variation with X will still have to follow $X^{2/3}$. Suppose the pressure is given by

$$p(X) = -\frac{1}{2}p_1 X^{2/3} \qquad (10.14)$$

then the similarity functions $M(s)$ and $\tilde{M}(\tilde{s})$ both satisfy the same equation,

$$M''' + MM'' + \frac{1}{2}(p_1 - M'^2) = 0. \qquad (10.15)$$

Borgas & Pedley simplified this equation by supposing $p_1 = \text{sgn}(p_1)d^4$ and applying a simple stretching transformation, $z = ds$, $\tilde{z} = d\tilde{s}$ then both functions M and $\tilde{M}$ could be expressed in terms of a single function N,

$$M(s) = dN(z), \quad \tilde{M}(\tilde{s}) = dN(\tilde{z}), \qquad (10.16)$$

where N satisfied

$$N''' + NN'' + \frac{1}{2}(\text{sgn}(p_1) - N'^2) = 0. \qquad (10.17)$$

In the case $\text{sgn}(p_1) = +1$ this is a Falkner–Skan equation. In the case $\text{sgn}(p_1) = -1$ it is not a normal Falkner–Skan equation but it is the sign change in the pressure gradient which allows non-unique solutions to be found. Since our interest is in flow through a channel expansion consider from now that $\text{sgn}(p_1) = -1$ and μ_L, μ_U to be positive.

The boundary conditions at the wall on (10.17) are $N(0) = N'(0) = 0$. Borgas & Pedley then considered solutions which satisfied

$$N''(0) = \tau, \qquad (10.18)$$

and showed that if a solution existed for large z then

$$N(z) \sim \frac{1}{2}\zeta(\tau)z^2 + \gamma(\tau) + o(z) \quad \text{as} \quad z \to \infty. \qquad (10.19)$$

Of course, this means that the stream-function far from the wall would behave according to

$$\psi \sim \frac{1}{4}\eta^2[\sqrt{\frac{8}{3}}d^3\zeta] + \frac{1}{2}\eta[2d^2\gamma X^{1/3}] + o(\eta) \quad \text{as} \quad \eta \to \infty, \qquad (10.20)$$

with a similar expansion for $\tilde{\Psi}_0$. Equating with the correct outer expansions gives

$$\zeta(\tau) = [\sqrt{\frac{8}{3}}d^3]^{-1}, \quad \gamma(\tau) = \frac{1}{2d^2}(A_0 - \mu_L), \qquad (10.21)$$

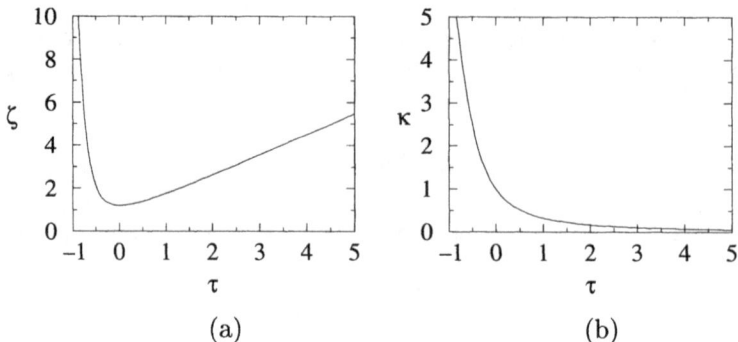

FIG. 10.10. (a) Plot of ζ, (b) Plot of κ.

and

$$\zeta(\tilde{\tau}) = [\sqrt{\frac{8}{3}}d^3]^{-1}, \quad \gamma(\tilde{\tau}) = -\frac{1}{2d^2}(A_0 + \mu_U). \tag{10.22}$$

Now of course the conclusion reached in chapter 8 was that $\tau = \tilde{\tau}$ and $A_0 = (\mu_L - \mu_U)/2$ so that when $\mu_L = \mu_U = \mu$ the displacement function would vanish with $A_0 = 0$. However that is not a necessary deduction from (10.21) and (10.22), rather we only need

$$\zeta(\tau) = \zeta(\tilde{\tau}), \tag{10.23}$$

so the question of uniqueness turns on whether the function ζ^{-1} is multiple valued or not.

Borgas & Pedley showed that the function $\zeta(\tau)$ did have a minimum and so its inverse would be multiple valued and that provides the possibility of multiple solutions for the system. The function $\zeta(\tau)$ is shown in Figure 10.10 (a) and has a minimum at $\tau = 0$ with $\zeta(0) = \zeta_0 = 1.198....$

First, it is necessary to eliminate d and A_0 from (10.21) and (10.22) and similarly to Borgas & Pedley, define a new function

$$\kappa(\tau) = -(\frac{3}{8})^{1/3}\gamma(\tau)\zeta(\tau)^{-2/3}, \tag{10.24}$$

so that the lower and upper wall velocity gradients, τ and $\tilde{\tau}$ must satisfy

$$\zeta(\tau) = \zeta(\tilde{\tau}) \quad \text{and} \quad \kappa(\tau) + \kappa(\tilde{\tau}) = 2\mu. \tag{10.25}$$

The behaviour of κ is also vital in showing that there is a pitchfork bifurcation as μ increases. The function κ is monotonic and decreasing as τ increases with $\kappa(0) = \kappa_0 = 0.992....$ If $\mu < \kappa_0$ then there can be only one solution which must have $\tau = \tilde{\tau} > 0$ and consequently $A_0 = 0$, that is the flow is symmetric.

If, however, $\mu > \kappa_0$ then there is still the symmetric solution with $\tau = \tilde{\tau}$ and $\kappa(\tau) = \mu$ only since $\tau, \tilde{\tau}$ would be negative, there would be backflow on each wall. There is another possibility because of the minimum of $\zeta(\tau)$ at $\tau = 0$. When $\mu > \kappa_0$ the condition (10.23) can be satisfied by values $\tau > 0$, $\tilde{\tau} < 0$ and vice versa. Thus the point $\mu_c = \kappa_0$ will be a bifurcation point. The nature of the bifurcation can be seen by looking near the bifurcation point and taking

$$\zeta = \zeta_0 + \zeta_2 \tau^2, \quad \kappa = \kappa_0 + \kappa_1 \tau + \kappa_2 \tau^2, \tag{10.26}$$

notably with $\zeta_2 > 0$, $\kappa_2 >)$, whence $\tilde{\tau} = -\tau$ and

$$\tau - \tilde{\tau} = \pm 2\sqrt{\frac{\mu - \mu_c}{\kappa_2}}. \tag{10.27}$$

The variable $\tau - \tilde{\tau}$ was used by Borgas & Pedley as a measure of asymmetry since it vanished when the flow was symmetric and it showed that there was indeed a super-critical pitchfork bifurcation at the heart of their model problem.

It is of course very difficult to relate this problem to more realistic situations since when backflow occurs, it occurs everywhere along the wall and the sharp right angled expansion at $X = 0$ is difficult to fit into an asymptotic theory which implicitly takes the wall slope to be bounded. In their work Borgas & Pedley indicated how this latter problem might be tackled, but the first problem remains.

If this bifurcation is the source of the Coanda effect it also means that the heuristic explanation of the Coanda effect in terms of a cross channel pressure difference needs at the very least, some modification. Perhaps it is the case that trying to argue about pressure variation on the basis of Bernouilli's equation for a channel flow is incorrect: it is certainly true that the explanation for the bifurcation calculated by Borgas & Pedley hinges around the pressure gradient being adverse while the flow is locally accelerating, entirely at variance with an argument incorporating Bernouilli's equation. It is also possible that the pitchfork bifurcation identified by Borgas & Pedley is not the only such bifurcation and that the bifurcation which we have called the Coanda effect occurs at a lower Reynolds number than that predicted for Borgas & Pedley's case and has a different physical basis.

There are other features which can be noted. If a channel of the type used by Borgas & Pedley were to have upstream width H and expand to $(2D+1)H$ after a distance LH then the scaling introduced in (8.102) would still be applicable so that the bifurcation would be predicted to occur at

$$R_c = \mu_c^3 L D^{-3}, \tag{10.28}$$

and this provides a partial answer for the behaviour of the bifurcation point as the expansion magnitude decreases. In this model problem the critical point at which the bifurcation occurs is the same point at which flow reversal occurs, something which is not in accordance with observation or calculation of solutions of the full Navier–Stokes equations where there is stable separated symmetric flow for a range of Reynolds numbers before the bifurcation occurs.

10.5 Summary

The Coanda effect is the simplest bifurcation associated with separated channel flow but as yet we do not have a thorough understanding of its origin or its behaviour as channel geometry varies. Application of the simplest interactive boundary layer model, that for wall indentations too small to provoke a cross channel pressure variation, is not completely satisfactory but does show that the non-linear boundary layer equations can support non-unique solutions. My own attempts to compute non-unique solutions from the non-linear boundary layer equations when there is a cross channel variation have so far failed, the solution either remaining symmetric or there being difficulty in obtaining converged solutions. Thus we still cannot resolve how this bifurcation and separation are related although it seems likely that the Coanda effect will at some time be fully explained using interactive boundary layer theory.

APPENDIX A

PROBLEMS AND COMPUTER PROGRAMS

Examples suitable for teaching or projects have not been included at the end of each chapter but some support material is collected in this appendix and other material such as computer program source code, is available on the World Wide Web at

http://www.sjc.ox.ac.uk/scr/sobey/iblt

A powerful way to understand many of the theoretical constructs discussed in this book is to use them to compute examples where physical intuition can be tested and brought to bear on calculated predictions. The material suggested is of two different types. First, I have tried to suggest mini-projects (some not so mini) relating to material for each chapter. Some of these can be computed very easily, others need more thought and some may have no easy outcome. The second type of material is source code such as generated data for this book. In many cases I have recomputed the methods of our predecessors using present computing power, a power undreamed of last century or early this century. What does come out of recomputing classical results is amazement that our predecessors were able to calculate so much! I am also making available some of my own codes for finite difference solution of the (unsteady) Navier–Stokes equations. There are many other sources for solvers and those I provide are just to illustrate that it is nowadays fairly straightforward to develop codes for two-dimensional laminar flow in particular geometries, such as channel flow for small to moderate Reynolds numbers. Calculations for larger Reynolds numbers remain at the cutting edge of what can be achieved. The examples I provide use fairly robust numerical techniques such as multigrid or conjugate gradient but are nevertheless easy to understand.

The structure of each section which follows is, first, to describe some problems and secondly to outline some software available at my web site. The web site will be updated through feedback from readers and the web site is the definitive appendix.

A.1 Chapter 1 – Introduction

A.1.1 *Software*

In this area there are example implementations of two and three level multigrid and conjugate gradient for a simple Poisson equation on a unit square. Although general methods such as multigrid and conjugate gradient are not discussed in

this book, they are vital to solution of the Navier–Stokes equations and some further details and notes are given in the web site. The programs are:
(i) mg2.c two level multigrid solver
(ii) mg3.c three level multigrid solver
(iii) cg.c conjugate gradient solver

A.2   Chapter 2 – Flat plate

A.2.1   *Problems*

A.2.1.1   Can you extend Imai's work for the drag to improve the far wake asymptotics - what effect does the wake have on the drag coefficient – is it just that Imai's formula with $x = 1$ is correct until you include triple deck correction?

A.2.1.2   The method used by Meksyn for the far wake (page 64) seemed to be somewhat ad-hoc. Can the methods be developed in a formal asymptotic framework?

A.2.1.3   Dean's method (page 47 *ff*) unexpectedly predicts the correct second term for the drag on a semi-infinite plate. Can you develop the method in a formal asymptotic context and examine how higher order terms come out. Does the method also lead to unknown constants in further terms?

A.2.2   *Software*

(i) b10.c velocity across Blasius boundary layer, results in b10.out
(ii) body.m contour potential flow streamlines around parabolic body
(iii) b12.c limiting values of f_2 and $c_{4,2}$ as a_2 varies, see Figure 2.5
(iv) composite.c calculate composite expansion for Goldstein's near wake

A.3   Chapter 3 & 4 – Triple deck

A.3.1   *Problems*

A.3.1.1   The transformation used by Dennis and Dunwoody (1966) has the attractive property that the leading and trailing edge points would not appear in a finite difference formulation. Does such a scheme provide enough accuracy to resolve the asymptotic structure of the drag coefficient? The computational domain has one less boundary than that used by McLachlan (1991a) and if a potential solution was used for $\xi > \xi_{M-1}$ to relate stream-function values on the boundary, $\psi_{M,j}$, to the values $\psi_{M-1,k}$, $k = 0, \cdots, N$ as advocated by Fornberg, can this method be even more able to resolve fine second and third order variation in the drag coefficient than the computations of McLachlan? McLachlan had some trouble with mesh refinement near the trailing edge, with this transformation refinement at both the leading and trailing edge should be straightforward.

A.3.1.2   The potential flow region for flow about a plate seems to revolve around a number of loose ends. Suppose the plate is located on $0 \leq x \leq 1, y = 0$. Over the main body of the plate the outer potential should be like

$$w \sim z - i\beta_0 \sqrt{\frac{2}{R}} z^{1/2},$$

the triple deck introduces a correction to this but not of order $z^{1/2}$, rather of order $(z-1)^{1/3}$ so far downstream the potential flow is still describing a body of parabolic extent rather than one where the flow returns to uniform velocity over the whole wake. Could this be rectified by assuming a distribution of sources $m(x,R)$ over $0 \le x \le 1$ where m was then expanded as an asymptotic sequence for large R? Obviously from continuity the sources would have to become sinks towards the trailing edge.

A.3.1.3 Talke and Berger (1970) used parabolic coordinates centred on the trailing edge to find a series truncation solution near the trailing edge using Goldstein's solution for the wake as an outer boundary condition. How would their results be modified if a series truncation solution near the trailing edge (in other words in a region of order $R^{-3/4} \times R^{-3/4}$) was matched to the near field triple deck solution rather than Goldstein's near wake? See also Berger and Scalise (1995).

A.3.2 *Software*

(i) newton.f program to solve trailing edge triple deck problem. The program requires a matrix inversion routine such as might be obtained from Press et al. (1986)
(ii) tdeck.m contour streamlines from output of triple deck program *newton.f*

A.4 Chapter 5 & 6 – Separation

A.4.1 *Problems*

A.4.1.1 Batchelor has conjectured that the correct limit for the Navier–Stokes equations as $t \to \infty$ and then $R \to \infty$ of a recirculation region is a constant vorticity region of finite extent. Normally computations are based around letting $t \to \infty$ to obtain a steady solution and then allowing the Reynolds number to increase. Is it possible to compute solutions with the limits interchanged, so $R \to \infty$ and then $t \to \infty$? One way this might be approached would be to take a viscous calculation at finite Reynolds number as the starting point for a time dependent inviscid (but rotational) calculation. How would a flow with a distribution of vorticity from a viscous calculation subsequently develop if viscosity were suddenly switched off? That is, from an initial distribution of vorticity $\omega_0(x,y)$ and corresponding stream function $\psi_0(x,y)$ solve

$$\omega_t + \mathbf{u} \cdot \nabla \omega = 0, \quad \nabla^2 \psi = -\omega, \tag{A.1}$$

with appropriate boundary conditions giving $\psi = 0$, $\omega_n = 0$ (n being normal derivative) on a body and uniform far field. It is difficult to see that how if this problem started with a recirculating region (such as from a viscous calculation of the steady vortex behind a cylinder) it would evolve by convection of vorticity alone to one where there were regions of constant vorticity. It would seem more plausible that the existing vorticity would be swept away into the wake and it would not be possible to achieve a Prandtl–Batchelor flow by this means.

A.4.1.2   In Brown and Stewartson (1969) it is claimed that the logarithmic terms cancel so that the u velocity is finite on the separation line (page 57 of their paper). It is very unusual for these normally meticulous authors not to reference a detailed justification of a remark such as this. Can you either track down a suitable reference or show this result yourself?

A.4.1.3   Modify Leigh's method using the initial profile

$$u(0, Y) = \begin{cases} \sin kY, & kY < \pi/2, \\ 1, & kY \geq \pi/2. \end{cases} \qquad (A.2)$$

and find a value k such that the separation point is the same as predicted by Leigh for Howarth's starting profile,

$$\alpha_s = 0.9585. \qquad (A.3)$$

Then use Terrill's method to calculate the wall shear for a linearly decreasing velocity profile and compare the results from the two methods. Does Terrill's method predict the same separation point? What is your estimate of the singularity power q, where $\sigma \sim (\alpha_s - \alpha)^q$ near separation?

A.4.1.4   Leigh and Terrill each use a semi-implicit method to solve a set of non-linear equations. Apply Newton's method to solve Terrill's discrete non-linear equations. Is the computation more efficient?

A.4.1.5   Implement Kawaguti's method to solve the Navier–Stokes equations. In Kawaguti & Jain the outer boundary condition was relaxed to have just the stream-function and vorticity zero. What effect does this have on the computed solution at $R_d = 40$?

A.4.2 Software

(i) `solvebrod.m` calculate first n Brodetsky coefficients $A_1, \ldots, A_n$ for given separation angle
(ii) `brodgen.f` calculate flow using Brodetsky coefficients calculated by *solvebrod.m*
(iii) `lcwcyl.f` Woods' method for cylinder with constant pressure on free streamline
(iv) `lcwcyl2.f` Woods' extended model for cylinder with variable pressure on free streamline
(v) `pohlhausen.f` Pohlhausen's method to solve integral form of boundary layer equation
with either Heimenz observations, potential flow or linear variation for flow at

the boundary layer edge
(vi) fskan.f determine existence of Falkner–Skan solutions by varying $f''(0)$ for fixed β
(vii) fskan1.f determine existence of Falkner–Skan solutions by varying β for fixed $f''(0)$
(viii) burgers.f Burgers' method for either Heimenz or potential outer flow
(ix) leigh.f Leigh's method for calculating solution of boundary layer equation
(x) terrill.f Terrill's method for calculating solution of boundary layer equation
(xi) thom.f Thom's method to calculate solution of Navier–Stokes equations for flow about a cylinder

A.5 Chapter 7 – Prediction of separation from a cylinder

A.5.1 Problems

A.5.1.1 Woods' model for variable pressure behind the cylinder results in a pressure gradient singularity strength

$$\kappa = \kappa(\alpha_s, C_p),$$

given by (6.100). Can this, instead of Brodetsky's singularity strength (which is the case $C_p = 0$), be combined with Sychev's hypothesis in any rational way?

A.5.1.2 We conjecture that the reattachment point for a rotational corner flow might be an interesting example for studying reattachment generally and particularly for comparison with the model in Peregrine (1985): is this true?

A.5.2 Software

(i) sw.f Stewartson & William's free interaction solution
(i) riley.m Riley's method to calculate slender Sadoviskii vortex shape

A.6 Chapter 8 – Channel flow

A.6.1 Problems

A.6.1.1 Use numerical quadrature of the analytic solution (8.125) to compute the coefficient for corner vorticity in (8.139) for the rectangular finite volume in Figure 8.25, (b).

A.6.1.2 The spectral method of Burggraf and Duck (1982) would seem to have ideal properties for calculating interactive boundary layer solutions for channels with periodic disturbances. Implement such a scheme and compare its resolution and computational requirement with the simple interactive boundary layer solver provided on the website.

A.6.1.3 One almost curious aspect of finite volume methods is to use differences between volume averaged values to represent derivatives or more precisely, fluxes. If Weinbaum's solution is used to calculate the average value of

ψ at the corner and to calculate the exact flux across the boundary of the finite volume, by how much does the difference between the average values of ψ at the corner and at say $(0, h)$ differ from the calculated flux across the line $y = h/2, -h/2 < x < h/2$?

A.6.2 Software

(i) coord.c generate conformal mesh for channel, uses X window interface
(ii) unsteady12.c time marching Navier–Stokes solver for channel using conformal mesh generated by coord.c. Uses X window interface and provides streamline contours as solution evolves. Uses multigrid to solve stream-function-vorticity equation. steady12.c is a steady version
(iii) channel.f 'triple deck' solution for channel with no cross channel pressure variation
(iv) lcw.f Implementation of Woods' method to calculate free streamline solution in a channel for various wall conditions
(v) corner.m contours near a corner using Weinbaum's solution

A.7   Chapter 9 – Upstream influence

A.7.1 Problems

A.7.1.1   We have noted that two-dimensional composite expansions have been little used. Can you firstly use the suggested composite expansion to generate the stream-function through a channel and contour lines of constant ψ? Can you develop an improved two-dimensional composite expansion?

A.7.1.2   One of the unsatisfactory features of predictions for channel flow is that of separation on both walls which comes from the $\epsilon = o(R^{-1/7})$ theory of chapter 8. We have shown one example where allowance of a cross channel pressure gradient results in separation being absent from the solution on the opposite flat wall in Figure 9.7. If we try to compute the equivalent flow for the channel with lower wall defined by (8.103) but allowing the pressure to vary across the channel then the computation is fraught with difficulty. Convergence is extremely slow and may be affected by an instability which develops as the number of iterations increases. An example of a flow calculated after 150 iterations using 51 mesh points in the X-direction and 101 across each boundary layer (so that a 609 × 609 matrix has to be inverted each Newton step) is shown in Figure A.1. The separation region on the flat wall is still evident although it has moved slightly downstream relative to that in the furrowed wall.

If we look at computation or observation of steady flow, separation on the opposite wall is not usually present. However, separation on the opposite flat wall is observed, see Sobey (1985) Figure 8 (a)), and can be computed, see Figure A.2, during the acceleration phase of unsteady flow. Were the observations to only be during the deceleration phase then the extra adverse pressure gradient from the deceleration could aid the occurrence of separation but these observations and calculations are for flows which have been accelerating. Is there any relation

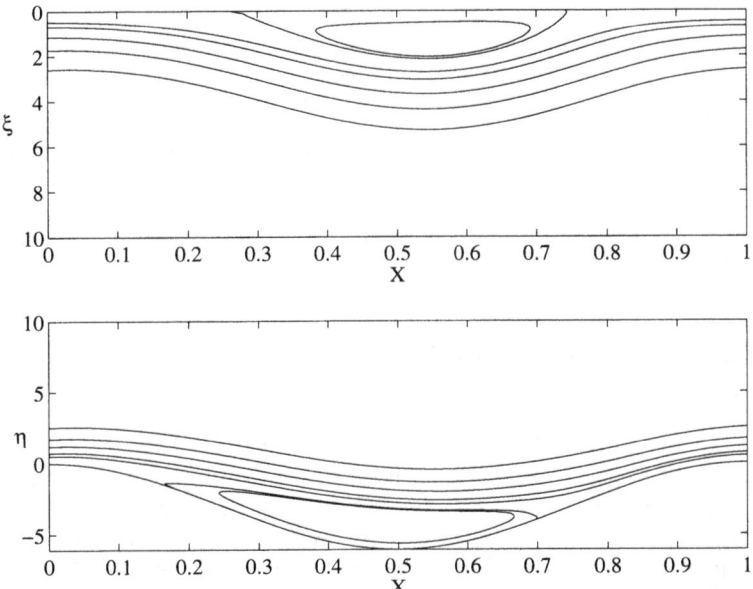

FIG. A.1. Solution of two layer interactive equations for $\mu = 6.08$ after 150 iterations.

between the prediction from steady theory of separation on the flat wall and the unsteady solutions. Does the extra acceleration of the fluid result in the fluid behaving as if it were at a much greater Reynolds number? To complicate matters the calculation of separation on the flat wall during an acceleration does not happen for all Strouhal numbers, there seems to be a band on values, greater than quasi-steady flow but not so large that viscous effects dominate for which this second vortex can be computed.

Are there a number of artificial effects happening in the interactive boundary layer computations: are the computations properly mesh converged, are they properly converged for a fixed mesh, is the simple method used to solve the interactive equations inadequate to resolve these questions and a more sophisticated method (for instance a spectral method) more appropriate for computing solutions to the interactive equations when applied to periodically varying channels?

A.7.2 *Software*

(i) `deck2.f` 'triple deck' solution allowing cross channel pressure variation
(ii) `freeint.f` free interaction development of separation in a parallel channel

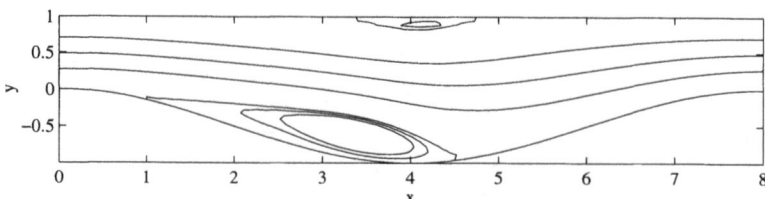

FIG. A.2. Computed solution of unsteady Navier–Stokes equations at instant of peak flow, $R = 7200$ for oscillatory flux.

A.8 Chapter 10 – Coanda effect

A.8.1 Problems

A.8.1.1 Riley (1987) gives one way of computing a Prandtl–Batchelor flow which might be applied to a channel flow if one assumes that there is potential flow through the channel with vorticity being confined to finite regions on the walls. If the channel has a symmetric expansion is it possible to compute an asymmetric Prandtl–Batchelor flow or are all solutions necessarily symmetric? Is it possible to compute asymmetric exterior Prandtl–Batchelor flows where the geometry and upstream outer flow is symmetric?

A.8.1.2 Can the $\epsilon \sim R^{-1/7}$ theory be applied in a symmetric channel in the same way as the $\epsilon \ll R^{-1/7}$ theory was applied by Borgas & Pedley; that is, as the limit of an asymmetric channel becoming symmetric? My experience with computed solutions of the two coupled non-linear boundary layer equations in a symmetric geometry is that when they do converge, they converge to a symmetric solution. Is it possible to compute asymmetric solutions by starting from an asymmetric initial condition or by introducing some form of psuedo time marching, for instance by solving

$$u_t + uu_X + vu_\eta = -P_L'(X) + u_{\eta\eta}, \quad \tilde{u}_t + \tilde{u}\tilde{u}_X + \tilde{v}\tilde{u}_\xi = -P_U'(X) + \tilde{u}_{\xi\xi}, \quad (A.4)$$

with P_L and P_U appropriately related? This could be done by either an explicit scheme or an implicit scheme using a suitable modification for the Newton iteration matrix.

A.8.2 Software

(i) bpgen.f generate solutions to Borgas & Pedley's variant of the Falkner–Skan equation (10.17)
(ii) channel.f time marching solution of Navier–Stokes equations for flow thorough a sudden channel expansion using a rectangular grid. Time stepping of vorticity can be by central, upwind, Lax–Wendroff or Quickest algorithms. Stream function-vorticity equation solved using multigrid

BIBLIOGRAPHY

Acker A. (1998). On the existence of convex classical solutions to a generalised Prandtl–Batchelor free boundary problem. *Zeitschrift für angewandte Mathematik und Physik*, **49**:1–30.

Ackerberg R. (1970a). Boundary-layer separation at a free streamline. Part 1. Two-dimensional flow. *Journal of Fluid Mechanics*, **44**:211–235.

Ackerberg R. (1970b). Boundary-layer separation at a free streamline. Part 2. Numerical results. *Journal of Fluid Mechanics*, **46**:727–736.

Acrivos A., Leal L., Snowden D., and Pan F. (1968). Further experiments on steady separated flows past bluff objects. *Journal of Fluid Mechanics*, **34**:25–48.

Acrivos A., Snowden D., Grove A., and Petersen E. (1965). The steady separated flow past a circular cylinder at large Reynolds numbers. *Journal of Fluid Mechanics*, **21**:737–760.

Alden H. (1948). Second approximation to the laminar boundary layer over a flat plate. *Journal of Mathematics and Physics*, **27**:91–104.

Allen D.N.d.G. and Southwell R. (1955). Relaxation methods applied to determine the motion, in two dimensions, of a viscous fluid past a fixed cylinder. *Quarterly Journal of Mechanics and Applied Mathematics*, **VIII**:129–145.

Apelt C. (1958). The steady flow of a viscous fluid past a circular cylinder at Reynolds numbers 40 and 44. *Aeronautical Research Council, Reports & Memoranda*, **3175**.

Bairstow L. (1925). Skin friction. *Journal of the Royal Aeronautical Society*, **29**:3–14.

Bairstow L., Cave B., and Lang E. (1922). The two-dimensional slow motion of fluids. *Proceedings of the Royal Society*, **100**:394–413.

Bairstow L., Cave B., and Lang E. (1923). The resistance of a cylinder moving in a viscous fluid. *Philosophical Magazine and Journal of Science*, **223**:383–432.

Batchelor G. (1956a). On steady laminar flow with closed streamlines at large Reynolds number. *Journal of Fluid Mechanics*, **1**:177–190.

Batchelor G. (1956b). A proposal concerning laminar wakes behind bluff bodies at large Reynolds number. *Journal of Fluid Mechanics*, **1**:388–398.

Batchelor G. (1967). An Introduction to Fluid Mechanics. Cambridge University Press.

Bender C. and Orszag S. (1978). Advanced Mathematical Methods for Scientists and Engineers. McGraw-Hill.

Benjamin T. (1967). Internal waves of permanent form in fluids of great depth. *Journal of Fluid Mechanics*, **29**:559–592.

Berger S. and Scalise D. (1995). An approximate theory for velocity profiles in the near wake of a flat plate. *Zeitschrift für angewandte Mathematik und Physik*, **46**:5612–5626.

Birkoff G., Goldstine H., and Zarantonello E. (1953-54). Calculation of plane cavity flows past curved obstacles. *Rendiconti del Seminario Matematico*, **13**:205–224. Università e Politecnico di Torino.

Birkoff G., Young G., and Zarantonello E. (1951). Numerical methods in conformal mapping. *Symposia in Applied Mathematics*, **4**:117–140. Proceedings, American Mathematical Society Symposium on Fluid Dynamics at University of Maryland. Edited by Martin, H.H., published by McGraw-Hill 1953.

Birkoff G. and Zarantonello E. (1957). Jets, Wakes and Cavities. Academic Press.

Blasius v.H. (1908). Grenzschichten in Flüssigkeiten mit kleiner Reibung. *Zeitschrift für angewandte Mathematik und Physik*, **56**:1–37.

Bogolepov V. (1985). Flow structure near the trailing edge of a plate. *Journal of Applied Mechanics and Technical Physics*, **26**:386–390. English translation from *Zhurnal Prikladnoi Mekhaniki i Teckhnicheskoi Fiziki*, No 3, pp. 95-99, 1985.

Borgas M. and Pedley T. (1990). Non-uniqueness and bifurcation in annular and planar channel flows. *Journal of Fluid Mechanics*, **214**:229–250.

Brillouin M. (1911). Les surfaces de glissement D'Helmholtz et la resistance des fluids. *Annales de Chimie et de Physique*, **Series 8, XXIII**:145–230.

Brodetsky S. (1923). Discontinuous fluid motion past circular and elliptic cylinders. *Proceedings of the Royal Society*, **A 102**:542–553.

Brown S. and Stewartson K. (1969). Laminar separation. *Annual Review of Fluid Mechanics*, **1**:45–72.

Brown S. and Stewartson K. (1970). Trailing edge stall. *Journal of Fluid Mechanics*, **42**:561.

Bull T. (1956). The tensile strength of liquids under dynamic loading. *Philosophical Magazine and Journal of Science*, **8**:153–165.

Bunyakin A., Chernyshenko S., and Stepanov G. (1996). Inviscid Batchelor-model flow past an airfoil with a vortex trapped in a cavity. *Journal of Fluid Mechanics*, **323**:367–376.

Bunyakin A., Chernyshenko S., and Stepanov G. (1998). High Reynolds number Batchelor-model asymptotics of a flow past an airfoil with a vortex trapped in a cavity. *Journal of Fluid Mechanics*, **358**:283–297.

Burgers J. (1922). Stationary streaming caused by a body in a fluid with friction. *Koninklijke Akademie van Wetenschappen te Amsterdam*, **XXIII Part 2**:1082–1107.

Burgers J. (1930). On the application of Oseen's theory to the determination of the friction experienced by an infinitely thin flat plate. *Koninklijke Akademie van Wetenschappen te Amsterdam*, **XXXIII**:605–613.

Burggraf O. and Duck P. (1982). Spectral computation of triple-deck flows. In Cebeci T., editor, *Numerical and Physical Aspects of Aerodynamic Flows*, pages 145–158. Springer-Verlag.

Carrier G. and Lin C. (1948). On the nature of the boundary layer near the leading edge of a flat plate. *Quarterly Journal of Applied Mathematics*, **6**.

Catherall D. and Mangler K. (1966). The integration of the two-dimensional laminar boundary layer equations past the point of vanishing skin friction. *Journal of Fluid Mechanics*, **26**:163–182.

Catherall D., Stewartson K., and Williams P. (1965). Viscous flow past a flat plate with uniform injection. *Proceedings of the Royal Society*, **A 284**:370–396.

Chen H. and Patel V. (1987). Laminar flow at the trailing edge of a flat plate. *Journal of the American Institute for Aeronautics and Astronautics*, **25**:920–928.

Cheng S.I. (1969). Accuracy of difference formulation of Navier–Stokes equations. *Physics of Fluids*, **12**:34–41. Supplement II.

Cherdron W., Durst F., and Whitelaw J. (1978). Asymmetric flows and instabilities in symmetric ducts with sudden expansions. *Journal of Fluid Mechanics*, **84**:13–31.

Chernyshenko S. (1988). The asymptotic form of the stationary separated circumfluence of a body at high Reynolds number. *Journal of Applied Mathematics and Mechanics*, **52**:746–753.

Chernyshenko S. (1993). Stratified Sadoviskii flow in a channel. *Journal of Fluid Mechanics*, **250**:423–431.

Chernyshenko S. and Castro I. (1993). High Reynolds number asymptotics of the steady flow through a row of bluff bodies. *Journal of Fluid Mechanics*, **257**:421–449.

Childress S. (1966). Solutions of Euler's equations containing finite eddies. *Physics of Fluids*, **9**:860–872.

Cliffe K. and Greenfield A. (1982). Some comments on laminar flow in symmetric two-dimensional channels. Technical Report TP 939, AERE, Harwell.

Cliffe K., Jackson C., and Greenfield A. (1982). Finite element solutions for flow in a symmetric channel with a smooth expansion. Technical Report E 10608, AERE, Harwell.

Coppel W. (1960). On a differential equation of boundary-layer theory. *Philosophical Transactions of the Royal Society*, **A 253**:101–136.

Copson E. (1965). *Asymptotic Expansions*. Cambridge University Press.

Coutanceau M. and Bouard R. (1977). Experimental determination of the main features of the viscous flow in the wake of a circular cylinder in uniform translation. Part I Steady flow. *Journal of Fluid Mechanics*, **79**:231–262.

Curle N. (1962). *The Laminar Boundary Layer Equations*. Oxford University Press.

Curle N. and Skan S. (1957). Approximate methods for predicting separation properties of laminar boundary layers. *The Aeronautical Quarterly*, **9**:257–268.

Daniels P. (1974). Numerical and asymptotic solutions for the supersonic flow near the trailing edge of a flat plate. *Quarterly Journal of Mechanics and Applied Mathematics*, **27**:175–191.

Daniels P. (1976). A numerical and asymptotic investigation of boundary-layer wake evolution. *Journal of the Institute of hematics and its Applications*, **17**:367–386.

Davis R. (1967). Laminar incompressible flow past a semi-infinite flat plate. *Journal of Fluid Mechanics*, **27**:691–704.

Davis R. and Werle M. (1982). Progress on interacting boundary-layer computations at high Reynolds number. In Cebeci T., editor, *Numerical and Physical Aspects of Aerodynamic Flows*, pages 187–210. Springer'–Verlag.

Dean W. (1954). On the steady motion of a viscous liquid past a flat plate. *Mathematika*, **1**:143–156.

Dean W. and Montagnon P. (1949). On the steady motions of a viscous liquid in a corner. *Proceedings of the Royal Society*, **45**:389–394.

Dennis S. and Chang G.Z. (1969). Numerical integration of the Navier–Stokes equations for steady two-dimensional flow. *Physics of Fluids*, **12**:88–93. Supplement II.

Dennis S. and Chang G.Z. (1970). Numerical solutions for steady flow past a circular cylinder at Reynolds numbers up to 100. *Journal of Fluid Mechanics*, **42**:471–489.

Dennis S. and Dunwoody J. (1966). The steady flow of a viscous fluid past a flat plate. *Journal of Fluid Mechanics*, **24**:577–595.

Dennis S. and Smith F. (1980). Steady flow through a channel with a symmetrical constriction in the form of a step. *Proceedings of the Royal Society*, **A 372**:393–414.

Dias F., Elcrat A., and Trefethen L. (1987). Ideal jet flow in two dimensions. *Journal of Fluid Mechanics*, **185**:275–288.

Dimpopoulos H. and Hanratty T. (1968). Velocity gradients at the wall for flow around a cylinder for Reynolds numbers between 60 and 360. *Journal of Fluid Mechanics*, **33**:303–319.

Drazin P. (1992). *Nonlinear Systems*. Cambridge University Press.

Eckhaus W. (1973). *Matched asymptotic expansions and singular perturbations*. North-Holland.

Edwards D. (1997). Viscous boundary-layer effects in nearly inviscid cylindrical flows. *Nonlinearity*, **10**:277–290.

Erdelyi A. (1955). *Asymptotic Expansions*. Dover.

Fage A. (1928). The air flow around a circular cylinder in the region where the boundary layer separates from the surface. *Aeronautical Research Council, Reports & Memoranda*, **1179**.

Fage A. (1929). The air-flow around a circular cylinder in the region where the boundary layer separates from the surface. *Philosophical Magazine and Journal of Science*, **Series 7, 7**:253–273.

Fage A. and Falkner V. (1931). Further experiments on the flow around a circular cylinder. *Aeronautical Research Council, Reports & Memoranda*, **1369**.

Falkner V. and Skan S. (1931). Solutions of the boundary-layer equations. *Philosophical Magazine and Journal of Science*, **12**:865–896.

Fearn R., Mullin T., and Cliffe K. (1990). Nonlinear flow phenomena in a symmetric sudden expansion. *Journal of Fluid Mechanics*, **211**:595–608.

Feynman R. and Lagerstrom P. (1956). Remarks on high Reynolds number flow in finite domains. *Proceedings of the 9th International Conference of Applied Mechanics*, **3**:342–343.

Filon L. (1926). The forces on a cylinder in a stream of viscous fluid. *Proceedings of the Royal Society*, **A 113**:7–27.

Floryan J. and Czechowski L. (1995). On the numerical treatment of corner singularity in the vorticity field. *Journal of Computational Physics*, **118**:222–228.

Fornberg B. (1980). A numerical study of steady viscous flow past a circular cylinder. *Journal of Fluid Mechanics*, **98**:819–855.

Fornberg B. (1985). Steady viscous flow past a circular cylinder up to Reynolds number 600. *Journal of Computational Physics*, **61**:297–320.

Fornberg B. (1994). Steady incompressible flow past a row of circular cylinders. *Journal of Fluid Mechanics*, **225**:655–671.

Gadd G. (1957). A theoretical investigation of laminar separation in supersonic flow. *Journal of the Aeronautical Society*, **24**:759–771.

Georgescu A. (1995). Asymptotic Treatment of Differential Equations. Chapman Hall.

Gilbarg D. (1960). Jets and cavities. *Handbuch der Physik*, **9**:311–445. Edited by S. Flügge.

Glendinning P. (1995). Stability, instability and chaos. Cambridge University Press.

Goldberg A. and Cheng S.I. (1961). An anomaly in the application of Poincaré–Lighthill–Kuo and parabolic coordinates to the trailing edge problem. *Journal of Mathematics and Mechanics*, **10**:529–535.

Goldstein S. (1930). Concerning some solutions of the boundary layer equations in hydrodynamics. *Proceedings of the Cambridge Philosophical Society*, **26**:1–30.

Goldstein S. (1933a). On the two-dimensional steady flow of a viscous fluid behind a solid body - I. *Proceedings of the Royal Society*, **A 142**:545–562.

Goldstein S. (1933b). On the two-dimensional steady flow of a viscous fluid behind a solid body - II. *Proceedings of the Royal Society*, **A 142**:563–573.

Goldstein S., editor (1938). Modern Developments in Fluid Dynamics. Clarendon Press, republished by Dover, 1965.

Goldstein S. (1948). On laminar boundary-layer flow near a position of separation. *Quarterly Journal of Mechanics and Applied Mathematics*, **1**:43–69.

Goldstein S. (1955). Some developments of boundary layer theory in hydrodynamics. University of Maryland, Lecture series No. 33.

Goldstein S. (1960). Lectures on Fluid Mechanics. Interscience.

Goldstein S. (1965). On backward boundary layers and flow in converging passages. *Journal of Fluid Mechanics*, **21**:33–45.

Grove A., Petersen E., and Acrivos A. (1964a). Velocity distribution in the laminar wake of a parallel flat plate. *Physics of Fluids*, **7**:1071. Supplement II.

Grove A., Shair F., Petersen E., and Acrivos A. (1964b). An experimental investigation of the steady separated flow past a circular cylinder. *Journal of Fluid Mechanics*, **19**:60–85.

Gurevich M. (1965). Theory of Jets in Ideal Fluids. Academic Press.

Hakkinen R. and Rott N. (1965). Similar solutions for merging shear flows II. *Journal of the American Institute for Aeronautics and Astronautics*, pages 1553–1554.

Hartree D. (1937). On an equation occurring in Falkner and Skan's approximate treatment of the equations of the boundary layer. *Proceedings of the Cambridge Philosophical Society*, **33**:223–239.

Hartree D. and Womersley J. (1937). A method for numerical or mechanical solution of certain types of partial differential equations. *Proceedings of the Royal Society*, **A 161**:353–366.

Heimenz K. (1911). Die Grenzschict an einem in den gleichförmigen Flüssigkeitsstrom eingetauchten geraden Kreiszylinder. *Dinglers Polytechnische Journal*, **326**:321–324,344–348,357–362,372–376,391–393,407–410.

Helmholtz H. (1868). On discontinuous motion of fluids. *Philosophical Magazine and Journal of Science*, **36**:337–346. Translated from *Monatsbericht der koniglich preussischen Akademie der Wissenschaften zu Berlin*, 1868, p215.

Hinch E. (1991). Perturbation Methods. Cambridge University Press.

Homann v.F. (1936a). Der Einfluß großer Zähigkeit bei der Strömung un den Zylinder und um die Kugel. *Zeitschrift für angewandte Mathematik und Mechanik*, **16**:153–164.

Homann v.F. (1936b). Einfluß großer Zähigkeit bei Strömung um Zylinder. *Forschung auf dem Gebiete des Ingenieurwesens*, **7**:1–10.

Howarth L. (1938). On the solution of the laminar boundary layer equations. *Proceedings of the Royal Society*, **A 164**:547–579.

Il'ichev K. and Postolovskii S. (1972). Calculation of the breakaway flow of a flat stream of a nonviscous liquid around bodies. *Izvestiia Akademii Nauk SSSR*

Mekhanika Zhidkosti i Gaza, **7**:72–82. English translation, *Fluid Dynamics* 1974 part 2, pp252-260.

Illingworth C. (1949). Steady flow in the laminar boundary layer of a gas. *Proceedings of the Royal Society*, **A 199**:533–558.

Imai I. (1951). On the asymptotic behaviour of viscous fluid flow at great distance from a cylindrical body, with special reference to Filon's paradox. *Proceedings of the Royal Society*, **A 208**:487–516.

Imai I. (1953). Discontinuous potential flow as the limiting form of the viscous flow for vanishing viscosity. *Journal of the Physical Society of Japan*, **8**:399–402.

Imai I. (1957). Second approximation to the laminar boundary-layer over a flat plate. *Journal of the Aeronautical Society*, **24**:155–156.

Jain P. and Sankara Rao K. (1969). Numerical solution of unsteady viscous incompressible fluid flow past a circular cylinder. *Physics of Fluids*, **12**:57–64. Supplement II.

Jobe C. and Burggraf O. (1974). The numerical solution of the asymptotic equations of trailing edge flow. *Proceedings of the Royal Society*, **A 340**:91–111.

Jones C. (1948). On a solution of the laminar boundary-layer equation near a position of separation. *Quarterly Journal of Mechanics and Applied Mathematics*, **1**:385–407.

Kaplun S. (1956). The role of coordinate systems in boundary-layer theory. *Zeitschrift für angewandte Mathematik und Physik*, **V**:111–135.

Kaplun S. (1967). Fluid Mechanics and Singular Perturbations. Academic Press. Edited by Lagerstrom, P.A., Howard, L.N. and Liu, C-S.

Karman T.v. (1921). Über Laminare und Turbulente Reibung. *Zeitschrift für angewandte Mathematik und Mechanik*, **1**:233–252.

Karman T.v. and Millikan C. (1934). On the theory of laminar boundary layers involving separation. *National Advisory Committee on Aeronautics, Washington, Report*, **504**.

Kawaguti M. (1953a). Discontinuous flow past a circular cylinder. *Journal of the Physical Society of Japan*, **8**:403–399.

Kawaguti M. (1953b). Numerical solution of the Navier–Stokes equations for the flow around a circular cylinder at Reynolds number 40. *Journal of the Physical Society of Japan*, **8**:747–757.

Kawaguti M. (1959). Note on Allen and Southwall's paper 'Relaxation methods applied to determine the motion, in two dimensions, of a viscous fluid past a fixed cylinder'. *Quarterly Journal of Mechanics and Applied Mathematics*, **12**:261–263.

Kawaguti M. and Jain P. (1966). Numerical study of viscous flow past a circular cylinder. *Journal of the Physical Society of Japan*, **21**:2055–2062.

Kim S.C. (1998). On Prandtl–Batchelor theory of a cylindrical eddy: Asymptotic study. *SIAM Journal of Applied Mathematics*, **58**:1394–1413.

Kirchoff G. (1869). Zur Theorie freier Flüssigkeitsstrahlen. *Journal fur Mathematik*, **LXX**:289–298.

Kiya M. and Arie M. (1977). An inviscid bluff-body wake model which includes the far-wake displacement effect. *Journal of Fluid Mechanics*, **81**:593–607.

Kovásznay L. (1949). Hot-wire investigation of the wake behind cylinders at low Reynolds numbers. *Proceedings of the Royal Society*, **A 198**:174–190.

Kuo Y. (1953). On the flow of an incompressible viscous fluid past a flat plate at moderate Reynolds numbers. *J. Math. Phys.*, **32**:83–101.

Kythe P. (1998). Computational Conformal Mapping. Birkhäuser, Boston.

Lagerstrom P. (1975). Solutions of the Navier–Stokes equation at large Reynolds number. *SIAM Journal of Applied Mathematics*, **28**:202–214.

Lagerstrom P. (1977). Solutions of the Navier-Stokes equations at large Reynolds number. In Rom J., editor, *International Symposium on Modern Developments in Fluid Dynamics*, pages 364–376. SIAM.

Lamb H. (1911). On the uniform motion of a sphere through a viscous fluid. *Philosophical Magazine and Journal of Science*, **Series 6, XXI**:112–121.

Lamb H. (1932). Hydrodynamics. Cambridge University Press, republished by Dover.

Leal L. (1973). steady separated flow in a linearly decelerated free stream. *Journal of Fluid Mechanics*, **59**:513–535.

Leigh D. (1955). The laminar boundary-layer equation: A method of solution by means of an automatic computer. *Proceedings of the Cambridge Philosophical Society*, **51**:320–332.

Levi-Civita T. (1907). Scie e Leggi di resistenza. *Rendconte de Circolo Matematico di Palermo*, **XXIII**:1–37.

Libby P. (1965). Eigenvalues and norms arising in perturbations about the Blasius solution. *Journal of the American Institute for Aeronautics and Astronautics*, **3**:2164–2165.

Libby P. and Fox H. (1963). Some perturbation solutions in laminar boundary layer theory, Part I. *Journal of Fluid Mechanics*, **17**:433–449.

Libby P. and Liu T. (1967). Further solutions of the Falkner-Skan equation. *Journal of the American Institute for Aeronautics and Astronautics*, **5**:1040–1042.

Lighthill M. (1949). A note on cusped cavities. *Aeronautical Research Council, Reports & Memoranda*, **2328**.

Lighthill M. (1953a). On boundary layers and upstream influence. I. A comparison between subsonic and supersonic flows. *Proceedings of the Royal Society*, **A 217**:344–356.

Lighthill M. (1953b). On boundary layers and upstream influence. II. Supersonic flows without separation. *Proceedings of the Royal Society*, **A 217**:478–507.

Lighthill M. (1958). An Introduction to Fourier Transforms and Generalised Functions. Cambridge University Press.

Linke v.W. (1931). Neue Messungen zur Aerodynamik des Zylinders, insbesondere seines reinen Reibungswiderstandes. *Physikalische Zeitschrift*, **XXXII**:900–914.
Lugt H. and Schwiderski E. (1965a). Flows around dihedral angles I. Eigenmotion analysis. *Proceedings of the Royal Society*, **A 285**:382–399.
Lugt H. and Schwiderski E. (1965b). Flows around dihedral angles II. Analysis of regular and singular motions. *Proceedings of the Royal Society*, **A 285**:400–412.
Lugt H. and Schwiderski E. (1965c). Flows around dihedral angles III. On symmetric compressible flows around a flat plate. *Proceedings of the Royal Society*, **A 285**:413–422.
Maccoll J. (1930). Modern aerodynamical research in Germany. *Journal of the Royal Aeronautical Society*, **34**:649–679.
McLachlan R. (1990). Separated viscous flows via multigrid. Ph.D. thesis, California Institute of Technology, Pasadena.
McLachlan R. (1991a). The boundary layer on a flat plate. *Physics of Fluids A*, **3**:341–348.
McLachlan R. (1991b). A steady separated viscous corner flow. *Journal of Fluid Mechanics*, **231**:1–34.
Meksyn D. (1951). Motion in the wake of a thin plate at zero incidence. *Proceedings of the Royal Society*, **A 207**:370–380.
Meksyn D. (1961). New Methods in Laminar Boundary Layer Theory. Pergamon.
Messiter A. (1970). Boundary-layer flow near the trailing edge of a flat plate. *SIAM Journal of Applied Mathematics*, **18**:241–257.
Messiter A. (1983). Boundary-Layer interaction theory. *Transactions of the American Society of Mechanical Engineers*, **50**:1104–1113.
Messiter A. and Enlow R. (1973). A model for laminar boundary-layer flow near a separation point. *SIAM Journal of Applied Mathematics*, **25**:655–670.
Messiter A., Hough G., and Feo A. (1973). Base pressure in laminar supersonic flow. *Journal of Fluid Mechanics*, **60**:605–624.
Meyer R. (1983). A view of the triple deck. *SIAM Journal of Applied Mathematics*, **43**:639–663.
Millikan C. (1936). A theoretical calculation of the laminar boundary layer around an elliptic cylinder, and its comparison with experiment. *Journal of the Aeronautical Society*, **3**:91–94.
Milne-Thompson L. (1938). Theoretical Hydrodynamics. Macmillan. Fifth edition, 1974, republished by Dover, 1996.
Moffatt H. (1964). Viscous and resistive eddies near a sharp corner. *Journal of Fluid Mechanics*, **18**:1–18.
Moore D., Saffman P., and Tanveer S. (1988). The calculation of some Batchelor flows: The Sadoviskii vortex and rotational corner flow. *Physics of Fluids*, **31**:978–990.

Morton K. (1996). Numerical Solution of Convection–Diffusion Problems. Chapman Hall.

Motz H. (1946). The treatment of singularities of partial differential equations by relaxation methods. *Quarterly Journal of Applied Mathematics*, **4**:371–377.

Murray J. (1964). Incompressible viscous flow past a semi-infinite flat plate. *Journal of Fluid Mechanics*, **21**:337–344.

Murray J. (1965). Incompressible viscous flow past a semi-infinite flat plate. *Journal of Fluid Mechanics*, **21**:337–344.

Murray J. (1967). A simple method for determining asymptotic forms of Navier-Stokes solutions for a class of large Reynolds number flows. *Journal of Mathematics and Physics*, **46**:1–20.

Neiland V. (1969). Theory of laminar boundary layers separation in supersonic flow. *Izvestiia Akademii Nauk SSSR Mekhanika Zhidkosti i Gaza*, **4**:53–57. English translation, *Fluid Dynamics* 1972 part 4, pp33-35.

Newton I. (1686). Philosophae Naturalis Principia Mathematica. English translation by A. Motte, 1729, republished with commentary by Florian Cajori, University of California Press, 1934.

Oleinik O. and Samokhin V. (1999). Mathematical Models in Boundary Layer Theory. Chapman & Hall/CRC.

O'Malley K., Fitt A., Jones T., Ockendon J., and Wilmott P. (1991). Models for high-Reynolds-number flow down a step. *Journal of Fluid Mechanics*, **222**:139–155.

Ono H. (1975). Algebraic solitary waves in stratified fluids. *Journal of the Physical Society of Japan*, **39**:1082–1091.

Oseen C. (1910). Über die Stokes'sche Formel, und über eine verwandte Aufgabe in der Hydrodynamik. *Arkiv för Matematik, Astronomi och Fysik*, **6**:1–20.

Panton R. (1996). Incompressible Flow. John Wiley & Sons. Second edition.

Parkinson G. and Jandali T. (1970). A wake source model for bluff body potential flow. *Journal of Fluid Mechanics*, **40**:577–594.

Payne R. (1958). Calculation of unsteady viscous flow past a circular cylinder. *Journal of Fluid Mechanics*, **4**:81–86.

Peregrine D. (1985). A note on the steady high-Reynolds-number flow about a circular cylinder. *Journal of Fluid Mechanics*, **157**:493–500.

Piercy N. and Winny H. (1933). The skin friction of flat plates to Oseen's approximation. *Proceedings of the Royal Society*, **140**:543–561.

Pierrehumbert R. (1980). A family of steady translating vortex pairs with distribution of vorticity. *Journal of Fluid Mechanics*, **99**:129–144.

Plotkin A. and Flügge-Lotz I. (1968). A numerical solution for the laminar wake behind a finite flat plate. *Transactions of the American Society of Mechanical Engineers, Journal of Applied Mechanics*, **35**:625–630.

Pohlhausen K. (1921). Zur näherungsweisen Integration der Differentialgleichung der laminaren Grenzschicht. *Zeitschrift für angewandte Mathematik und Mechanik*, 1:252–268.

Prandtl L. (1904). Über Flüssigkeitsbewegung bei sehr kleiner Reibung. In *Proceedings, Third International Congress for Mathematics*, pages 484–491. Also republished in Prandtl, L. *Gesammelte Abhandlungen zur angewandten Mechanik, Hydro und Aeromechanik, volume 2: Grenzschlichten und Widerstand. Turbulenz und Wirbelbildung. Gasdynamik* pages 576-584, Springer-Verlag 1961.

Prandtl L. and Tietjens O. (1934). Applied Hydro and Aeromechanics. Engineering Society Monograph, republished by Dover, 1957.

Press W., Flannery B., Teukolsky S., and Vetterling W. (1986). Numerical Recipies. Cambridge University Press.

Proudman I. and Pearson J. (1957). Expansion at small Reynolds numbers for the flow past a sphere and a circular cylinder. *Journal of Fluid Mechanics*, 2:237–262.

Pullin D. (1981). The nonlinear behaviour of a constant vorticity layer at a wall. *Journal of Fluid Mechanics*, 108:401–421.

Pullin D. (1984). A constant-vorticity Riabouchinsky free-streamline flow. *Quarterly Journal of Mechanics and Applied Mathematics*, 37:619–631.

Ragab S. and Nayfeh A. (1982). A comparison of the second-order triple-deck theory with interacting boundary layers. In Cebeci T., editor, *Numerical and Physical Aspects of Aerodynamic Flows*, pages 237–254. Springer'–Verlag.

Rayleigh L. (1876a). Notes on hydrodynamics. *Philosophical Magazine and Journal of Science*, **Series 5,** 2:441–447.

Rayleigh L. (1876b). On the resistance of fluids. *Philosophical Magazine and Journal of Science*, **Series 5,** 2:430–441.

Rayleigh L. (1911). Hydrodynamic notes. *Philosophical Magazine and Journal of Science*, **XX1**:177–195.

Reyhner T. and Flügge-Lotz I. (1968). The interaction of a shock wave with a laminar boundary layer. *International Journal of Non-linear Mechanics*, 3:173–199.

Riabouchinsky D. (1920). On steady fluid motions with free surfaces. *Proceedings of the London Mathematical Society*, 19:206–215.

Riley N. (1987). Inviscid separated flows of finite extent. *Journal of Engineering Mathematics*, 21:349–361.

Rogers D. (1969). Further similar laminar-flow solutions. *Journal of the American Institute for Aeronautics and Astronautics*, 7:976–978.

Rogers D. (1992). Laminar Flow Analysis. Cambridge University Press.

Rosenhead L., editor (1963). Laminar Boundary Layers. Oxford University Press.

Roshko A. (1955). On the wake and drag of bluff bodies. *Journal of the Aeronautical Society*, **1955**:124–132.

Roshko A. (1961). Experiments on the flow past a circular cylinder at very high Reynolds number. *Journal of Fluid Mechanics*, **10**:345–356.

Rothmayer A. and Smith F. (1998a). Free interactions and breakaway separation. In Johnson R., editor, *The Handbook of Fluid Dynamics*. Springer-Verlag.

Rothmayer A. and Smith F. (1998b). Incompressible triple-deck theory. In Johnson R., editor, *The Handbook of Fluid Dynamics*. Springer-Verlag.

Rothmayer A. and Smith F. (1998c). Numerical solution of two-dimensional steady triple-deck problems. In Johnson R., editor, *The Handbook of Fluid Dynamics*. Springer.

Rubin S. (1982). A Review of marching procedures for parabolized Navier–Stokes equations. In Cebeci T., editor, *Numerical and Physical Aspects of Aerodynamic Flows*, pages 145–158. Springer-Verlag.

Sadoviskii V. (1971). Vortex regions in a potential stream with a jump of Bernouilli's constant at the boundary. *Journal of Applied Mathematics and Mechanics*, **35**:729–735. Translated from *Prikladnaia Matematika i Mekhanika*, volume 35, pages 773-779, 1971.

Saffman P. (1981). Dynamics of vorticity. *Journal of Fluid Mechanics*, **106**:49–58.

Saffman P. (1992). Vortex Dynamics. Cambridge University Press.

Saffman P. and Tanveer S. (1982). The touching pair of equal and opposite uniform vortices. *Physics of Fluids*, **25**:1929–1930.

Saffman P. and Tanveer S. (1984). Prandtl–Batchelor flow past a flat plate with a forward-facing step. *Journal of Fluid Mechanics*, **143**:351–365.

Schlichting H. (1979). Boundary Layer Theory. 7th edition, McGraw-Hill.

Schmieden C. (1929). Die unstetige Strömung um einen Kreiszylinder. *Ingeniur Archiv*, **1**:104–109.

Schmieden C. (1932). Über die Eindeutigkeit der Lösungen in der Theorie der unstetigen Strömungen. *Ingeniur Archiv*, **3**:356–370.

Schmieden C. (1934). Über die Eindeutigkeit der Lösungen in der Theorie der unstetigen Strömungen II. *Ingeniur Archiv*, **5**:373–375.

Schneider L. and Denny V. (1971). Evolution of the laminar wake behind a flat plate and its upstream influence. *Journal of the American Institute for Aeronautics and Astronautics*, **9**:655–660.

Smith A. and Cuttler D. (1963). Solution of the incompressible boundary-layer equations. *Journal of the American Institute for Aeronautics and Astronautics*, **1**:2062–2071.

Smith F. (1972). Theoretical and experimental study of flow past a porous surface with strong blowing and two related problems. Oxford University. D.Phil. Thesis.

Smith F. (1974). Boundary layer Flow near a discontinuity in wall conditions. *Journal of the Institute of hematics and its Applications*, **13**:127–145.

Smith F. (1976a). Flow through constricted or dilated pipes and channels: Part 1. *Quarterly Journal of Mechanics and Applied Mathematics*, **XXIX**:343–364.

Smith F. (1976b). Flow through constricted or dilated pipes and channels: Part 2. *Quarterly Journal of Mechanics and Applied Mathematics*, **XXIX**:365–376.

Smith F. (1977a). The laminar separation of an incompressible fluid streaming past a smooth surface. *Proceedings of the Royal Society*, **A 356**:443–463.

Smith F. (1977b). Upstream interactions in channel flow. *Journal of Fluid Mechanics*, **79**:631–655.

Smith F. (1978). Flow through symmetrically constricted tubes. *Journal of the Institute of hematics and its Applications*, **21**.

Smith F. (1979a). Laminar flow of an incompressible fluid past a bluff body: the separation, reattachment, eddy properties and drag. *Journal of Fluid Mechanics*, **92**:171–205.

Smith F. (1979b). The separating flow through a severely constricted symmetric tube. *Journal of Fluid Mechanics*, **90**:725–754.

Smith F. (1981). Comparisons and comments concerning recent calculations for flow past a circular cylinder. *Journal of Fluid Mechanics*, **113**:407–410.

Smith F. (1982). On the high Reynolds number theory of laminar flows. *Institute of Mathematics and Applications Journal of Applied Mathematics*, **28**:207–281.

Smith F. (1985). A structure for laminar flow past a bluff body at high Reynolds number. *Journal of Fluid Mechanics*, **155**:175–191.

Smith F. (1986a). Concerning inviscid solutions for large scale separated flows. *Journal of Engineering Mathematics*, **20**:271–292.

Smith F. (1986b). Steady and unsteady boundary-layer separation. *Annual Review of Fluid Mechanics*, **18**:197–220.

Smith F., Brighton P., Jackson P., and Hunt J. (1981). On boundary-layer flow past two-dimensional obstacles. *Journal of Fluid Mechanics*, **113**:123–152.

Smith F. and Daniels P. (1981). Removal of Goldstein's singularity at separation, in flow past obstacles in wall layers. *Journal of Fluid Mechanics*, **110**:1–37.

Smith F. and Duck P. (1980). On the severe non-symmetric constriction, curving or cornering of channel flows. *Journal of Fluid Mechanics*, **90**:727–753.

Smith J. (1986c). Vortex flows in aerodynamics. *Annual Review of Fluid Mechanics*, **18**:221–242.

Smith R. (1938). Laminar boundary layer based on a minimum theorem. *Journal of the Aeronautical Society*, **5**:266–272.

Sobey I. (1980). On flow through a furrowed channel Part I. Calculated flow patterns. *Journal of Fluid Mechanics*, **96**:1–26.

Sobey I. (1985). Observation of waves during oscillatory channel flow. *Journal of Fluid Mechanics*, **151**:395–426.

Sobey I. and Drazin P. (1986). Bifurcations of two-dimensional channel flows. *Journal of Fluid Mechanics*, **171**:263–287.

Son J. and Hanratty T. (1969a). Numerical solution for the flow around a cylinder at Reynolds number of 40, 200 and 500. *Journal of Fluid Mechanics*, **35**:369–386.

Son J. and Hanratty T. (1969b). Velocity gradients at the wall for flow around a cylinder at Reynolds numbers from 5×10^3 to 10^5. *Journal of Fluid Mechanics*, **35**:335–368.

Southwell R. and Vaisey G. (1946). Relaxation methods applied to engineering problems. XII Fluid motion characterised by 'free' streamlines. *Philosophical Transactions of the Royal Society*, **A 240**:117–161.

Squire H. (1934). On the laminar flow of a viscous fluid with vanishing viscosity. *Philosophical Magazine and Journal of Science*, Series 7, **17**:1150–1160.

Stewartson K. (1949). Correlated incompressible and compressible boundary layers. *Proceedings of the Royal Society*, **A 200**:84–100.

Stewartson K. (1954). Further solutions of the Falkner-Skan equation. *Proceedings of the Cambridge Philosophical Society*, **50**:454–465.

Stewartson K. (1957). On asymptotic expansions in the theory of boundary layers. *J. Maths Physics*, **36**:173–191.

Stewartson K. (1958). On Goldstein's theory of laminar separation. *Quarterly Journal of Mechanics and Applied Mathematics*, **11**:399–410.

Stewartson K. (1964). The theory of laminar boundary layers in compressible fluids. Oxford University Press.

Stewartson K. (1965). The boundary layer. Lewis. Inaugural Lecture, University College, London.

Stewartson K. (1968). On the flow near the trailing edge of a flat plate. *Proceedings of the Royal Society*, **A 306**:275–290.

Stewartson K. (1969). On the flow near the trailing edge of a flat plate II. *Mathematika*, **16**:106–121.

Stewartson K. (1970a). Is the singularity at separation removable? *Journal of Fluid Mechanics*, **44**:347–364.

Stewartson K. (1970b). On laminar boundary layers near corners. *Quarterly Journal of Mechanics and Applied Mathematics*, **23**:137–152.

Stewartson K. (1974). Multistructured boundary layers on flat plates and related bodies. *Advances in Applied Mechanics*, **14**:145–239.

Stewartson K. (1977). On the asymptotic theory of separated and unseparated fluid motions. In Rom J., editor, *International Symposium on Modern Developments in Fluid Dynamics*, pages 305–322. SIAM.

Stewartson K. (1981). d'Alembert's paradox. *SIAM Review*, **23**:308–342.

Stewartson K. and Williams P. (1969). Self induced separation. *Proceedings of the Royal Society*, **A 312**:181–206.

Stewartson K. and Williams P. (1973). Self induced separation II. *Mathematika*, **20**:98.

Stokes G. (1850). On the effect of internal friction of fluids on the motion of pendulums. *Proceedings of the Cambridge Philosophical Society*, **9**:8. Reprinted in *Mathematical and Physical Papers*, volume 3, Cambridge University Press 1901. See particularly Article I, section IV, pages 55-67.

Sychev V. (1972). Laminar Separation. *Izvestiia Akademii Nauk SSSR Mekhanika Zhidkosti i Gaza*, **7**:47–59. English translation, *Fluid Dynamics* 1974 part 3, pp407-417.

Sychev V., Ruban A., Sychev V., and Korolev G. (1998). Asymptotic theory of separated flows. Cambridge University Press. Translation by Maroko, E.V., English translation edited by Messiter, A.F. and Van Dyke, M., of 'Asimptoticheskaia teoriia otrynvnykh techenii' 1987.

Takami H. and Keller H. (1969). Steady two-dimensional viscous flow of an incompressible fluid past a circular cylinder. *Physics of Fluids*, **12**:51–56. Supplement II.

Talke F. and Berger S. (1970). The flat plate trailing edge problem. *Journal of Fluid Mechanics*, pages 161–189.

Taneda S. (1956). Experimental Investigation of the wakes behind cylinders and plates at low Reynolds numbers. *Journal of the Physical Society of Japan*, **11**:302–307.

Tani I. (1977). History of boundary-layer theory. *Annual Review of Fluid Mechanics*, **9**:87–111.

Terrill R. (1960). Laminar boundary-layer flow near separation with and without suction. *Philosophical Transactions of the Royal Society*, **A 253**:55–100.

Thom A. (1928a). The boundary layer of the front portion of a cylinder. *Aeronautical Research Council, Reports & Memoranda*, **1176**.

Thom A. (1928b). An investigation of fluid flow in two dimensions. *Aeronautical Research Council, Reports & Memoranda*, **1194**.

Thom A. (1930). The pressure on the front generator of a cylinder. *Aeronautical Research Council, Reports & Memoranda*, **1389**.

Thom A. (1933). The flow past circular cylinders at low speeds. *Proceedings of the Royal Society*, **A 141**:651–669.

Thom A. and Orr A. (1931). The solution of a torsion problem for circular shafts of varying radius. *Proceedings of the Royal Society*, **A 131**:30–37.

Thoman D. and Szewczyk A. (1969). Time dependent viscous flow over a circular cylinder. *Physics of Fluids*, **12**:76–86. Supplement II.

Thomee V. and Vasudeva-Murthy A. (1998). A numerical method for the Benjamin-Ono equation. *BIT Numerical Mathematics*, **38**:597–611.

Thwaites B. (1949). Approximate calculation of the laminar boundary layer. *The Aeronautical Quarterly*, **1**:245–280.

Thwaites B. (1960). Incompressible Aerodynamics. Oxford University Press, republished by Dover(1987).

Tillett J. (1968). On the laminar flow in a free jet of liquid at high Reynolds numbers. *Journal of Fluid Mechanics*, **32**:273–292.

Tollmien W. (1931). Grenzlichttheorie. *Handbuch der Experimental Physik*, **4**:269.

Tuck E. (1971). Transmission of water waves through small apertures. *Journal of Fluid Mechanics*, **49**:65–74.

Turfus C. (1993). Prandtl–Batchelor flow past a plate at normal incidence in a channel - inviscid analysis. *Journal of Fluid Mechanics*, **249**:59–72.

Uchibori Y. and Sobey I. (1992). Dependence on numerical algorithm of bifurcation structure for flow through a symmetric expansion. Technical Report NA-92-14, Oxford University Computing Laboratory.

Underwood R. (1969). Calculation of incompressible flow past a circular cylinder at moderate Reynolds numbers. *Journal of Fluid Mechanics*, **37**:95–114.

Van de Vooren A. and Dijkstra D. (1970). The Navier–Stokes solution for laminar flow past a semi-infinite flat plate. *Journal of Engineering Mathematics*, **4**:9–27.

Van Dyke M. (1962a). Higher approximations in boundary layer theory. Part I General results. *Journal of Fluid Mechanics*, **14**:161–177.

Van Dyke M. (1962b). Higher approximations in boundary layer theory. Part II Application to leading edges. *Journal of Fluid Mechanics*, **14**:481–495.

Van Dyke M. (1964). Perturbation Methods in Fluid Mechanics. Parabolic Press.

Van Dyke M. (1994). Nineteenth-century roots of the boundary-layer idea. *SIAM Review*, **36**:415–424.

Veldman A. and van de Vooren A. (1974). Drag of a finite plate. *Proceedings of the 4th International Conference on Numerical Fluid Dynamics*, pages 423–430. Boulder, Colorado. Lecture Notes in Physics, 35, Springer-Verlag.

Villat H. (1914). Sur la validité des solutions de certains problémes d'Hydrodynamique. *J. Maths. Pure Appl.*, **6th Series, 10**:231–290.

Weinbaum S. (1968). On the singular points in the laminar two-dimensional near wake flow field. *Journal of Fluid Mechanics*, **33**:38–63.

Weyl H. (1942). On the differential equations of the simplest boundary-layer problem. *Annals of Mathematics*, **43**:381–407.

Williams J. (1977). Incompressible boundary-layer separation. *Annual Review of Fluid Mechanics*, **9**:113–144.

Wilmott P. and Fitt A. (1992). A composite cavity model for axisymmetric high Reynolds number separates flow. I Modelling and analysis. *Journal of Engineering Mathematics*, **26**:539–555.

Wood W. (1957). Boundary layers whose streamlines are closed. *Journal of Fluid Mechanics*, **2**:77–87.

Woods L. (1954a). Compressible subsonic flow in two-dimensional channels with mixed boundary conditions. *Quarterly Journal of Mechanics and Applied Mathematics*, **VII**:263–282.

Woods L. (1954b). A note on the numerical solution of fourth-order differential equations. *The Aeronautical Quarterly*, **V**:176–184.

Woods L. (1955). Two-dimensional flow of a compressible fluid past given curved obstacles with infinite wakes. *Proceedings of the Royal Society*, **A 227**:367–386.

Woods L. (1956). Generalised airfoil theory. *Proceedings of the Royal Society*, **A 238**:358–388.

Woods L. (1961). The Theory of Subsonic Plane Flow. Cambridge University Press.

Wu T.Y.T. (1962a). A wake model for free streamline flow theory. Part I. Full & partially developed wake flows & cavity flows past an oblique flat plate. *Journal of Fluid Mechanics*, **13**:161–181.

Wu T.Y.T. (1962b). A wake model for free streamline flow theory. Part I Fully and partially developed open wake flows and cavity flows past an oblique flat plate. *Journal of Fluid Mechanics*, **13**:161–181.

Wu T.Y.T. (1972). Cavity and wake flows. *Annual Review of Fluid Mechanics*.

Young A. (1989). Boundary Layers. BSP.

Zarantonello E. (1952). A constructive theory for the equations of flows with free boundaries. *Collectanea Mathematica*, 5:175–225. Consejo Superior de Investigaciones Cientificas, Universidad de Barcelona.

Zdravkovich M. (1997). Flow Around Circular Cylinders. Oxford University Press.

AUTHOR INDEX

Acker, A., 212
Acrivos, A., 120, 219
　see Grove et al.(1964), 59, 120
Alden, H.L., 42
Allen, D.N. de G., 191
Apelt, C.J., 192
Arie, M., 123

Bairstow, L., 13, 69, 154
Batchelor, G.K., 3, 113, 119, 168, 212, 219, 301
Bender, C.M., 6
Benjamin, T.B., 215
Berger, S.A., 96, 301
Birkoff, G., 122, 133
Blasius, H., 25, 30
Bogolepov, V.V., 257
Borgas, M.S., 290, 292
Bouard, R., 116, 121, 189, 190
Brighton, P.W.M.
　see Smith et al.(1981), 96
Brillouin, M., 130
Brodetsky, S., 130, 166, 168
Brown, S.N., 176, 207, 302
Bull, T.H., 123
Bunyakin, A.V., 214
Burgers, J.M., 164
Burggraf, O.R., 87, 91, 98, 104, 303

Carrier, G.F., 94
Castro, I., 214
Catherall, D., 194
Cave, B.M.
　see Bairstow et al.(1923), 13, 69
Chang, G-Z., 189, 190, 193
Chen, H.C., 94
Cheng, S-I., 194
Cherdron, W., 288, 292
Chernyshenko, S.I., 214, 219
　see Bunyakin et al.(1996, 1998), 214
Childress, S., 212
Cliffe, K.A., 289, 291, 292
Coanda, H., 284
Coppel, W.A., 159
Copson, E., 6
Coutanceau, M., 116, 121, 189, 190
Curle, N., 13
Czechowski, L., 261

Daniels, P.G., 87, 103, 204, 276

Davis, R.T., 49, 98
Dean, W.R., 47, 253, 300
Dennis, S.C.R., 92, 189, 190, 193, 281, 300
Denny, V.E., 93
Dias, F., 123
Dimopoulos, H.G., 121
Drazin, P.G., 285, 288, 289, 292
Duck, P.W., 98, 104, 275, 276, 282, 303
Dunwoody, J., 92, 300
Durst, F., 288
　see Cherdron et al.(1978), 292

Eckhaus, W., 6
Edwards, D., 212
Elcrat, A.R.
　see Dias et al.(1987), 123
Enlow, R.L., 77, 205
Erdelyi, A., 6

Fage, A., 118
Falkner, V.M., 118, 157, 293
Fearn, R.M., 289
Feynman, R., 212
Filon, L., 46, 49
Fitt, A.D., 214
　see O'Malley et al.(1991), 214
Flügge-Lotz, I., 93, 199
Flannery, B.P.
　see Press et al.(1986), 301
Floryan, J.M., 261
Fornberg, B., 95, 169, 210, 211, 218
Fox, H., 35

Gadd, G.E., 199
Georgescu, A., 6
Gilbarg, D., 122
Goldstein, S., 13, 25, 42, 49, 56, 59, 62, 155, 157, 161, 169, 170, 194
Goldstine, H.H., 133
Greenfield, A.C., 291, 292
　see Cliffe et al.(1982), 291
Grove, A.S., 59, 120
　see Acrivos et al.(1965), 120
Gurevich, M.I., 122, 129

Hakkinen, R.J., 51
Hanratty, T.J., 121
Hartree, D.R., 159, 169, 177
Heimenz, K., 117, 151

Helmholtz, H., 123, 168
Hinch, J., 6
Homann, von F., 119
Howarth, L., 162, 169, 177
Hunt, J.C.R.
 see Smith et al.(1981), 96

Il'ichev, K.P., 123
Imai, I., 25, 45, 49, 59, 93, 167, 300

Jackson, C.P.
 see Cliffe et al.(1982), 291
Jackson, P.S.
 see Smith et al.(1981), 96
Jain, P., 192, 194
Jobe, C.E., 87, 91, 98
Jones, C.W., 175
Jones, T.V.
 see O'Malley et al.(1991), 214

Kaplun, S., 6, 31, 176
Karman, Th. von, 16, 61, 151, 163
Kawaguti, M., 167, 190, 192
Keller, H.B., 194
Kim, S.C., 212
Kirchoff, G., 125
Kiya, M., 123
Korolev, G.L.
 see Sychev et al.(1998), 98, 205
Kovásznay, L.S.G., 120
Kuo, Y.H., 70
Kythe, P.K., 123

Lagerstrom, P.A., 10, 212, 219
Lamb, H., 11, 13, 69
Lang, E.D.
 see Bairstow et al.(1923), 13, 69
Leal, L.G., 217
 see Acrivos et al.(1968), 120
Leigh, D.C.F., 177
Levi-Civita, T., 129
Libby, P.A., 35, 160
Lighthill, M.J., 76, 169, 199, 232, 250, 291
Lin, C.C., 94
Linke, von W., 119
Liu, T.M., 160
Lugt, H.J., 253

Maccol, J.W., 154
Mangler, K.W., 194
McLachlan, R.I., 94, 169, 217, 300
Meksyn, D., 13, 59, 62, 300
Melnik, R.E., 91
Messiter, A.F., 26, 76, 77, 97, 205
Meyer, R.E., 96

Milne-Thompson, L.M., 122
Moffatt, H.K., 253
Montagnon, P.E., 253
Moore, D.W., 212, 214
Morton, K.W., 192, 210
Motz, H., 259
Mullin, T., 289
Murray, J., 43, 44

Nayfeh, A.H., 103
Neiland, V.Ya., 77, 199
Newton, I., 17

O'Malley, K., 214
Ockendon, J.R.
 see O'Malley et al.(1991), 214
Oleinik, O.A., 13
Ono, H., 215
Orr, A., 187
Orszag, S.A., 6
Oseen, C.W., 11

Pan, F.
 see Acrivos et al.(1968), 120
Panton, R.L., 3
Patel, V.C., 94
Payne, R.B., 192
Pearson, J.R.A., 12
Pedley, T.J., 290, 292
Peregrine, D.H., 218, 219, 303
Petersen, E.E.
 see Acrivos et al.(1965), 120
 see Grove et al.(1964), 59, 120
Piercy, N.A.V., 13, 70
Pierrehumbert, R.T., 215
Plotkin, A., 93
Pohlhausen, K., 16, 151
Postolovskii, S.N., 123
Prandtl, L., 13, 25, 168
Press, W.H., 301
Proudman, I., 12
Pullin, D.I., 214

Ragab, S.A., 103
Rayleigh, Lord, 126, 253
Reyhner, T.A., 199
Riabouchinsky, D.P., 122, 168
Riley, N., 214, 216, 303
Rogers, D.F., 13, 158
Rosenhead, H., 13
Roshko, A., 120
Rothmayer, A.P., 96–98
Rott, N., 51
Ruban, A.I.
 see Sychev et al.(1998), 98, 205

Rubin, S.G., 98

Sadoviskii, V.S., 214
Saffman, P.G., 212, 214, 215
  see Moore et al.(1988), 214
Samokhin, V.N., 13
Sankara Rao, K., 194
Scalise, D.T., 96
Schlichting, H., 13
Schmieden, C., 133, 166
Schneider, L.I., 93
Schwiderski, E.W., 253
Shair, F.H.
  see Grove et al.(1964), 120
Skan, S.W., 157, 293
Smith, F.T., vii, 77, 87, 96–98, 107, 121,
    204, 206, 208, 211, 215, 218, 219,
    224, 227–230, 257, 263, 266, 273,
    275–277, 280–282
Smith, J.H.B, 212
Smith, R.H., 15
Snowden, D.D.
  see Acrivos et al.(1965), 120
  see Acrivos et al.(1968), 120
Sobey, I.J., 247, 288, 289, 291, 292, 304
Son, J.S., 121
Southwell, R.V., 169, 191
Squire, H.B., 166, 167
Stepanov, G.Yu.
  see Bunyakin et al.(1996, 1998), 214
Stewartson, K., 13, 26, 59, 62, 64, 66, 76,
    97, 159, 175, 199, 201, 202, 204,
    207, 257, 273, 302
Stokes, G.G., 11
Sychev, V.V., 77, 204
  see Sychev et al.(1998), 98, 205
Sychev, Vl.V., 98, 205
Szewczyk, A.A., 194

Takami, H., 194
Talke, F.E., 301
Tani, I., 13
Tanveer, S., 214, 215
  see Moore et al.(1988), 214
Terrill, R.M., 176, 180
Teulosky, S.A.
  see Press et al.(1986), 301
Thom, A., 119, 177, 185, 187
Thoman, D.C., 194
Thomee, V., 215
Thwaites, B., 122, 163, 168
Tietjens, O.G., 13
Tillett, J.P.K., 234, 280
Tollmien, W., 49, 56, 59
Topakoglu, H.E., 257

Trefethen. L.N.
  see Dias et al.(1987), 123
Tuck, E.O., 7, 8
Turfus, C., 214

Uchibori, Y., 291
Underwood, R.L., 193

Vaisey, G., 169
van de Vooren, A.I., 91
Van Dyke, M., 6, 7, 12, 13, 29
Vasudeva-Murthy, A.S., 215
Veldman, A.E.P., 91
Vetterling, W.T.
  see Press et al.(1986), 301
Villat, H, 130

Weinbaum, S., 253, 255–257
Werle, M.J., 98
Weyl, H., 29
Whitelaw, J.H., 288
  see Cherdron et al.(1978), 292
Williams, J.C., 114
Williams, P.G., 76, 199, 201, 273
Wilmott, P., 214
Winny, H.F., 13, 70
Womersley, J.R., 177
Wood, W.W., 212
Woods, L.C., 122, 128, 139, 142, 147, 192,
    234
Wu, T. Y-T., 112, 149

Young, A.D., 13
Young, D.M., 133

Zarantonello, E.H., 122, 133
Zdravkovich, M.M., 115
Zhukovskii, N.E., 129
  see Gurevich(1965), 129

SUBJECT INDEX

Airy function, 232, 267
asymmetric channels
   $R^{-1} \ll \epsilon \ll R^{-1/7}$, 228
   $\epsilon \sim R^{-1/7}$, 263
asymptotic methods, 5–9
  basic boundary layer, 13
  expansion, 6
  matched asymptotic analysis, 7
  order $o()$, $O()$, 6

Benjamin–Ono equation, 215
Bernouilli's equation, 4, 19, 78, 112, 117, 128, 232, 297
bifurcation, 284
  Hopf, 288
  pitchfork, 286
  variation with R, 297
biharmonic equation, 11
Blasius boundary layer, 15
  displacement thickness, 31
  equation, 29
  expansion far from plate, 30
  power series solution, 29–30
    radius of convergence, 30
Borda's mouthpiece, 123
boundary conditions, 41
  wall, 5
boundary layer
  x-momentum, 14, 27
  y-momentum, 15, 27
  adverse pressure gradient, 162
  asymptotic expansion, 26
  basic theory, 13–17
  Blasius, 15
  constants
    α_0, 29, 39, 41, 47, 50, 51, 59
    β_0, 31, 32, 35, 38, 39, 43, 44, 46, 59
  cylinder
    Heimenz calculation, 151
    Pohlhausen calculation, 151
    using Heimenz observed u_e, 152
    using potential velocity, 152, 163
  displacement thickness, 16
  Falkner–Skan, 15, 157
  flat plate, 25
  form factor, 17
  free stream velocity, 15
  Goldstein's singularity, 169
  momentum integral, 16–17, 163
  momentum thickness, 16
  numerical solution, 176–184
    Heimenz outer flow, 183
    Leigh, 177
    potential outer flow, 182
    Terrill, 180
    Thom, 177
  thickness, 27
  Thwaites' method, 163
  transverse scaling, 27
  variable pressure gradient, 149
Brodetsky
  cylinder radius, 133, 136
  drag prediction, 138
  mapping, 130
  numerical solution, 133
  singularity at separation, 136
  Smith's calculation of separation, 209
  Squire's prediction of separation, 166
Brodetsky coefficients, computed values, 134
Burgers' prediction of separation, 164

Cauchy principal value, 86, 99, 204, 207, 214, 215
cavitation bubble, 112
channel flow
  asymmetric channel, 228, 263
    $R^{-1} \ll \epsilon \ll R^{-1/7}$, 228
    computed solution, 276
  asymmetric channels
    $\epsilon \sim R^{-1/7}$, 263
  bifurcation, 288
    observation, 289
  composite solution, 247
  conformal transformation, 250
  core region, 228, 264
  core velocity scaling, 226
  corner, 252
  criterion for separation, 247
  displacement function, 225, 230
  expansion in core, 225
  interactive solution
    computed examples, 246
  linearised theory, 231
  lower wall layer, 228
  numerical solution
    triple deck, 106
  Prandtl transformation, 229, 231, 265
  Prandtl–Batchelor flow, 282

pressure, 223
Reynolds number, 224
separation
    critical Reynolds number, 252
    free interaction, 271
singularity
    free interaction, 271
smooth expansion, 291
sudden expansion, 290
symmetric channel, 233, 277
    $R^{-1} \ll \epsilon \ll 1$, 233
    $\epsilon = O(1), \sigma = R^{-1/3}$, 277
    fine indentation, 278
    moderate indentation, 280
    severe indentation, 281
symmetric vs asymmetric flow, 226
symmetry breaking, 288
triple deck
    computational technique, 106
upper wall layer, 229
upstream influence
    asymmetric channel, 263
    displacement function, 269
    Fourier transform solution, 269
    symmetric channel, 277
wall layer, 226, 234
wall slope, 228
wall vorticity, 249
Coanda effect, 284
combined boundary layer – free streamline
        models, 164
complex potential
    definition, 10
    outer plate solution, 32
composite expansion, 8
composite solution
    channel flow, 247
    near wake of plate, 56
computed Navier–Stokes solutions
    channel, 250
        bifurcation, 290
        comparison with interactive solution,
            251
        conformal transformation, 250
    cylinder, 184, 210
        Allen & Southwell, 191
        Apelt, 192
        Dennis & Chang, 193
        Fornberg, 210
        Kawaguti, 190
        Payne, 192
        Thom, 185
    flat plate, 91
constants
    numerical values, 59

continuity equation, 4
convolution integral, 269
coordinates
    Reynolds number stretched, 35
corner flow, 252
    discretisation near, 259
    finite volume approximation, 259
    interactive approximation, 257
    potential solution, 253
    Stokes singularity, 256
    Stokes solution, 254
    vorticity estimate, 261
cylinder
    boundary layer
        Heimenz calculation, 151
        Pohlhausen calculation, 151
    Burgers' prediction of separation, 164
    computed Navier–Stokes solutions, 184
        Allen & Southwell, 191
        Apelt, 192
        coordinate transformation, 186, 190
        Dennis & Chang, 193
        Kawaguti, 190
        Payne, 192
        Thom, 185
    flow visualisation, 121
    free streamline flow, 130
    Kuwaguti's prediction of separation, 167
    laminar flow zone, 116
    pressure measurement, 117
    Reynolds number, 115
    separation
        Smith's theory, 208
    separation angle, 206, 208
    separation length, 115, 120, 122
    smooth separation, 132
    Squire's prediction of separation, 166
    vortex shedding, 120
    wake size, 218

d'Alembert's paradox, 20, 123, 139
Dean's approximation for plate drag, 47
displacement function, 84, 200, 207, 267,
        294
    asymptotic properties, 87
    channel flow, 225, 230
displacement thickness, 16, 194
    Blasius boundary layer, 31
doublet, 13
drag, 17–20
    Brodetsky's model, 138
    d'Alembert's paradox, 20
    finite plate, 69
    Imai's formula, 46
    semi-infinite plate, 45

Subject Index

drag coefficient
  Dean's approximation, 47
  definition, 18
  finite plate, 69
  free streamline flow, 128
  Imai's expansion, 46
  momentum thickness relation, 19
  plate, 62, 73
  section of plate, 47
  triple deck modification, 90

Euler equations, 9
exchange of stability, 285

Falkner–Skan
  similarity variable, 158
Falkner–Skan boundary layer, 15
Falkner–Skan equation, 157–161, 294
  Harteee's solution, 159
  non-uniqueness, 159
far field
  vorticity decay, 41
far wake
  higher order expansion, 62
  logarithms in expansion, 66
  Meksyn's model, 64
  Tollmien's model, 60
  transverse velocity, 63
fine indentation, 278
finite plate
  wake, 49
finite volume approximation, 100, 259
flat plate
  rectangular coordinates, 26
flow visualisation
  Homann, 119
  Thom, 119
  vortex length, 120
  wall stress measurement, 121
form factor, boundary layer, 17
Fourier transform, 105, 231, 267, 280
free interaction
  channel, 270
  expansions near singularity, 272
free streamline
  re-entrant jet, 123
free streamline theory, 122–149
  Ω map, 127
  auxiliary plane, 128
  Brillouin, 129
  Brodetsky's method, 130
    numerical solution, 133
  channel flow, 234–245
  drag, 138
  far field, 126

  force on body, 128
  general formulation, 127
  Helmhotz, 123
  Kirchoff, 125
  Levi-Civita, 129
  Rayleigh, 126
  Schmieden, 133
  Schwarz–Christoffel transformation, 128
  smooth separation, 132
  trajectories, 127
  Woods' method, 139
    channel flow, 234
free streamline, at separation
  Brodetsky, 136
  Woods, 144
friction coefficient
  definition, 19

Goldstein
  boundary layer
    variable pressure gradient, 154
  inner wake, 51–54
  near wake, 56
  outer wake, 54–55
  singularity
    attempts to resolve, 194
    boundary layer expansion, 174
    Catherall & Mangler, 194
    form, 179
    Stewartson's attempt to resolve, 202
    Stewartson's modified expansion, 175
    sublayer expansion, 171
  singularity at separation, 169

Heimenz
  boundary layer calculation, 151
  boundary layer using his observations, 183
Hilbert integral, 86, 99, 201, 204, 207, 214, 215
  computation, 102
Hopf bifurcation, 288
Howarth
  adverse pressure gradient, 162

integral form of boundary layer, 163
irrotational, definition, 4

Karman integral method, 117, 163
kinematic viscosity, 3
Kuo's wake model, 70

Levi-Civita
  mapping, 129

matched asymptotic expansion

basic boundary layer, 14
matched asymptotic expansions, 7
　Van Dyke's criterion, 9
Meksyn, see far wake
moderate indentation, 280
momentum flux, 68
momentum flux definition, 18
momentum thickness, 16
　drag coefficient relation, 19

Navier–Stokes equations, 4
　channel flow, 223
　computed solutions, 184
　cylindrical coordinates, 149
　definition, 3
　parabolic coordinates, 37
Newton iteration, 99, 101, 185, 211, 216
numerical solution
　asymmetric channel, 276
　Blasius boundary layer, 31
　boundary layer on cylinder, 176
　boundary layer, adverse pressure gradient
　　Leigh, 177
　　Terrill, 180
　channel
　　Navier–Stokes, 250
　Coada effect, 290
　cylinder
　　Navier–Stokes, 184, 210
　　Thom, 185
　Falkner–Skan equation, 157–161
　free interaction
　　channel, 270
　　plate, 201
　free streamline for cylinder
　　Brodetsky, 133
　　Woods, 142
　Goldstein near wake, 56
　pressure on cylinder, 188
　Sadoviskii vortex, 216
　triple deck, 88, 97–107
　Woods' free streamline method
　　channel flow, 238

oblique plate
　free streamline flow, 126
order $o()$, $O()$, 6
Oseen's approximation, 11–13
　Burgers' prediction of separation, 164
　finite plate, 69

parabolic coordinates
　expansion in rectangular coordinates, 45
　Navier–Stokes equations, 37

plate expansion, 38
　semi-infinite plate, 36
pitchfork bifurcation, 286
plate
　boundary layer equation, 27
　computed solutions, 91
　displacement surface, 32
　drag, 69
　drag coefficient, 62, 73
　drag on section, 45
　far field expansion, 26
　finite length, 49
　Kuo's wake model, 70
　leading order solution, 28
　length, 50
　logarithms in expansion, 42
　near wake, 50
　Oseen flow, 69
　parabolic coordinates, 36
　potential body shape, 78
　power series failure, 39
　second order far field, 31
　second order inner expansion, 34
　semi-infinite, 26
　stream-function expansion, 26
　wake
　　centreline velocity, 53, 61
Pohlhausen
　boundary layer calculation, 151
Pohlhausen's method, 162
Poisseuille flow, 223
　stream-function, 228
potential flow, 9–10
　definition, 4
Prandtl transformation, 229, 231, 265
Prandtl–Batchelor flow, 168, 212, 282
　Childress, 213
　pressure coefficient, 213
　rotational corner vortex, 217
　Sadoviskii vortex, 214
　vortex sheet strength, 216
Prandtl–Glauert equation, 199
pressure coefficient, 113, 117, 188, 209, 213
pressure gradient at separation, 208
pressure gradient, adverse, 162, 177, 295
pressure measurement
　Fage & Falkner, 118
　Heimenz, 117
　Thom, 119
pressure scale
　Navier–Stokes, 3
　Stokes flow, 10
pressure–displacement relation, 86, 99, 201, 204, 207

radius of convergence

Subject Index

Blasius power series, 30
Rayleigh layer, 112
re-entrant jet, 123
rectangular coordinates
    flat plate, 26
Reynolds number
    channel flow, 224
    cylinder, 115
    definition, 3, 289
Riabouchinsky vortex, 122, 214
Richardson extrapolation, 178
rotational corner vortex, 217

Sadoviskii vortex, 214
Schwarz–Christoffel transformation, 123, 128, 235, 243
self induced separation, 199
    numerical solution, 201
semi-infinite plate
    drag on section, 45
separation, 1, 2, 111–114
    angle on cylinder, 206
    Burgers' prediction, 164
    channel
        critical Reynolds number, 252
        criterion for channel flow, 247
    cylinder
        separation angle, 208
    free interaction
        compressible, 199
    free interaction in channel, 271
    free streamline singularity
        Brodetsky, 136
        Woods, 144
    Goldstein singularity
        boundary layer expansion, 174
        Stewartson's modified expansion, 175
        sublayer expansion, 171
    Kawaguti's prediction, 167
    length behind cylinder, 122
    length for cylinder, 115
    observation, 115–122
    self induced, 199
    Smith's theory for cylinder, 208
    Squire's prediction, 166
    Stewartson's attempt to resolve Goldstein's singularity, 202
    Sychev's hypothesis, 204
        Smith's numerical solution, 206
    triple deck around, 207
series truncation method, 49
severe indentation, 281
similarity solution
    boundary layer, 28
    Falkner–Skan, 158

similarity variable, 15, 63, 293
    definition, 28
singularity
    free interaction in channel, 271
    expansions near, 272
    free streamline, at separation, 136, 144, 239
        Brodetsky, 134
        Sychev's hypothesis, 204
    Goldstein's at separation, 169
        attempts to resolve, 194
    separation
        Goldstein boundary layer expansion, 174
        Goldstein sublayer expansion, 171
        Stewartson's modified expansion, 175
    Woods' modified wake, 149
slip velocity, 5
smooth separation, 132, 133, 167, 205
spectral method
    see triple deck, 104
Squire's prediction of separation, 166
Stewartson
    attempt to resolve Goldstein's singularity, 202
Stokes doublet, 13
Stokes flow, 10–11
    pressure scale, 10
Stokes paradox, 12, 69
stream-function
    definition, 4
    Poisseuille flow, 228
stream-function - vorticity, 5
Strouhal number
    definition, 3
Sychev's hypothesis, 204
    Smith's numerical solution, 206
symmetric channel
    $R^{-1} \ll \epsilon \ll 1$, 233
    $\epsilon = O(1)$, $\sigma = R^{-1/3}$, 277

Thwaites
    integral boundary layer method, 163
    using potential velocity, 163
trailing edge
    outer potential, 78
    pressure perturbation, 78
    singularity, 52, 56
    viscous region, 82
triple deck
    Y, W, Z scales, definition, 82
    around separation, 207
    comparison with computed solutions, 91
    computation technique for channels, 106
    computed in sublayer coordinates, 103

computed solution, 88
displacement function, 84
drag, 90
formulation, 82
fundamental problem, 97
history, 76
inner deck, 86
middle deck, 83
numerical solution, 98
outer deck, 85
pressure–displacement relation, 86
reduction of parametric dependence, 97
spectral method solution, 104
turbulence
  laminar flow zone, 116

upstream influence, 263–283
  asymmetric channel, 263
  boundary conditions, 275
  computed solution, 270
  displacement function, 269
  fine indentation, 278
  free interaction singularity, 271
  linearised theory, 266
    Fourier transform solution, 269
  moderate indentation, 280
  separation distance, 273
  severe indentation, 281
  symmetric channel, 277

velocity potential, 10
viscosity, 3, 9
vorticity, 4
  definition, 4
  far field decay, 41
  transport equation, 4

wake
  composite near wake, 56
  constants
    $\gamma_0, \mu_0, \lambda_0$, 52, 59
    $\gamma_1, \mu_1, \lambda_1$, 56, 59
  far wake, 59
  finite plate, 49
  Goldstein inner, 50–54
  Goldstein outer, 50, 54–55
  higher order expansion, 62
  Kuo's model, 70
  plate
    centreline velocity, 61, 68
  structure behind cylinder, 218
  transverse velocity, 63, 70
  Woods' non-diverging, 146
wall stress measurement, 121
Woods' free streamline solution

channel flow, 234–245
  expansion near separation, 239
  fundamental solution, 237
  mapping, 235
  numerical solution, 238
  singularity at separation, 239
cylinder, 139–149
  expansion near separation, 142
  far field expansion, 144
  fundamental solution, 139
  iterative solution, 142
  modified singularity, 149
  non-diverging wake, 146
  numerical method, 142
  singularity at separation, 144

The manufacturer's authorised representative in the EU for product safety is
Oxford University Press España S.A. of el Parque Empresarial San Fernando de
Henares, Avenida de Castilla, 2 – 28830 Madrid (www.oup.es/en or product.
safety@oup.com). OUP España S.A. also acts as importer into Spain of products
made by the manufacturer.

www.ingramcontent.com/pod-product-compliance
Lightning Source LLC
LaVergne TN
LVHW022003060526
838200LV00003B/80